THE MYTH OF
AMERICA'S DECLINE

THE MYTH OF AMERICA'S DECLINE

Leading the World Economy into the 1990s

HENRY R. NAU

OXFORD UNIVERSITY PRESS
New York Oxford

Oxford University Press

Oxford   New York   Toronto
Delhi   Bombay   Calcutta   Madras   Karachi
Petaling Jaya   Singapore   Hong Kong   Tokyo
Nairobi   Dar es Salaam   Cape Town
Melbourne   Auckland

and associated companies in
Berlin   Ibadan

First published in 1990 by Oxford University Press, Inc.,
200 Madison Avenue, New York, New York 10016

First issued as an Oxford University Press paperback, 1992

Oxford is a registered trademark of Oxford University Press

Library of Congress Cataloging-in-Publication Data
Nau, Henry R., 1941–
The myth of America's decline : leading the world economy
into the 1990s / Henry R. Nau.
p. cm.   Includes bibliographical references.
ISBN 0-19-506001-6
1. United States — Foreign economic relations.   2. United States —
Economic policy — 1945–   3. International finance.   I. Title.
HF1455.N394   1990
337.73 — dc20   90-6760

ISBN 0-19-507272-3 (PBK.)

2 4 6 8 10 9 7 5 3 1

Printed in the United States of America

To my daughter,
Kimberly Alison Nau,
whose future
in freedom is hers
to choose.

PREFACE

How can America be in decline and yet win the Cold War in 1989 and the Persian Gulf war in 1991? If that question intrigues you, you need to read this book. On the other hand, has America triumphed so completely in the world that it can safely retreat to its own shores and tend to its own domestic agenda? If you believe America can "come home," you also need to read this book. For America faces both unprecedented opportunities and challenges in the decade ahead. It leads a community of Western industrial nations that for the first time in history is entirely democratic and free market-oriented. It has the landmark opportunity to expand this community to include former communist and authoritarian states in Eastern Europe, the former (after the failed coup in August 1991) Soviet Union, and much of the developing world. But it also faces the threat of growing ethnic and economic conflicts around the world that could snuff out reforms in the East and fragment markets once again in the West. Much depends on what the United States learns from its relatively successful leadership during the postwar period and whether it can maintain a proper balance between this leadership abroad and its quest for a more perfect and just society at home.

When the cloth version of this book was completed in the spring of 1989, there were only faint hints of the dramatic events that would eliminate communist regimes in Eastern Europe and the Soviet Union and repel Iraqi aggression in Kuwait. Nevertheless, the book anticipated these events by trying to understand the longer-term and deeper sources of American political and economic strength in the postwar period. This strength derived from critical policy choices the United States made in the late 1940s to promote open societies and liberalize economic markets among Western countries and thereby to overcome the ethnic and economic conflicts that had ravaged relations among the countries of Western Europe and the Atlantic for more than a century. Temporarily, Vietnam and inflation eroded America's strength. But in the 1980s, America struggled back, to a renewed sense of political purpose and an improved economic position in global markets. While these gains were always partial and reversible, they were real, and they provided the basis for the unprecedented political leadership which the United States exercised in the Persian Gulf war—even more impressive in this author's judgment than the superb American military performance—and for the strong resurgence of America's manufacturing productivity and exports that occurred in the last half of the 1980s.

How can judgments differ so sharply between those who see America on the rise and those who see it in decline? The differences have to do with what analysts emphasize and how they interpret history. Three differences distinguish this study from decline arguments.

First, this study is concerned not only with the rise and fall of relative power among nations but also with the purposes for which nations exercise power. If these purposes increasingly converge, as they have among the industrial countries since World War II, relative power becomes less important. A decline in relative American power may actually translate into an overall increase in America's position if the world community increasingly reflects purposes compatible with America's interests. In a world community that is moving toward greater democracy and more open markets, America needs less power to protect and promote its national interests.

Decline arguments miss this point completely. They read contemporary events through the prism of old nineteenth century balance-of-power politics. They assume that political ideologies and institutions among countries always differ and can never converge significantly. Thus, peace is possible in international affairs only in two circumstances: if power is unequally distributed such that one set of national interests dominates over other sets, or if power is equally distributed and perfectly balanced such that no one set of national interests dominates. Because American economic power has relatively declined since 1945 and the world has never succeeded in perfectly balancing power, the decline perspective expects more national conflict in the 1990s. Countries such as the United States and Japan will inevitably clash with one another to protect their national and economic security and to equilibrate power.

The most astonishing feature of the contemporary world, however, is that serious national conflicts have not increased. Instead, there has been a broad convergence of national interests and institutions over the past forty-five years toward democracy and freer markets. This movement was apparent initially among industrialized countries with Germany, Japan, and then Spain, Portugal, and others joining the ranks of free nations, but it is now also potentially a possibility with former communist states and much of the developing world as well. However vigorously the United States, Japan, and Europe quarrel about burden-sharing and trade issues, their relationships today are a far cry from the military rivalries of the 1930s or the nineteenth century. In ten or twenty years, relationships between the United States and the Soviet Union (or whatever replaces it) and between Eastern Europe and Western Europe may achieve a similar, less threatening dimension. Indeed, already in 1991, a broad convergence among the major nations of the world, especially between East and West, enabled the member states of the United Nations to form a coherent coalition behind U.S. leadership in the Persian Gulf crisis. More than one hundred nations trusted the United States to use force on their behalf to liberate Kuwait. America's allies, if they did not contribute military forces, ultimately paid for the entire cost of the war. This could not have happened if those nations had not shared common external goals with the United States or if they had had no confidence in America's domestic institutions and political leadership.

None of this is to say that the world today is made over in America's image. Of course, the democracies of Japan and the United States, of France and Czechoslovakia, of Mexico and India are different. It trivializes the changes taking place in these countries to argue that they are simply reflections of American democracy or American culture. But they are still democracies, and from research on wars over the past two centuries, there is persuasive historical evidence that democracies have not fought against one another. This fact suggests that America's national security may be best advanced by policies that ensure that Germany and Japan continue to grow democratically, not by policies that drive them into premature military commitments or provoke destructive trade wars. It also suggests that no American objective is more vital than the success of democratic reforms in Eastern Europe, the new Soviet Union, and modernizing countries in the developing world.

A second difference between the perspective in this book and decline arguments has to do with whether policy choices can change circumstances or must inevitably adapt to them. In decline perspectives, policy choices are determined largely by circumstances. For example, the decline school starts from the premise that power has shifted away from the United States and will lead to growing conflict in the contemporary world. As America becomes entangled in more foreign conflicts with fewer relative resources, it faces the danger of "imperial overstretch." Hence the decline school counsels America to disengage from the world, stop leading the West politically and militarily, and put more emphasis on its own national economic interests and domestic agenda. Decline theorists end up urging America to abandon the very policies of freer trade and national outreach that helped to construct the politically converging and prosperous economic world that we see today.

For today's world of congealing community among industrial nations did not just happen. It was at least partially constructed through the policies of open societies and open markets which the United States and its Western allies pursued after World War II. After all, the Soviet Union too was a great power in the communist world in 1945. If relative power were all that mattered, the Soviet Union should have succeeded in unifying the Eastern bloc. Yet it failed to develop common purposes and shared prosperity with its allies. Bad luck? Wrong side of history? Perhaps! But more likely, it simply pursued less effective policies—policies that did nothing to try to reconcile and heal traditional national and ethnic differences or to share markets, technology, and resources to build efficient integrated economies. Today, as a result, ethnic tensions and crumbled economies litter the landscape of Eastern Europe and the Soviet Union.

The United States and its allies, by contrast, deliberately opened their politics, societies, and economic markets to one another. The United States shared some of its technology and wealth to nurture prosperity and democratic societies in postwar Germany, Japan, and other allied countries. Although it lost relative economic power, the United States gained friends who share more of its political purposes and therefore pose less of a threat to its fundamental national interests. In a world of receding threat, it should be possible to depend more, not less, on cooperation with allies such as foreign investment from Japan and to work on domestic and international agendas simultaneously.

The third difference between this study and perspectives emphasizing America's decline has to do with the assessment of America's domestic agenda and political self-image. Decline perspectives argue that the 1980s was a lost decade for America—a time of unrestrained consumption and waste. They cite America's status as the world's largest debtor nation, its educational deficiencies, drug problems, racial inequalities, decaying infrastructure, savings and loan bailouts, and banking difficulties as palpable evidence that the nation is in decline. They see America increasingly vulnerable to foreign investors, imports, and influence. In the face of such general weakness, they urge America to pull in its horns, focus on its domestic problems, and get tougher with allies it no longer needs to defend against the Soviet Union.

The perspective in this book does not deny the domestic challenges America faces. But it tries to understand how the United States has met these challenges in the past and how it is that, despite continuing shortcomings in American society, the rest of the world displays an affinity for and confidence in America that does not square with the extreme negative self-image of the decline school. For, with all its faults, what other country (especially if we think of America's principal economic competitors, Germany and Japan) has done more internally over the past forty-five years to address racial discrimination, to employ women and young people, to benefit consumers (that is, people rather than companies), to offer opportunities to immigrants, to push back the frontiers of technology and higher education, to develop leisure and entertainment industries, to foster an active and voluntarist citizenry at all levels of political and social life (creating, in a sense, the problems we have today with multiple, fragmented special interest groups), and to uphold the cause of human rights and freedom around the world?

America has earned a better self-image also in the economic area. As this study details, and as is now evident from America's renewed export competitiveness, much was accomplished in the 1980s. America's share of the Gross National Product of the industrial world bounced back and is now the same in 1991 as it was in 1970 (after declining throughout the 1970s). Our manufacturing sector constitutes the same share of U.S. GNP in 1991 as it did in 1950, and America's manufacturing productivity grew faster in the 1980s—3.6 percent per year—than it did in the 1950s and 1960s. In the last half of the 1980s, American manufacturing productivity grew as fast as that of Japan and twice as fast as that of Germany. Our absolute level of productivity is still the highest in the world, with Japan's level equaling only 80 percent of the U.S. level. Once the value of the dollar declined, American manufactured exports soared, growing at a compounded rate of 15 percent per year in the late 1980s.

Surely on this basis, America can compete without closing its markets. What it cannot do is continue to tie its own hands behind its back with foolish economic policies. The rebuilding of the American economy in the 1980s was only partial. While inflation and the slow growth of output and manufacturing productivity were overcome, fiscal policy went out of control and produced a cumulative massing of public, corporate, and consumer debt that seriously weakened the U.S. financial economy and now, in late 1991, restrains the recovery of the American economy. This fiscal and broader debt problem remains the single biggest obstacle to America's future. It is more than an economic problem; it is a test of America's capacity to live up to its own ideals and to shape its domestic and international identity in the new world community.

Whether America meets this test or not depends on how Americans think about themselves and their future opportunities at home and abroad. That is why the debate about America's rise or decline is so critical. These ideas will shape today's policy choices, which in turn will shape tomorrow's world circumstances.

The federal budget is in many ways an expression of the ideas of a nation. How people or countries spend their money tells us who they are. Budget expenditures generally reflect three aspects of a society: self-worth, self-maintenance, and self-development. Self-worth can be measured crudely by what a nation spends to express and defend itself in the world. Only a nation that feels good about itself is likely to be active in international affairs. Defense and aid expenditures by the United States, which comprise 25–30 percent of the federal budget, went up significantly in the 1980s and probably had something to do with the successful containment and subsequent transformation of aggressive Soviet military policies.

Self-maintenance may be measured by so-called entitlement expenditures, about 45 percent of the current federal budget. These expenditures constitute spending for medicare, social security, pensions, agriculture, and industry. Such spending maintains a certain quality of life, particularly for the middle class. These expenditures grew dramatically over the past decade and are funded in large part by payroll taxes which are regressive—that is, taxes on poorer income groups, such as many blue-collar workers, that benefit the middle class, such as a large number of the elderly.

Self-development may be measured by expenditures in the infrastructure, such as education and transportation, and by expenditures in social welfare, such as medicaid and welfare payments. Outlays for self-development seek to improve the overall potential of the society and particularly its poorer citizens. These outlays (about 15 percent of the current federal budget) fell relatively during the 1980s. The remaining 10–15 percent of the federal budget is taken up by interest payments on the national debt, the part of the budget that exploded in the 1980s because of the fiscal deficits.

America needs a vigorous debate, hopefully in the context of the presidential elections in 1992, to adjust its spending priorities to shape the emerging new world community. The perspective in this study suggests shifting priorities from self-worth and self-maintenance activities toward self-development objectives. Self-maintenance, which involves spending scarce government resources on citizens irrespective of income, borders on self-indulgence. Well off middle-class Americans have to ask themselves if they really need small social security checks or medicare reimbursements just because they are "entitled" to them. These resources could be used for self-development. Advances in education and infrastructure contribute more directly to self-worth and strengthen America's influence abroad as America's defense needs recede in the more friendly world community we inhabit. With less need to project military power abroad, America can increase its influence through a more vigorous and wholesome domestic identity, and thereby continue to nurture the convergence of social and political values with other societies, further reducing threats and hence defense needs.

Throughout its history, America has been tempted to pursue alternately the twin illusions of utopia at home (isolationism) and permanent peace abroad (internationalism). Realistically, it can attain neither. But in the early 1990s, by comparison with any other time in the twentieth century and probably in the history of the American republic, the United States has achieved a substantial amount of both justice at home and peace abroad. This book contends that it did so largely because it was willing to share power at home and abroad and cultivate diverse groups that came to share increasingly its basic political and economic values. Common sense dictates that America continue this kind of leadership abroad to bring the former communist and emerging developing nations into the new world community, even as it enhances its self-image by improving aspects of America's own domestic society.

Washington DC   Henry R. Nau
October 1991

CONTENTS

THE MYTH OF
AMERICA'S DECLINE

INTRODUCTION

Leadership or Decline:
America's Choice

Recent headlines declare that the American era is over and with it the most prosperous international economic period in human history. The reason for this passing era, we are told, is the decline of American power. Postwar prosperity was a fortuitous consequence of unparalleled American preeminence after World War II. This preeminence enabled the United States to project its liberal economic ideology onto the rest of the Western world and to integrate Western markets and defend American foreign policy interests against world communism. America's power, however, inevitably declined; economic rivals caught up (Kennedy 1987). Japan, in particular, pioneered a renaissance in Asia and the Pacific that progressively eclipsed the Eurocentric world of the postwar period and championed a new, interventionist economic ideology that challenged the liberal policies of the American era. Economic competition strained Western military alliances. With domestic reform in Eastern Europe and the Soviet Union, the communist threat receded. The United States became both less willing and less able to bear the burdens of global leadership, and Europe and Japan hesitated to assume global roles. The Western world drifted toward conflict with less unified purpose and more fragmented power.

America's politics, we are also told, hastened the decline of American power. Political interest groups that seek to redistribute growth rather than to create it proliferate as dominant societies evolve (Olson 1982). Under the influence of these groups, domestic consumption outpaces investment and production; the burdens of foreign leadership, especially military burdens, grow; and the benefits of technological innovation spread to other nations (Gilpin 1981, 1987). In the 1980s American politics gridlocked and America went on a consumption binge, borrowing from abroad to become, by 1986, the world's largest debtor. Going into the 1990s, America seemed once again to be an ordinary power and the postwar period but another episode in the perpetual cycle of dominant and declining powers, rising and falling prosperity.

But is this all true? Is the American era over or has it simply been replaced by a Western era in which noncommunist and now reforming communist countries increasingly share American purposes and seek to emulate Western institutions of pluralist democracy and free market economies? Moreover, has American power declined significantly, when the U.S. share of the Gross National Product in the industrial world is the same today as it was in 1970 (after adjustments for exchange rate fluctuations) and when American economic policies in the 1980s have produced growth rates and manufacturing productivity increases in the United States characteristic of the boom years of early postwar recovery? Whether current policies can sustain or further enhance America's power is doubtful, but isn't it therefore more relevant to debate these policies and to recognize that America may not need as much power today in a world that more broadly shares some of its basic purposes than to conclude that American power and politics have irretrievably declined?

This book argues that America's dominance and interest-group politics explain the ups and downs of American influence in the postwar period less persuasively than America's choices of national purpose and economic policy. When America's strategic and economic policies inspired a renewal and reconstruction of free societies in Western Europe and the Far East in the 1940s and when America's choices challenged totalitarian societies and accelerated technological change in the 1980s, America's influence spread and actually contributed in some measure to the dramatic turnaround that is now taking place in the communist world. In other periods, such as the 1970s, when America questioned its self-worth and chose inefficient economic policies, its influence waned.

The debate about America's choices, therefore, is far more important for America's influence in the future than the historical conclusion that power inevitably shifts and produces conflict. America's choices can recognize and seek to reinforce the remarkable convergence in the world today toward freer political and more competitive economic institutions or conclude that the United States and new powers, such as Japan and perhaps a reunited Germany, have fundamentally different interests and must inevitably collide.

A Choice-Oriented Perspective

To advance the debate about America's future, this study develops a choice-oriented perspective on historical events, which emphasizes the role of national (or human) purpose and policy choices on world economic and security affairs. In four areas this perspective contrasts markedly with conventional, or what I call structuralist, approaches that stress the constraints on human choice of power balances and political interest groups. First, it argues that international politics and economics are about not only cycles of power and wealth, as structuralist views contend, but also a search for shared political community in which human societies, most recently organized in national units, freely choose alternative forms of domestic political society and then compete to establish, among themselves, a world community that satisfies higher goals of political and human development. National societies today disagree about the basic principles for organizing political life among human beings. Some societies reserve significant rights and privileges to private individuals and take public decisions on the basis of a broad political franchise; others control most aspects of private life and take public decisions on the basis of a privileged elite. These broad distinctions between democratic and totalitarian societies motivate security and economic activities in international life and give history and historical cycles a political or moral, and not just material, dimension.

From this perspective the postwar period of American power does not represent just another cycle in the rise and fall of great powers. This period stands out as an extraordinary era of political and economic development, particularly, but by no means exclusively, among Western countries. By 1990 the entire industrial world, including the fascist powers of World War II — Germany, Japan, Italy, and Spain — shared to some degree basic democratic values and institutions. In addition, unlike previous periods of dominant-power rule (e.g., that of the Netherlands in the seventeenth century and of Great Britain in the nineteenth century), the American era witnessed the political decolonization and economic advancement of the developing world. Moreover, many of the newly independent nations were free after 1945 to choose non-Western forms of government and domestic economy, although by 1990 more and more of them were experimenting voluntarily with their own versions of democratic and economic liberalization.

Not all of this convergence of basic political view in the postwar world was due to America's leadership or to a simple projection of America's political culture through imperial power. But it did happen on America's watch, and unless it was accidental, it had something to do with the national purpose and economic policies America projected. Throughout the postwar period, America, together with its Western allies, championed principles of free societies at home and competitive markets abroad. These principles, embodied in various concepts of national purpose in postwar American policy, not all of which were equally effective, provided a standard for the political and economic organization of the postwar world community. In recent years, the world community has increasingly and voluntarily affirmed this standard even as America's power has relatively declined. This study develops the concept of national purpose and international political community to get at this critical role of human and political standards in postwar economic events and to escape the material and moral cynicism of perspectives emphasizing primarily power and wealth.

Second, this study emphasizes the role of economic policy choices in shaping international outcomes of wealth and power. The cycle of dominance and decline of great powers is not just a consequence of uneven patterns of growth and technological change, circumstances that themselves have no explanation (see, for example, Kennedy 1987). It is also a consequence of policy choices that spur growth, change technology, and alter circumstances. After World War II, the United States and other industrial countries made specific economic policy choices that radically altered prevailing circumstances of soaring inflation, rapid nationalization of industrial assets, and bilaterally managed trade. They created the liberal, postwar, Bretton Woods economic system that emphasized stable prices, flexible domestic markets, and freer trade (permitting fixed

exchange rates, which became the hallmark of this system). Under these policies, for the next two decades world industrial production and trade grew faster
and inequalities between industrial and developing countries narrowed more
sharply than in any previous historical period since the industrial revolution.
With the backlog of wartime demand, circumstances were favorable, to be
sure, but they were not that favorable. As one study concludes, "it was unprecedented for the world economy in 1939–45 to absorb the costs of a hegemonic
war at the beginning of an [economic] upswing and then to continue sustained
growth for another two decades after the war" (Goldstein 1988, p. 343).

America abandoned these Bretton Woods policies in the late 1960s. Well
before the first oil crisis, inflationary policies, spreading regulations, and nas-
cent protectionism slowed economic performance. Policy choices then shifted
once again in the late 1970s and early 1980s, reversing inflation and intervention but exacerbating protectionism. Each time as the world's leading economy,
America influenced the choices of other countries, including a growing number
of developing countries. According to this study, these policy choices had a
much greater impact on world economic performance than relative power balances or unexplained favorable or unfavorable circumstances.

A third difference between this study and conventional power or structuralist views concerns the mechanisms by which policies alter circumstances in the
world economy. The choice-oriented perspective in this study emphasizes the
role of nonbargaining, competitive mechanisms of policy influence, such as
international markets. Structuralist arguments, by contrast, emphasize bargaining and institutional mechanisms of change, such as international conferences and organizations. According to the perspective of this study, nations
act unilaterally as well as collectively to influence policy change in the world
community. They exert leverage on one another through independent actions in
complex and interdependent world markets, as well as through collective decisions in traditional diplomatic arenas such as summit conferences and international negotiations.

As the world community has become more diverse and pluralistic, markets
have become more important and valuable mechanisms of influence. They provide greater flexibility for narrowing differences among a growing number of
countries. National leaders do not have to negotiate and compromise domestic
interests directly at summit conferences in the glare of media lights but can move
their policies toward one another indirectly in response to market forces which
they can claim are outside their control or immediate responsibility. Structuralist
arguments, by contrast, raise unrealistic expectations that international conferences or summit meetings can always resolve outstanding differences. These
perspectives thus frequently contribute either to frustration, when international
cooperation fails, or to unsound compromises, when the need to cooperate takes
precedence over the substantive content of economic policy.

Markets work as a mechanism for narrowing policy differences in a pluralistic and complex international economy, however, only if they are circumscribed by a broad consensus among participating nations to keep markets
open (i.e., avoid protectionism), maintain stable domestic prices (i.e., prevent

wildly fluctuating exchange rates), and encourage flexible domestic markets (i.e., reallocate resources to meet international competition). Thus, as a fourth area of difference with structuralist arguments, the choice-oriented perspective in this study stresses the key role of policy ideas that cut across national perspectives, political parties, and special interest groups and that ultimately shape or fail to shape the consensus that creates international markets and institutions and permits them to function efficiently. These ideas originate in the major participating nations in what I call in this study the cocoon of nongovernmental organizations that surrounds domestic bureaucratic and international institutional decision-making processes. Ideas compete in this cocoon to shape social views. These views then permeate, through political coalitions and elections, more immediate policymaking processes at both the domestic and international level to influence choices about national purpose and economic efficiency. This larger consensus in the cocoon of nongovernmental organizations enables highly pluralistic, democratic policymaking institutions to work. Without it domestic as well as international policymaking would grind to a halt, stalemated by unconstrained rivalries and a special-interest-group perspective that sees all policy choices and outcomes as a matter of winners and losers and no solutions that serve the interests of all groups.

From the choice-oriented perspective, ideas do not emerge in a vacuum. They are influenced by circumstances and interests and to affect policy, they have to meet a certain test of public acceptance and succeed in mobilizing political and ultimately bureaucratic forces. Moreover, translated into policies, ideas have to yield results. They have to meet certain performance tests established by limited economic resources and by irreducible elements of military and social power. Nevertheless, ideas are not simply rationalizations of interests or the consequence of external circumstances. Some ideas make this claim. For example, Marxism claims that ideas are the product of class interests. Structuralist arguments claim that certain policy options are unrealistic or ideological because they conflict with what are perceived to be irreversible circumstances. But these claims are themselves arguments on behalf of certain policy ideas. They simply interpret history to preclude ideas that originate independently of class or circumstances. Such claims have to be tested critically against alternative interpretations of historical events and debated in the cocoon of nongovernmental organizations to shape the consensus that influences the next round of policymaking.

In that sense, this book, as well as those that adopt structuralist arguments, is very much a part of the debate about America's future in the world political economy. The interpretations this study draws from the historical record lead to quite different recommendations for American policy in the 1990s. Whereas perspectives emphasizing the decline of American power and the gridlock of American politics counsel paring down America's role and burdens in the world community, the perspective in this book urges a more assertive and self-confident role for America that recognizes the widespread appeal of America's purposes in the world today and that embraces policies to continue to pay for that world both in terms of providing security for major allies, such as Germany and Japan, and in terms of extending the liberal, prosperous, postwar economic system to include developing countries. The costs of such a role are not beyond America's power or politics; they are determined by America's purposes and policies—in short, by America's choices and America's priorities.

Postwar History from a Choice-Oriented Perspective

The preceding differences between the choice-oriented perspective and more commonplace structuralist arguments emerge from a detailed examination in this book of the postwar evolution of American foreign economic policy. Subsequent chapters examine this postwar experience in four separate phases: a first phase, from 1942 to 1947, when the United States, despite considerable efforts, failed to launch a successful program to reconstruct the postwar world; a second phase, from 1947 to 1967, when the United States, initially through the Marshall Plan, led the Western world toward rapid and stable growth; a third phase, from 1967 to 1979, when America's purposes and economic policies faltered and world growth and trade slowed; and a fourth phase, from 1979 to the present, when America's purposes became more self-confident and its economic policies restored noninflationary growth but also exacerbated protectionism and international debt, leaving America's power increasingly constrained and its politics gridlocked.

In each of these phases, America's definition of itself—that is, its national purpose—and the specific economic policy choices it made had more to do with the variation of outcomes than the decline of American power or the increasing fragmentation of American politics. In the first phase, from 1942 to 1947, America's power was more dominant and its politics more unified than in any subsequent phase, yet its purpose and policies failed to inspire the strengthening of democratic institutions or the launching of economic reconstruction in central Europe. The early U.S. view of world community through world law and commerce embodied in United Nations (U.N.) institutions ignored fundamental differences in domestic political community between free and totalitarian societies, and the economic policy rules agreed to at the major U.N. economic conference in Bretton Woods, New Hampshire, in July 1944 failed to prescribe the liberal trade policies or the flexible, noninterventionist domestic policies necessary to create open and efficient international markets. As a result, from 1945 to 1947, the world drifted toward political conflict and the world economy stagnated.

In the second phase, from 1947 to 1967, the United States found a more compelling purpose and initiated more efficient economic policies to launch postwar peace and reconstruction. This happened despite the greater division in American politics as the Republicans took over both houses of Congress in 1946. America's new purpose of containment recognized and reacted to the struggle going on in central Europe between differing ideas of domestic political community, and the Marshall Plan encouraged new economic policies to integrate and use Western resources more efficiently, by liberalizing trade, reducing inflation, and removing excessive domestic regulations that restricted the free movement of labor, capital, and products in European markets. This period, in fact, gave rise to what I call in this book the Bretton Woods policy triad of postwar economic rearmament (labeled Bretton Woods because of the association of the triad with the larger postwar system, not the specific 1944 Bretton Woods agreements). This triad consists of the three elements essential to establish and maintain open, stable, and growing world markets:

1. moderate Keynesian fiscal and monetary policies to ensure *stable domestic prices* and thereby stable international prices or exchange rates.
2. *reduction of trade barriers* and foreign exchange restrictions to promote freer trade, comparative advantage, and, hence, competitive and more efficient use of domestic and international resources.
3. *limits on noneconomic motivated intervention by governments in the workings of the domestic market* so as to facilitate a flexible flow of labor where it is needed, the movement of capital resources into new and more innovative enterprises, and a stream of new products uninhibited by arbitrary regulation of individual sectors — all of which helps to control balance-of-payments imbalances through a reallocation of domestic resources, instead of through inflation, frequent exchange rate changes, restrictions on trade and foreign exchange, or unconditional external borrowing.

This triad emerged in the United States already in 1946 and subsequently in all the other industrial countries under the influence of Marshall Plan programs and institutions. From the perspective of this study, the shift in Western economic policies after 1947 was decisive for the impressive economic performance of the initial Bretton Woods economic system from 1947 to 1967.

In the third phase of postwar U.S. policy, from 1967 to 1979, the United States and other Western countries abandoned key elements of the postwar policy triad and world markets accordingly became more unstable, less open, and less efficient. New ideas of detente in Europe and the failed performance of containment in Vietnam challenged America's sense of national purpose and weakened the resolve of the United States and other Western countries to pursue efficient economic policies. New economic policy ideas emerged that encouraged more expansive Keynesian fiscal and monetary policies, regardless of the consequences for domestic price stability and the fixed value of the dollar, and that sanctioned increasing government intervention in individual domestic markets to achieve social equality, environmental, and other noneconomic objectives. Runaway fiscal and monetary policies, price controls, and creeping protectionism all preceded the first oil crisis and contributed to the shocks and vulnerabilities that afflicted the U.S. and world economies in the 1970s. Policy weakened American power, and politics was not a primary cause of the policy shifts. Contrary to what interest group perspectives predict, the Bretton Woods consensus collapsed first in fiscal and monetary policy areas where domestic interest groups were less active, whereas it persisted in the area of freer trade where domestic interest groups and political fragmentation were much stronger.

In the fourth phase of postwar policy, from 1980 to the present, America struggled back to find a renewed sense of self-worth and to reinstate more efficient policies of lower inflation, less government intervention, and more liberal trade. Despite the widely proclaimed decline of American power and politics, Reagan administration policies went a long way to restore the self-confidence of American foreign policy and to usher in a new era of low inflation and revived productivity and growth in the industrial world. But Reagan policies did so at the cost of large budget and trade deficits and a widely fluctuating dollar that distorted world financial markets, particularly for heavily indebted developing countries, and burdened world trade markets, where governments threatened increasingly to restrict and manage trade. The industrial world appeared to be on the verge of abandoning the last element of the Bretton Woods policy triad—freer trade—even as the first two—more stable prices and more flexible and productive markets—were being partially restored. The United States became the world's largest debtor, lowered world interest rates and the dollar to accommodate increasing debt and protectionist pressures, and missed a golden opportunity to reinforce market-oriented domestic and trade policy reforms abroad by continuing massive budget deficits and raising trading barriers at home. Many of the developing countries all but dropped out of an expanding world economy, suffering for the first time in the postwar period substantial losses in real income.

In the end, America's purposes and policies came up short, not because America lacked power or a sufficiently unified political base, but because President Reagan and his administration failed in 1985, when they were at the top of their political game, to follow through on their purposes and policies. They drew back from decisive budget cuts, particularly in middle-class entitlement programs and poorly managed defense programs, and surrendered leadership in both budget and trade policy to a more nationalistic and divided U.S. Congress.

Choices for the Future

This brief summary of the argument that unfolds in the rest of this study suggests that the key question for America in the future is not whether its power has declined but what purposes it seeks to achieve in the world community and what specific economic policies it intends to follow. The debate about these purposes and policies is more important than the debate about American power or burden sharing with the allies. It is also more important than the repeated calls for the reform of American politics or bureaucratic reorganization of the government. In this sense, the contemporary American crisis is not about power and politics or institutions and resources. It is about the intellectual and political soul of America, and the kind of political community America wants at home and in the world and is ready to pay for.

Studies that focus primarily on power and politics exaggerate the costs of continuing American leadership in the world economy. They overlook the more homogeneous political world in which America seeks to exert its power today, and they pay too little attention to the content of American economic policies that can enhance American power and to the market mechanisms by which America can less visibly and less expensively assert its power.

America's purposes are more widely shared in the world today than they were in 1947 or 1967. Accordingly, to influence this world, America does not need as much power as it did 20 or 40 years ago. It can maintain its leadership by being more of a global pope than a global power, supporting basic political values of individual freedom and economic initiative throughout the world less by its military legions than by its diplomacy and domestic example. In this sense, some adjustment of America's military presence in the world community is called for.

How America adjusts its role, however, is crucial. It can do so from the standpoint of a self-confident global leader that continues to articulate a clear vision of shared political community, or it can do so from the standpoint of a declining society that has become self-indulgent, preoccupied with narrow economic interests, and insistent that its allies share more of the burden for no other reason than because America's power has declined. Even with its reduced power, America remains the only military and economic superpower in the world. It has an unparalleled opportunity, therefore, and indeed a special responsibility, to continue to set the political tone for the world community, affirming the basic values of human freedom and economic competition that lie at the heart of American society and now also a larger part of world society. This responsibility is even more important as societies in Eastern Europe, the Soviet Union and the developing world seek to reform. Recent efforts to liberalize political systems in these societies draw their inspiration from Western values and experience. If America and the West now came to doubt their own values, the world would lose its political anchor for peaceful change. For without a clear sense of continuing differences between Eastern and Western societies, the West can neither maintain its own unity nor help inspire the East to create a more humane political society.

How America plays its role will also influence its allies. If America asks Europe and Japan to share greater burdens simply to compensate for declining American power, the allies will rely less and less on American power. If, on the other hand, America appeals to its allies to share in preserving and enhancing a more congenial world, one that modestly and carefully nurtures Western values of liberty and economic opportunity also in reforming countries, the allies may respond with the best traditions and policies of their own societies. For if American citizens tend to react more quickly (perhaps overreact) to political (i.e., totalitarian) threats to democratic society, citizens in the allied countries, particularly in Europe, may be more sensitive to economic and social threats. Thus, Europe and Japan have special responsibilities to motivate the industrial world to open its societies and markets increasingly to the rest of the world.

The most important prerequisite for an open and growing world economy, therefore, is that the United States, it allies, and a growing number of reformist countries retain and develop a sense of common purpose and shared political community. For without this common purpose, the citizens of these societies will not be ready to accept the domestic changes that are required to create and maintain open economic markets and transparent political systems. Instead, they will conclude that they do not understand each other internally, that they have nothing in common politically, and that they cannot find "fair" rules to govern their relations economically.

America must above all recognize and preserve the policies that have created the more hospitable world of the late 1980s. Since 1945, America has provided the primary security for Germany and Japan, while these countries have developed more open and democratic societies. America should continue to provide this security, both because it may be less expensive to do so in the 1990s if the superpowers agree to reduce overall arms and because Germany and Japan, as relatively young democracies, should be allowed to decide at their own pace when they can assume greater military responsibilities, particularly in the case of a reunited Germany. Otherwise, America will withdraw from these countries now to achieve modest savings in its budget only to return later at much greater cost when disturbances, either within Germany and Japan or within their regions, create a far less congenial world for Western interests.

On the other hand, Europe and Japan have greater economic obligations, especially toward reforming countries. Europe needs to adjust internally (e.g., agriculture) so as to accommodate continued and even greater openness toward the outside world, and Japan has to use its global economic and financial power for more explicit social and political objectives, both at home to create a more emancipated domestic society and abroad to emphasize standards of human and political development that go beyond Japanese commercial interests.

These choices to project national purposes of free and open Western societies, especially toward reforming societies, are fully consistent with internal choices to use American and Western resources more efficiently. The most critical internal adjustment for the United States is to reduce its budget deficit, not because the deficit has been a calamity from the beginning, as some critics contend, but because, at approximately 3 percent of GNP in 1989, it is unjustified economically as the U.S. economy nears full employment and because it represents, politically, an incapacity on the part of the United States to lead other countries to reform their economic policies. Europe has to adjust by reducing excessive government regulation in labor, financial, and critical-product markets such as agriculture and telecommunications. Germany, especially a reunited Germany, has a special role to play. It is the balance wheel between the process of deepening integration in Western Europe and the challenge of widening integration with Eastern Europe. And it must ensure that this regional development does not come at the expense of American, Japanese, or developing-country partners. Finally, Japan has to attack the last layers of trade policy restrictions in the Japanese economy that derive from Japan's retail distribution system, land use policies, and other measures that excessively constrain domestic consumption.

These domestic policy adjustments in the allied countries would restore fully the Bretton Woods policy triad of sound macroeconomic, especially fiscal, policies, which exist in Europe but not yet in the United States; more flexible labor, capital, and product (e.g., agriculture) markets, which exist to a greater extent in the United States but lag in Europe; and freer trade, which continues, albeit falteringly, in Europe and the United States but remains suspect in Japan.

The substance of these policy adjustments is more important than the mechanisms by which they are achieved. As this study shows, the industrial nations do not necessarily need more extensive institutional mechanisms for international economic policy coordination. International cooperation under the glare of media lights often inhibits or substitutes for, rather than facilitates, domestic policy actions. Highly politicized processes of monetary and economic policy coordination, such as the Group of Five (G-5) and Group of Seven (G-7) meetings (G-5 includes France, Germany, Japan, the United Kingdom, and the United States, and G-7 adds Canada and Italy) initiated at the Plaza Hotel in September of 1985, work as long as policy focuses on easing monetary polices and lowering interest rates. These exercises, thus, tend to tilt world economic policies toward potential inflation. When the situation calls for tighter monetary policy and raising interest rates, political leaders are generally less enthusiastic about highly visible coordinating processes. They end up squabbling in public, as U.S. and German finance ministers did in October 1987 before the stock market crash, or they tone down public coordination of policies, as the G-7 did in 1988 and 1989.

Decisive policy actions taken at home and projected through competitive international markets may be a better mechanism for encouraging policy convergence, especially when the participants seek to emphasize price stability and medium-term structural and trade policy reforms. Reagan administration policies in 1981–1983 to disinflate the world economy, the European Economic Community's decision in 1985 to create a single European internal market, and Japan's decision in 1987 to restructure the Japanese economy all represent unilateral national or regional actions that have contributed significantly to real adjustments in the world economy. Only the Japanese restructuring can be said to have had anything to do with active financial and exchange rate coordination. Thus, this study recommends that active coordination of exchange rates and monetary policies be downplayed in the future so that industrial and reformist countries can concentrate on developing more effective mechanisms to encourage medium-term fiscal, microeconomic, and trade policy reforms around the Bretton Woods policy triad of restrained fiscal and monetary policies, appropriate deregulation of domestic markets, and incremental liberalization of trade.

Structuralist studies argue that the policy triad and market mechanisms are not enough to stabilize contemporary world markets, given new circumstances of highly mobilized and speculative capital flows. But capital flows, while many times larger than trade flows, are not unrelated to real economic developments in domestic markets. Much of the volatility of capital flows in the 1970s and 1980s, it can be argued, was due not to new circumstances but to prevailing policies of deregulating global financial markets while reregulating domestic and trade markets. More flexible money moved around in more inflexible product markets, exacerbating exchange rate and other instabilities. Restoring stable and flexible domestic markets, as called for by the Bretton Woods policy triad, may go a long way toward restoring stable international markets and reducing speculative capital flows.

The Plan of the Book

The perspective in this study, which stresses domestic political and economic policy choices, is sufficiently novel that it warrants conceptual development and contrast with structuralist approaches to world political economy that stress external constraints and circumstances. For this reason the next two chapters lay out the analytical basis and justification of the study. The reader who feels that he or she already sympathizes with, or understands, the choice-oriented perspective may wish to skip these chapters and go on to the historical interpretation of postwar U.S. foreign economic policy beginning in Chapter 3. On the other hand, the reader who is skeptical of the arguments outlined in this introduction may wish to examine the next two chapters with some care.

Chapters 3 and 4 begin the historical analysis. Chapter 3 explores U.S. policy choices in the period from 1941 to 1947, when the United States chose to project a Wilsonian vision of national purpose emphasizing law and open commerce but went on to compromise efficient economic policies in the Bretton Woods agreements of 1944 and to create economic rules that were completely incompatible with open and stable world markets. Chapter 4 describes U.S. efforts from 1947 on to redefine America's national purpose in terms of containment of Soviet Communism and to project, through the Marshall Plan, more efficient economic policy choices to integrate Western markets, sparking the unprecedented growth of world output and trade from 1947 to 1967.

Chapters 5 and 6 document the breakdown of the U.S. and allied consensus on containment and efficient markets and the shift toward detente and more inflationary and interventionist economic policies in the 1970s. Chapters 7 through 9 then explore the Reagan years and the groping and partial attempt to retrieve the noninflationary and market-oriented policies of earlier postwar decades.

Chapter 10 examines postwar U.S. policy choices in East–West economic relations and draws attention back to the underlying framework of shared political community that ultimately motivates and limits economic exchanges. Finally, the Conclusion of the study addresses the question of whether and how this framework of shared political community can be preserved and enhanced among Western and other nations that are seeking to reform their political and economic institutions, and what specific economic policy choices and adjustments this will require on the part of both industrial and reforming countries to preserve open, stable, and growing world markets.

I

National Choices and International Constraints in U.S. Foreign Economic Policy

1

Purpose, Policy, and Ideas: A Choice-Oriented Model of American Foreign Economic Policy

This chapter presents the basic ideas of a choice-oriented approach to American foreign economic policy. This approach emphasizes the role of public policy in shaping events in the world economy and contrasts significantly with the widespread structuralist approach, which sees circumstances and constraints that are largely outside the control of public policy as the determiners of policy outcomes.

Conceptual models lurk behind all policy debates. Critics of recent U.S. foreign economic policies charge that the Reagan administration was ideological and did not take into account imperatives of structural change in the international system, particularly the decline of American power. Yet these critics conceal an ideological or conceptual orientation of their own, one that emphasizes structural constraints on policy. As they see it, "limits set by domestic resource constraints, the international distribution of power, and international configurations of interest" ultimately defeated Reagan policies and ensured that these policies "reverted largely to lines established by the Nixon, Ford, and Carter administrations." "By and large," they conclude, "external limits and internal political processes curtailed the Reagan experiment" (Oye, Lieber, and Rothchild 1987; quotes from Chapter 1). According to this account, American choices of policy and purpose ultimately had to yield to real-world constraints of power and politics (see also Nye 1988, who appeals to "historical trends in world politics" pushing Reagan policies "back to the center").

The structuralist approach is widespread in debates about American policy. It views favorably policy developments in the 1970s that accommodated the purpose and content of American policy to the decline of American power and fragmentation of American politics. It does not emphasize, as the present study does, that American policy in the 1970s may have weakened American power and divided American politics; it concedes only that American choices may have been "a shade too pessimistic" (Oye, Lieber, and Rothchild 1987, p. 7).

17

Similarly, the structuralist approach is unable to account for significant improvements in the effectiveness of American policies in the 1980s. It has no explanation, at least in terms of U.S. policies, for the unprecedented progress in arms control negotiations, from the limitation of nuclear weapons (strategic arms limitation talks or SALT) to their substantial reduction (the agreement on intermediate-range nuclear forces or INF and the pending strategic arms reduction talks or START). It passes over similar improvements in economic policies, from the reversal of the worldwide trend toward higher and higher inflation in the 1970s to the strong, investment-led character of economic recovery in the United States in the 1980s (especially from 1983 to 1985 and again from 1987 on) and the dramatic turnaround of American manufacturing productivity, which grew at an unprecedented pace of 4.1 percent per year from 1981 to 1987 (see Chapter 8). To ignore these improvements, however partial, or to attribute them to secular forces outside the influence of public policy is to abandon all standards of accountability for American policy choices.

United States policies in the 1980s attempted to assert domestic power, both economic and military, to alter international structures. The policies succeeded in part, as this study documents, in the economic area; where they failed, they did so less because of external constraints than because of the failure of the administration to follow through on its policies when it had the requisite power and political support (the case certainly with the budget deficit in 1985) or because it failed to link its policies to clear and consistent purposes that might have rallied and sustained the necessary political support and mobilized American power more effectively — the case perhaps with its initial international economic policies and with the defense program. America's purposes and policies, in short, had at least as much to do with actual outcomes as did structural constraints of power and process.

A Choice-Oriented or Voluntarist Model

The conceptual orientation in this study offers an alternative framework for critiquing U.S. foreign economic policies. It develops domestic and international concepts that are more voluntarist and that rely less on structural determinants. Three concepts in particular are central. First, the concept of national purpose defines the domestic political community that a nation represents and inevitably projects into the international arena. On the basis of these national purposes, nations compete to establish the terms of shared political community in international society; they do not just seek to balance power or interests that in turn constrain or define political preferences. A second concept, international economic policy, defines the specific content of four main economic policies by which each country relates to the international economy — domestic economic policy, exchange rate policy, trade policy, and financial policies. A country chooses the content of these policies and thereby creates, maintains, or destroys the constraints of economic interdependence, rather than merely adjusting to these constraints. Finally, a third concept in this study is the notion of a cocoon of nongovernmental and quasi-governmental organizations that surrounds the official, bureaucratic policymaking process and develops the larger ideas that penetrate and ultimately bring together the many diverse political interests in the policy process.

Debate about the nature of shared political community and the content of economic policies a country pursues goes on primarily within the cocoon of nongovernmental organizations. Here consensus is achieved, allowing pluralist policy processes to supersede special interests or class conflict and to advance common interests of human liberty and material progress. Politics, in brief, is much more than the narrow constraints of domestic bureaucratic infighting or official bargaining at international institutions. It is also more than some concept of the state in which certain officials act autonomously on behalf of the common interest. It is, in fact, the broader debate among competing ideas that cuts across state, bureaucracy, society, and international borders and galvanizes social and political coalitions on a largely voluntarist basis to support certain common notions of shared political community and to choose certain common guidelines for international economic policy.

The voluntarist concepts of purpose, policy, and ideas emphasized in this study do not exist in a vacuum. National purposes compete to define international political community, and in that context international power is a factor influencing the choice of shared political community. Economic policies may be more or less efficient, and in that sense some economic choices will cope effectively with an environment of scarce resources and others will not. Ideas about political community and economic policy may be more or less convincing and some will make the passage through politics to build supporting coalitions and influence bureaucracies and others will not. The conceptual approach in this study does not ignore structural constraints (see Chapter 2). In extreme situations, such as bankruptcy, these constraints become overriding. In most situations, however, structural factors are themselves subject to change.

Differences between structuralist and voluntarist views are matters of degree and emphasis. As Robert Gilpin notes, structuralist studies of the cycles of dominant or hegemonic powers have "underemphasized the importance of motivating ideologies and domestic factors . . . " (1987, p. 91). Paul Kennedy, in his epic study of the rise and fall of great powers, admits that he downplays the personal and leadership aspects of the historical story (1987, pp. 27, 135, 192 and passim). These differences of degree, however, are frequently decisive.

This study concludes, for example, that Reagan policies have had a greater impact on reality than on perceptions. Economic policies moved back, at least partially, toward the efficient premises of the Bretton Woods policy triad of price stability, flexible markets, and freer trade. These policies produced improved performance, especially in inflation and growth. Structural realities moved in the direction of policies, rather than policies being forced back in the direction of structural constraints. But Reagan officials never understood or articulated a larger intellectual framework within which U.S. policies and purposes fit together. They failed to shape a persuasive rationale to mobilize sustained public support, both at home and abroad. As a result, perceptions did not grow in line with performance. By the end of the Reagan years, for example, the public perceived a loss of American competitiveness, even though

American manufacturing productivity, a primary measure of a country's capacity to compete, had grown faster in this period than at any time since 1945.

The structuralist critique, by contrast, contends that Reagan rhetoric has had more impact than Reagan policies (Schneider 1987; Gourevitch makes a similar argument with respect to conservative policy shifts in all major industrial countries; 1986, p. 184). The rhetoric, it is argued, has given everybody a good feeling, while the policies have been repudiated by structural constraints. The result, according to this view, is a personally popular president (less so after the Iran-Contra affair) with policies largely *and* necessarily retrieved from the 1970s.

Structuralist arguments emphasize repetitive patterns of the past (Ikenberry et al., 1988, p. 5). The perspective in this book, however, regards the post–World War II era as a truly remarkable and perhaps unique period of human history. Progress in this period on both political and economic grounds has been far superior to that of any previous historical period. Interpreting this period, therefore, calls for something more than recounting the rise and fall of dominant powers in the past. The rest of this chapter looks briefly at the economic performance of the postwar system, in comparison with previous historical periods, and then in more detail at the voluntarist concepts necessary to understand the postwar period more fully in terms other than the simple rise and decline of American dominance.

Postwar Performance

When we examine the postwar period we find that it is distinguished from earlier periods of hegemonic rule by the unprecedented spread of politically legitimate institutions and by the achievement of both higher and more equitable levels of economic growth.[1] The claim can be made, in purely political terms, that the postwar international economic system has been one of the most legitimate in recorded history. Over 100 countries achieved independence in this period, and democratic institutions, which require regular and popular political legitimation, have spread to all industrial and many developing countries. This record contrasts sharply with the colonization of the world and suppression of political choice that characterized British hegemony in the nineteenth century and Dutch rule in the seventeenth.

What further sets the postwar system apart from these earlier historical periods is that it also achieved faster and more equitable distribution of growth. The latter comes as a bit of a surprise because liberal economic systems are often thought theoretically to lead to greater economic inequalities. Yet the postwar record shows the opposite result and may suggest that, when liberal economic policies and democratic processes of political legitimization proceed in tandem, both economic growth and equality may be enhanced.

[1]On the importance of both legitimacy and growth as criteria of success, see Katzenstein 1985, p. 29.

Arthur Lewis identifies two periods in world history of what he calls extraordinary growth — 1853–1873 and 1951–1973 (Lewis 1984, p. 15). Although data for the eighteenth and nineteenth centuries are not always reliable and techniques of measurement of growth vary and are controversial, Table 1-1 displays several series of growth rates for world trade and industry from 1705 to 1985.

In every series the post–World War II period stands out. The two separate series of growth of world industrial production calculated by Goldstein (first and second columns in Table 1-1) show higher growth rates for the post-1940 period than for any previous period with the exception of the years 1833 to 1856, for which Goldstein calculated two widely diverging growth rates (3.1 and 6.9 percent) from overlapping data sets. Rostow shows a rate of industrial growth for 1948 to 1971 (third column in Table 1-1) that is almost 35 percent higher than in any earlier period. Further data in Table 1-1 show that trade liberalization contributed substantially to the postwar growth record. Rostow's trade series (fourth column) records a 7.21 percent annual rate of growth of trade in the first two decades after World War II, 30 percent higher than in any previous decade. Katzenstein's series (fifth column) shows that trade in the 1960s grew in value twice as much as in any previous historical decade (133 percent per decade compared to 61–63 percent in the 1950s, from 1835 to 1845 and from 1890 to 1913), a result that is adjusted only slightly (to 100.9 percent per decade) when trade growth is measured in volume rather than in value (sixth column in Table 1-1).

In the 1970s and 1980s trade growth by volume slowed sharply, as did economic growth. Thus, trade growth correlates quite closely with output growth in the postwar period, and it seems hard not to attribute a considerable part of the total growth performance to the liberalization and expansion of trade under the Bretton Woods system.

The most impressive feature of postwar growth is the spread of benefits that occurred. Writers have identified for years the spread, as well as concentration, effects associated with liberal international economic policies (Myrdal 1968; Gilpin 1975). Yet the view persists that income gaps widen under market forces. Postwar statistics refute this view. According to World Bank statistics shown in Table 1-2, average annual growth of GNP per capita (in constant 1980 dollars) from 1950 to 1980 was approximately the same in some 60 middle-income developing countries (per capita income greater than $370 in 1979) as in the industrial countries — 3.1 percent versus 3.2 percent, respectively. When comparative purchasing power is taken into account, real per capita income in the middle-income developing countries actually grew more than twice as fast as that in the industrial countries (World Bank 1981b, p. 7). By 1980, the absolute income gap between middle-income and industrial countries was considerably less in purchasing power terms than in real terms ($6270 compared to $9080 — see figures in parentheses at top of Table 1.2).

These are hardly trivial outcomes, and they suggest that the *relative* gap between the industrial and middle-income developing countries was closing during the first three decades after World War II. Confusion arises because in

Table 1-1 World Industry and Trade Growth Rates from Selected Series, 1705–1985

Goldstein World Industrial Production (average annual percent)				Rostow World Industry and Trade (average annual percent)					Katzenstein World Trade (percentage growth per decade)		
	Series 1		Series 2		Industry		Trade			Value	Volume
1720–1746	2.7	1732–1746	2.7	1705–1785	1.5	1720–1780	1.10	1750–1825*		10.1	
1747–1761	-0.1	1747–1774	0.9	1780–1830	2.6	1780–1830	1.37	1825–1835		30.2	
1762–1789	2.1	1775–1798	3.2	1820–1840	2.9	1820–1840	2.81	1835–1845		61.5	
1790–1813	2.6	1799–1832	2.8	1840–1860	3.5	1840–1860	4.84	1845–1855		59.8	
1814–1847	3.4	1833–1856	3.1(6.9)	1860–1870	2.9	1860–1870	5.53	1855–1865		52.7	
1848–1871	3.4	1857–1877	3.3	1870–1900	3.7	1870–1900	3.24	1865–1875		53.7	
1872–1892	3.5	1878–1901	3.8	1900–1913	4.2	1900–1913	3.75	1875–1884*		43.4	
1893–1916	3.7	1902–1924	1.8	1913–1929	2.7	1913–1929	0.72	1881/85–1886/90*		42.0	
1917–1939	2.4	1925–1965	3.6	1929–1938	2.0	1929–1938	-1.15	1890–1913*		63.5	
1940–1967	4.0	1966–1975	4.0	1938–1948	4.1	1938–1948	0.00	1928–1938		-55.1	
				1948–1971	5.6	1948–1971	7.21	1951–1960		61.2	66.0
								1961–1970		133.0	100.9
								1971–1980		467.0	26.1
								1981–1985*		-4.2	25.6

Sources: First and second columns: Selected data from tables on pages 212 and 216 from *Long Cycles* by Joshua S. Goldstein, © 1988. Reprinted by permission of Yale University Press. The two time series are based on different data sets for the periods 1740–1850 and 1850–1975 and represent a base dating scheme (first column) and a lagged series (second column). Figure in parentheses in second column is a second growth rate for this period calculated from overlapping data in the two historical data sets.

Third and fourth columns: Rostow (1978), p. 67 and Appendixes A and B.

Fifth column: Figures for 1750–1938 are assumed to be value figures and are taken from Katzenstein (1975), Table 1, p. 1024.

Fifth and sixth columns: Figures after 1951 are calculated from various issues of GATT, *International Trade*.

*For periods shorter or longer than 10 years, per-decade figures are computed on basis of adjusted annual growth rates.

Table 1-2 Three Decades of Progress: Income, Health, Education, 1950–1980*

	Industrial countries	Middle-income countries	Low-income countries	Nonmarket countries
Income				
GNP per Person				
(1980 dollars)				
1950	4,130	640	170	
1980	10,660 (8,960)†	1,580 (2,690)†	250 (730)†	
Average Annual Growth Rate of GNP per Person (percent)				
1950–1980	3.2	3.1	1.3	
Health				
Life Expectancy at Birth (years)				
1950	67	48	37	60
1979	74	61	51	72
Education				
Adult Literacy Rate (percentage)				
1950	95	48	22	97
1976	99	72	39	99

Source: Adapted from World Bank, *World Development Report 1981* (New York: Oxford University Press, 1981), p. 6. Reprinted by permission.

*Figures in this table do not include China.

†The figures in parentheses are GNP per person levels based on purchasing power parity conversion. For middle-income and low-income countries, these figures include oil importers only. See World Bank (1981b), p. 17.

the early years of such relative gains, the absolute gap between the two income groups continues to widen.[1] But this initial increase in the absolute gap is not the complete story. If the differential rates of growth from 1950 to 1980 were to be sustained, a quick calculation shows that within another 70 years the absolute per capita income gap between the two sets of countries would disappear completely. Starting from a much smaller base, middle-income countries experience widening absolute gaps in the early years and then rapidly diminishing gaps in the later years. Gains for these countries follow an exponential curve. Thus, sustaining the postwar liberal economic system for an additional 70 years would eliminate altogether real per capita income differences between some 60 developing countries and the industrial world. This calculation assumes, of course, that population growth rate differentials would remain the same as earlier in the two groups of countries, a doubtful assumption given the

[1]For example, in absolute terms, GNP per person in the industrial countries grew from $4,130 in 1950 to $8,960 in 1980 (in terms of purchasing power), or a total of $4,830, and it grew in middle-income developing countries in the same period by $2,050. Thus, the absolute gap between the two groups of countries increased by $2,780.

tendency of population to grow ever faster in many middle-income countries. Nevertheless, by any standards, this is a truly remarkable historical result.

Middle-income developing countries achieved these gains in good part by participating in the liberal postwar international economic system. The developing countries that joined the system in the 1960s—the so-called newly industrializing economies—are precisely the middle-income countries that grew fastest in the postwar period (Nau 1985b) and that account for much of the results recorded by middle-income countries in Table 1.2.

None of the preceding data assuages the fact that a majority of developing countries, or some 90 low-income countries with per capita incomes of less than $370 in 1979, did not grow as fast as industrial countries from 1950 to 1980. For them real per capita incomes rose by only 1.3 percent per year, or less than one half the rate in the industrial and middle-income countries. Even after adjustment for purchasing power, these countries still experienced a relative decline.

Yet under the postwar system low-income countries made historically unprecedented progress in meeting basic human needs that are a prerequisite for modern economic growth—education, health, and nutrition. As Table 1-2 shows, adult literacy rates in these countries rose from 22 to 39 percent over the period from 1950 to 1976, and life expectancy went up from 37 to 51 years during the period from 1950 to 1979. Child mortality rates also declined during this period from 29 to 19 deaths per thousands in low-income Africa and from 24 to 12 deaths per thousand in low-income Asia (World Bank 1981b, p. 108). It must also be borne in mind that the low-income countries, almost without exception, chose development strategies that waived participation in an efficient international economic system, even after some middle-income developing countries joined the system in the 1960s. Although some may wish to refrain from criticizing low-income countries for this choice, no one should blame the international system for the outcomes, at least not primarily or exclusively.

After 1980 developing countries as a whole kept pace with real per capita income growth in industrial countries (1.8 percent per year compared to 1.9 percent per year in industrial countries for the period 1980–1987; see World Bank 1988b, p. 37). But performance among developing countries varied widely. The manufacturing exporters grew much faster than industrial countries, at a pace of 4.6 percent per year; whereas the highly indebted countries (seventeen altogether by World Bank definition, mostly in Latin America) suffered a decline of real income per capita of 1.3 percent per year, and the Sub-Saharan African countries tumbled deeper into poverty with a steep decline of 2.9 percent per year. This performance reflects both the legacy of inefficient domestic policy choices made by developing (and industrial) countries in the 1970s and the external consequences of vigorous, albeit partial, policy adjustments made by the industrial countries in the 1980s. Despite the negative outcomes for certain countries, the developing economies that fared best during this period—for example, China, Korea, Hong Kong, Singapore—depended most on international trade. In addition, low-income countries continued to make gains in life expectancy and infant mortality rates (World Bank 1988b, p. 220). Thus, the open, liberal international economy, even in an unusually volatile decade, still seemed to be good for development, as long as industrial and developing countries pursued the requisite domestic economic policies.

Acknowledging the unprecedented achievements of the postwar industrial economic system by historical standards is not to say that more could not have been achieved or that it *should* not have been achieved. But to do more requires understanding how much one has already done. How did these achievements come about and how might the rate of improvement be accelerated in the future?

Accounts that seek to explain this extraordinary postwar performance in terms of preponderant American power fall short on two counts. They cannot explain the increasing homogeneity of basic political values and institutions within the noncommunist world and now also with some communist countries, which has grown stronger in the postwar period as American power has grown weaker. And they cannot explain the sharp drop in economic performance in the postwar system after 1967, when American power among the industrial countries declined only modestly from 1950 to 1965 and then not at all from 1965 to 1985 (see Tables 2-1 and 2-2). The concept of national purpose helps us to understand the first point; the concept of efficient economic policies, the second. And the concept of the cocoon of policymaking explains the process by which the content of national purpose and international economic policy changed during different periods of the postwar era leading to different performance outcomes. In the rest of this chapter, we examine each of these concepts in greater detail.

National Purpose

The concept of national purpose captures the essence of any political society. It concerns the way societies, including nations and groups of nations, choose (the voluntarist assumption in this study) to organize their political relationships with one another. These choices, about which philosophers have disagreed at least since the time of Aristotle, involve two basic issues: (1) What aspects of political life in any given group, country, or international system are going to be reserved for private decision making and initiative? (2) How are those aspects that remain public to be decided? The first issue concerns the central question of human and civil rights in political life; the second involves the question of whether the voting franchise for public decision making will extend to all or most individuals or only to a small, politically privileged elite.

Groups and nations in varying historical circumstances and economic conditions have made different choices on these issues. The national group, as we know it today in the West, emerged in the sixteenth and seventeenth centuries largely as a result of the decision to remove the matter of religious choice from the province of the universal Catholic Church and vest it in the hands of national sovereigns. In 1648, the Peace of Westphalia generally marked the end of the notion of a universal Christian community in which "the Holy Roman Emperor had claims over the sovereignty of all other rulers . . . " (Morse 1976, p. 25). After this period, sovereigns acquired full authority to decide the basic issues of private rights and public decision making within their territorial jurisdiction, including the religious preference of their subjects.

Roughly over the same period, a basic change also took place in the nature of the political franchise within Western states. In medieval society, property was the exclusive right of the feudal lords, a claim made legitimate by religious and other customs. In the modern society that emerged around the sixteenth and seventeenth centuries, property became the exclusive right of mutually independent sovereigns. As religious and other medieval customs weakened, sovereign monarchs had to work out a new basis of property rights with feudal lords and a rising merchant class. This legitimization crisis, as John Ruggie points out, gave rise to the "works we regard today as the modern classics in political theory and international legal thought" (Ruggie 1986, p. 144).

Political theory and subsequent practice produced a variety of answers to the question of how to reconcile private property and public authority. The answer that emerged in much of Europe and subsequently in the United States was the Lockian notion of natural individual private rights, the idea that certain rights, including that of property, existed prior to political society and that public authority existed to secure these rights, not to interfere with or confiscate them. Over time, the process of safeguarding private rights against public authority led to the evolution of democratic institutions and processes whereby not only private rights were safeguarded but public decisions of all kinds were made on the basis of an increasingly widening franchise. Two key features thus came to characterize the democratic approach to political society — the reservation of important political rights to private individuals and the making of public decisions by broadly representative participatory processes.

The modern nation-state is in fact the cumulative expression of choices individuals and societies have made over time on these issues of public and private rights and democratic responsibilities. Some of these choices reflect continued centralization of power in the hands of the state; others reflect democratic systems or authoritarian systems in evolution toward democracy; still others involve reversions from democratic institutions back to authoritarian systems. Out of these choices arise contemporary confrontations between totalitarian and democratic societies. The nation-state today is the principal expression as well as defense of varying group choices on the ageless question of political community. Contemporary states now include societies from non-Western as well as Western traditions.

In the absence of widespread agreement on these matters, nation-states compete to define the political nature and organization of international society. What actions in the world community are to be considered the sovereign domain of states, and what actions require international decision making and how will these decisions be made? Do states have an absolute and unconstrained authority over their citizens, or do certain universal human rights (e.g., minimal protection of the physical safety of the human being) exist independent of political society, both national and international, that must be protected by that society? On issues that are properly in the domain of national or international public authorities, how will public decisions be made—by law, by class, by vote of all individuals or only some, by war?

International society today has to address the same political issues that national societies have been deciding for four centuries or more. In this sense, international politics is no different than domestic politics. The content (human rights) and mechanisms (law, vote, war, etc.) of political life in each setting differ only in degree, not in kind. Politics is a seamless web, despite the commonplace notion in textbooks of international politics that anarchy is the distinguishing feature of international life (Waltz 1979). Before national societies formed, anarchy was also the distinguishing feature of domestic politics, and it becomes the distinguishing feature again when national society breaks down (i.e., when there is civil war). Domestic societies may appear to be more stable, but that is largely because the breakdown of community in the domestic setting has fewer consequences than the collapse of order in the international system.

In this study national purpose defines a society's approach to the two key issues of international politics—the *domestic* political content (private rights) and the international *diplomatic* mechanisms (international organizations, protocols, law, etc.) that independent nation-states share in coming together to form international society. In the postwar period, America's definition of national purpose has gone through various phases. Different emphasis has been placed on the domestic and diplomatic content of international society.

The historical chapters of this study (Chapters 3–10) detail the evolution of different U.S. conceptions of national purpose. Table 1-3 summarizes this evolution. Under isolationism prior to World War II, America placed exclusive emphasis on the domestic content of international politics to the exclusion of any active diplomatic role. America's domestic identity recoiled at the idea of any diplomatic involvement or relationship to the world community (Schneider 1987). During and immediately after World War II, America went to the opposite extreme and entertained a universal-utopian (Wilsonian) vision of world community that stressed diplomatic relationships structured around common functional problems (e.g., food, health, money, trade) and completely ignored domestic political differences.

Containment in 1947 brought the pendulum back again to a focus on domestic differences but with an active international diplomacy to defend and express America's and the free world's concept of individual political freedom against the totalitarian tenets of Communism. Detente carried diplomatic efforts one step further to build a global web of interdependence of political and economic relations between East and West in which societies with fundamentally different political systems might coexist in a nuclear world and might even, as Wilsonian advocates envisioned, resolve some of their political differences and demilitarize international life.

In the early 1980s, Reagan's policies emphasized once again the competition between different domestic conceptions of political community in East

Table 1-3 Varying Definitions of U.S. National Purpose

Dimensions of International Political Community	Isolationism Pre–World War II	Wilsonianism 1942–1946	Containment 1947–1967	Detente 1967–1979	Reagan Policies 1979–1987	Era of Liberalization 1988–?
Domestic content	Uniqueness of American politics	No basic differences	Freedom versus communism	Coexistence and possible long-term convergence	Freedom versus communism	Deepening liberalization in Western Europe and initial liberalization of Eastern societies toward more pluralist and competitive standards
Diplomatic construction	Noninvolvement	Functional problems	Military alliance and economic integration within the West	Web of political and economic interdependence capping military race and possibly demilitarizing international relations	Military and economic strength of the West leading to balanced arms reductions, including possible denuclearization of East–West relations	Reduction of military forces, including possible denuclearization of central Europe and integration of Eastern countries into the international economic system

and West but sought diplomatic accommodations that might limit rivalries, especially military ones, without any necessary expectations of resolving fundamental differences or demilitarizing international life (at least, in the case of nuclear arms, until Reykjavik). After 1985, the unprecedented domestic reform initiatives by Mikhail Gorbachev in the Soviet Union raised the Wilsonian prospect once again, albeit problematic and long-term, that political society in communist countries may evolve sufficiently toward Western standards that military tensions may be reduced and international cooperation organized primarily around functional problems of economic relations.

The varying definitions of America's national purpose supplied the motivation for America's policies in the different periods, both military and economic. From the perspective of this study, military and economic motivations follow conceptions of shared political community rather than leading them. Foreign policy is not just a response to material power or market pursuits; it advances a conception of political community for which power and wealth is desired (Krasner 1978, p. 10). Thus, national purpose goes beyond traditional realist concepts of national security and national interests, as well as pluralist concepts of interest groups and bureaucratic politics. It rejects the determinism of power (realists) or wealth (Marxism) as well as of technology (often emphasized by contemporary liberal perspectives). It insists on a political or moral dimension in international politics both in nonthreatening circumstances, where neorealist perspectives concede a role for nonmaterial influences (Krasner 1978, p. 41), *and* in circumstances of national threat and survival. From this perspective, national security does not justify the survival of any set of national purposes. As Robert Osgood wrote, "national security pursued without relation to ideal goals may lead to the sacrifice of individual freedom, social purposes and other transcendent values which make security worth having in the first place" (1953, p. 444). The question is always what political purposes states advocate when they have sufficient power to maneuver freely or when they face tests of conflict and survival.

The concept of national purpose raises a serious issue for contemporary academic analysis. Structuralist approaches deal with it in various ways. The purest realist or neorealist theories duck the issue altogether. They consciously expunge internal attributes of national actors, such as domestic political values, from their theoretical framework (Waltz 1979). This is done in the interest of parsimony of analysis and theorizing across centuries, if not millennia. But this approach, which underlies the fascination with long-term cycles of international power and prosperity, deprives human history of moral content. Every state has the right to arm and wrap itself in the moral mantle of national security whether it brutalizes its citizenry or not — that is, whatever its conception of the domestic content of international society. Hitler's Germany and the apartheid state of South Africa are treated in the same way as any other state.

From this perspective, states fight only about power and wealth. If morality enters the picture, it does so only in the form of prudence. As long as states do not aggress, they can treat their own citizens in any manner they choose. In

June 1989, China brutally suppressed student demonstrations in Tiananmen Square and invoked the doctrine of realism to declare to the international community that this action was none of the world's business.

Other structuralist studies pay attention to both circumstances — the material costs and benefits of rational behavior — and social norms (Gilpin 1981, p. xii). But by their own admission, these studies are "founded on a pessimism regarding moral progress and human possibilities" (Gilpin 1986, p. 304). "I am not even sure," Robert Gilpin writes, "that progress exists in the moral and international spheres" (1986, p. 321). Thus, moral purpose rises and falls in historical cycles just like power rises and falls, and one dominant power is just like any other. The pessimism about purpose in international politics leads to a preoccupation with process. Game theorists, for example, representing still another structuralist approach, model international relations exclusively in terms of conflict and study the prospects for cooperation irrespective of the particular substantive content of such cooperation (see Chapter 2).

The concept of national purpose in this study comes closest perhaps to Ernst Haas' concept of nationalism. "Nationalism," he argues, is " . . . a civil religion . . . that contains a set of core values that, whether for objectivist or subjectivist reasons, come to be accepted by the population of a state . . . " and "become the definers of selfhood" for that state (1986, p. 709). Such a civil religion, Haas argues, rationalizes political life within a given society; but it also opens up a Pandora's box in international politics because "various national selfhoods are arrived at by mutual exclusivity and outright hostility" (1986, p. 711). Haas considers this international hostility to derive from innate dogmas of various civil religions, rather than from an inevitable competition among them, to define the political terms of international society. He is therefore reluctant to consider any of the historical concepts of national "selfhood" that have been tried thus far as "authentic," "true," or "legitimate" (1986, p. 714). He hopes that all nations, whatever their national self-conception, will eventually "encounter difficulties related to technology, welfare commitments, and international economic interdependence" (i.e., structural constraints) that will cause previously accepted national self-conceptions to deteriorate and permit new choices to be made based on autarchy, regionalism, international regimes, national decentralization, or a new global identity involving class or religion (1986, p. 742). Haas seems to be putting his hopes on the processes of modernization (technology, economic interdependence) and a benevolent politics of socialism (welfare commitments) to define civil religions in the future that either will not conflict with one another or will be able to coexist at historically unprecedented levels of peace and cooperation.

The concept of national purpose in this study shares with Haas' notion of civil religion the idea that definitions of national selfhood change and thus affect the prospects of shared political community. But it questions why it is not possible now to pass some judgment on the relative merits of past choices of "civil religions" because such choices in the future must continue to contend with the same issues of how private rights and public decisions are apportioned in national and international society that have applied in the past. Although this study also would not lay the mantle of "true" on any particular definition of national purpose, it would argue that some definitions have been better than others in terms of their consequences for human progress. By labeling past conceptions of selfhood "nationalisms," Haas makes it easy to reject them all, while still hoping that technological and economic change will somehow make future definitions "better."

Judgments about national "civil religions" or purposes are difficult to make, to be sure, and this study, although it does make judgments, does so with great modesty. In a nuclear age, survival is also a moral value. Tension arises between the moral value of survival and the moral values (e.g., freedom) for which we seek to survive. Charles Krauthammer argues that, although individual self-sacrifice is morally justified, self-sacrifice by an entire civilization is senseless. He agrees with Jonathan Schell that survival must take precedence because the pursuit of all other values depends on it (Krauthammer 1984, p. 17; Schell 1982). In this sense, a civilization chooses survival in order to live on and fight for other values another day. But this discussion *in extremis* hardly means that our analytical frameworks, by which we judge more common events, should suspend all moral evaluation of the internal choices societies make that define their national purpose from within and that express this purpose in their foreign policy relations toward the outside. Although no society in a nuclear age can afford to press its moral claims to the point of threatening the survival of another society, and therefore its own survival, moral issues do not thereby cease to exist. Some societies (e.g., those that practice human genocide) do not have a *moral* claim to exist, although in a nuclear world, other societies may not have a moral right to destroy them. Our analysis of international relations must continue to address both the domestic and diplomatic dimensions of global political society—the former dealing with the moral nature of that society and the latter with its survival in a nuclear age.

In this study, therefore, the concept of national purpose has both analytical and prescriptive dimensions (as all concepts do; see Alker and Biersteker 1984). It is used to describe and differentiate various U.S. definitions of the domestic and diplomatic content of international politics over the postwar period (see Table 1-3) and ultimately to judge which of these definitions seemed to correlate best with the advancement of human rights (the domestic component) and participatory decision-making processes (the diplomatic component) in world politics. Seen from this perspective, America's security and foreign economic policies have always discriminated and must continue to discriminate between liberal and illiberal societies. Containment, which in its original form projected a vision of shared Western community based primarily on political and economic factors but backed by adequate military power, was by far the most successful postwar framework for advancing U.S. aims. Detente rescued the political and economic spirit of early containment from its subsequent and excessive militarization in Vietnam, but in the process detente also let the sinews of American military power atrophy. Reagan's policies restored military strength and put America on the side of democracy in both rightist (Philippines, Panama, Haiti, etc.) and leftist (Nicaragua, Cuba) third world countries.

But the administration's credibility suffered after Reykjavik, when it proposed the elimination of nuclear arms and thus the elimination of existing deterrence strategies to defend the West, and after the Iran–Contra affair, when it collaborated with a totalitarian and terrorist state.

Not every conception of national purpose fares as well as any other. The conception must mesh with the domestic and diplomatic realities of world affairs. If it does not, the country projecting its national purpose will find either that it extends this conception too far and arouses the political opposition of other societies with different conceptions of national purpose or that it fails to defend itself sufficiently and succumbs to the different national purposes of other societies. National purpose, in short, is not the same as ideology. It is a concept that recognizes the realities of power but does not concede that power is all that there is in international politics.

International Economic Policies

The shared political community that arises from the interrelationship of competing national purposes establishes the framework within which international economic activities occur. Four questions follow that influence the extent and efficiency of economic activities within that framework: (1) Are the scope and intensity of political community sufficient to support extensive economic exchanges? (2) Does the content of political community in any way limit the choice of economic policies within the economic system? (3) What are the various economic policy choices nations have to make to organize and operate an international economic system? (4) How efficient are these choices and what consequences do they have for economic growth and equity?

Political Framework for Economic Exchanges

This study argues that from 1945 to 1947 the political framework for postwar economic exchanges was insufficient to support extensive economic activity, either between East and West or within the Western world. In the winter of 1947–1948, however, such a framework emerged within the West and subsequent economic policy choices made by Western countries established the efficient policy premises of the Bretton Woods system. These choices had something to do with the unprecedented economic performance from 1947 to 1967, and policy choices made after 1967 contributed to the decline of that performance.

Economic exchanges require a certain amount of mutual trust or political community (Etzioni 1988). If political societies differ radically, economic exchanges will be limited, as has been the case since World War II in East–West relations. East–West trade has been limited both by the different emphasis that the domestic political systems in East and West placed on efficiency and growth (i.e., change) and by the strategic context in which the West has relied on a military technological lead to deter conventional military advantages of the East in central Europe (see Chapter 10).

Even among Western countries, a shared political community is necessary to support extensive economic interdependence, at least to the degree interdependence exists today. In the interwar period such a political community did not exist, and in 1945, although the United States helped establish the Bretton Woods system of global economic institutions, this system did not function, because the allies did not share sufficient common political purposes and the United States in particular lacked sufficient motivation to finance the system (see Chapter 3).

America was dominant in both periods, so power alone cannot explain the presence or absence of such incentives. The existence of an uncommon power, in fact, does not appear to be as much a prerequisite of a common economic community, and particularly a common currency, as confidence in the collective purposes of that community. The European Community (EC), for example, pursues common purposes and a common economic system today without one country that dominates in both economic and military areas.

The dollar became the world's key reserve currency after 1947 and remains so today, despite a relative decline of American power, because Western countries agreed then and continue to agree broadly today with American purposes in the world community. From this perspective the ECU (the common European Community monetary unit) and yen will become more meaningful reserve currencies in the world economy only to the extent the EC and Japan succeed in defining broader political purposes that inspire international confidence. A reserve currency, which serves as a store of value for international capital, is a direct expression of confidence in the political and economic policies of the issuing country.

Competitive and Command Economic Systems

The second question raised by the existing political framework is whether it constrains the type of economic system that emerges. In premodern societies, religion and other social mores severely circumscribed economic exchanges and limited them largely to reciprocal arrangements—gift giving and barter. The transformation of medieval society in the sixteenth and seventeenth centuries introduced alternative economic systems. Redistribution and mobilization systems organized the production, storage and distribution of goods through central institutions, for the benefit of either elites or the state, whereas market economies organized economic exchanges on the basis of openness and competition primarily for the benefit of individual consumers (Gilpin 1977, pp. 20–23).

After World War II, different political communities established in the East and West led to different economic systems. The Soviet Union, dominant in Eastern Europe, created a command economy resembling that of a redistribution or mobilization system. The United States, dominant in the West, created a market economy system. The type of political system clearly constrains some aspects of the economic system. A command economy, for example, can be efficient only if it enjoys unchallenged support from the population, as it did in Japan and Germany in the 1930s. However, such support has not prevailed in the Soviet Union and Eastern Europe since World War II.

Some writers believe that competitive or market systems are linked historically to the rise of free societies. In their account of *How the West Grew Rich*, Nathan Rosenberg and L. E. Birdzell, Jr., describe the connection as follows:

> Initially, the West's achievement of autonomy stemmed from a relaxation, or a weakening, of political and religious controls, giving other departments of social life the opportunity to experiment with change. Growth is, of course, a form of change, and growth is impossible when change is not permitted. And *successful* change requires a large measure of freedom to experiment. A grant of that kind of freedom costs a society's rulers their feeling of control, as if they were conceding to others the power to determine the society's future. The great majority of societies, past and present, have not allowed it. Nor have they escaped from poverty [1986, p. 34].

Relatively competitive political and economic systems did characterize the three periods of most rapid historical growth under Dutch, British, and American auspices (the mid-seventeenth century, the mid-nineteenth century, and the period immediately after 1945—see Table 1-1). And one may be inclined to agree with Rosenberg and Birdzell that it is doubtful whether a non-Western society "could be as innovative as the West without using the main substantive features of private ownership of the means of production and without curtailing central authority over the uses of the means of production so greatly as to make the feasibility of planning dubious" (1986, p. 30). Political leaders in the Soviet Union, Eastern Europe and, before 1989, in China may also agree, as they experimented themselves with private incentives and market decentralization to revive stalled economies.

Yet if political systems act as ultimate constraints on economic choices and growth, they probably do so no more directly than other structural factors. From the perspective of this study, such structural factors matter less than policy choices. Non-Western societies are free to choose competitive economic policies even under systems of public ownership of property. The shift to market competition may entail political consequences, as Soviet and Chinese leaders experienced in the late 1980s. But it is probably misleading to argue that choosing markets necessarily implies choosing political freedom. As Milton and Rose Friedman observe and the discussion in Chapter 10 confirms, "voluntary exchange may prevent a command economy from collapsing, may enable it to creak along and even achieve some progress . . . [but] it can do little to undermine the tyranny on which a predominantly command economy rests" (Friedman and Friedman 1980, p. 11).

From the perspective of this book, the intensity of political communities is more important for growth than their inherent structure (e.g., the definition of property rights). Totalitarian systems have intense political communities that enable them to establish single-purpose objectives and mobilize immense resources, through both internal repression and external expansion, to achieve their objectives. Economic policy in this centralized approach can be efficient as long as resources are not a constraint and the political system retains its internal legitimacy.

Industrial democracies in the West mobilize resources through competition rather than by central direction. The key concept in the competitive model is opportunity costs (Rhoads 1985, Chapter 2). The most efficient use of resources is decided not on the basis of elite command or consensual decision making by an entire society (e.g., the idealized model of Japan, Inc.), but on the basis of alternative calculations of costs and benefits by competitive actors. Competitive calculations ensure that the use of scarce resources will be optimal in general rather than just within a single, centrally determined objective. In a command economy, there is no mechanism by which alternative objectives can be raised and evaluated, except perhaps by an omniscient central bureaucracy. Economies operating on command principles therefore tend to feel resource constraints sooner than competitive economies, contributing also perhaps to the tendency of command systems to repress their own populations or expand abroad to secure additional resources.

No market economy is purely competitive, of course, any more than a command economy is totally centralized. The difference is a matter of degree, but this difference is also significant. Market economies exist in a context reflecting various influences — geography, infrastructure, institutions (public and private), and more immediate government policies. These influences range from those that are relatively immune to government influence (e.g., geography) to those that can be affected by government policies only over a period of a generation or more (e.g., infrastructure and institutions) to those that are more immediately affected by government policies (e.g., macroeconomic conditions). Table 1-4 identifies these influences.

Some studies collapse the various factors in Table 1-4 and see outcomes largely determined by long-term and deep-seated institutions or social coalitions relatively immune to government policy (e.g., Hall [1986], who equates market policies with institutions, and Ruggie [1986], whose concept of social purpose embraces social formations that endure for centuries). This study, by contrast, emphasizes the role of more immediate government policies that affect directly the price of goods and capital assets in international exchanges (see the last column of Table 1-4) and that seek to place broad limits on the behavior of governments in areas where institutional differences exist (see the third column of Table 1-4, e.g., limits on subsidies to state-owned firms).

If policy or institutional differences become too large, they inhibit significant economic exchanges. For example, if governments pursue diverging exchange rate policies, as they did in the 1930s and to a lesser extent in the 1970s and early 1980s, they discourage stable trade and investment flows. Or if countries differ too greatly in domestic traditions and institutions, as Eastern and Western countries have done in the postwar period or as some analysts now claim the United States and Japan do, they may be unable to accept common

Table 1-4 Factors Affecting International Economic Competition

Endowments	Infrastructure	Domestic Institutions	More Immediate Government Policies
Natural resources	Transportation Roads Ports Railroads Airlines	State/society relations Role of private vs. public actors in various activities Production R & D Consumption Finance	Domestic policies Macroeconomic Fiscal Monetary Overall regulatory (e.g., antitrust) Microeconomic* Labor regulations Sector-specific policies (e.g., for agriculture, financial services, etc.) Subsidies to industry
Population	Communications Telecommunications	State policymaking Degree of centralized vs. decentralized administration	
Climate	Energy and power		Exchange rate policies
Land		Society networks Market structures (e.g., competitive vs. oligopolistic) Corporate structures (e.g., large vs. small, management–labor–bank relations, etc.)	Trade policies Tariffs Nontariff barriers (e.g., health and safety standards)
Human capital Literacy/Education Nutrition Health			Financial market and official exchange reserve policies

*Microeconomic policies are also commonly referred to as structural policies. We use both terms in this study but the term *structural* should not be confused with the term *structuralist*, which refers to broad conceptual approaches to international economic relations discussed in Chapter 2.

rules on more immediate government policies to, for example, maintain convertible currencies or liberalize trade.

Efficient Bretton Woods Policy Choices

Thus, a third question affecting the character of the international economy emerging within a given political framework is what specific economic choices nations make with respect to policy and institutional areas where government influence is most immediate. These choices determine how open or closed and hence efficient the international economy will be. The Bretton Woods policy choices developed after 1947 set rules for immediate government policies that encouraged international competition and required domestic economic policies to adjust to this competition. The Bretton Woods rules, whose origins are examined in detail in Chapters 3 and 4, fixed exchange rates and called for lowering trade restrictions primarily on manufactured goods. In addition, these rules created only modest international financial resources, in the form of quotas in the International Monetary Fund (IMF), to finance international balance of payments deficits.

Under these rules, governments, facing limits on their ability to change exchange rates, raise trade barriers, or borrow on international markets, had to resort primarily to changes in domestic macroeconomic and microeconomic policies to correct external balance of payments deficits. If a country runs a deficit on its current account, it is either losing foreign exchange reserves or borrowing an equal amount on international markets through the capital account. The balance of payments, which includes the current and capital accounts, must by definition balance (Table 1-5). Unless the country's foreign exchange reserves are massive or financial resources (bank loans, foreign investment, aid, etc.) are easily and widely available in the international economy, the country with a current account deficit must eventually adjust other economic policies to reduce the deficit and its related borrowing. Since the Bretton Woods rules placed firm limits on the use of exchange rate and trade policies and the United States, which provided the principal financing for the system through the Marshall Plan, conditioned this financing on domestic

Table 1-5 Components of the Balance of Payments

---

1. The current account, which includes
   a. The merchandise trade account (i.e., all trade in goods)
   b. The services account (travel, banking, insurance, transportation, investment income, etc.)
   c. Unilateral transfers, principally private remittances and government transfers
2. The capital account, which includes all transactions involving short- and long-term financial claims or liabilities abroad, both by government and private sources.*

---

*In a fixed exchange rate system, the capital account excludes foreign exchange reserves, which appear in the official settlements account; in a clean floating rate system, exchange reserves remain constant and thus the capital account directly equals (with an opposite sign) the current account.

policy changes (see Chapter 4), the postwar system after 1947–1948 relied primarily on domestic economic policy adjustment to correct deficits or surpluses in the external balance of payments. Countries were asked to reallocate resources through domestic macroeconomic and microeconomic policies rather than through exchange rate changes, trade restrictions, or unconditional external borrowing.

The heart of the postwar system, therefore, was not finance or even fixed exchange rates, as is often assumed because of the identity of the system with the Bretton Woods institutions of the IMF and World Bank. Finance was relatively limited and conditional, and fixed exchange rates were possible only because of a more fundamental, underlying commitment to domestic economic policies that stabilized domestic prices, especially on the part of the United States, whose currency became the principal reserve asset. The heart of the system, as we see in Chapter 4, was relatively conservative fiscal and monetary policies, to preserve domestic price stability; incremental trade liberalization, to spur competition; and restrained microeconomic intervention, to permit flexible labor, capital, and product markets to reallocate resources as needed to alter the level of savings or investment in the economy.

These three premises constitute what I refer to in this study as the Bretton Woods policy triad: price stability, flexible domestic markets, and freer trade. The first premise to maintain price stability and the exchange rate of the dollar was a U.S. commitment to exchange dollars for gold at a fixed rate, namely $35 per ounce. But, given the inevitable limits on gold supplies — even the massive U.S. gold reserves after World War II — it was also a commitment to back the dollar with American goods and capital assets at a stable domestic price. As long as American dollars were worth their stated value in terms of gold *or* the equivalent in stably-priced American products, dollars retained a fixed relationship to real assets. Other countries assumed no obligation to back their currencies with gold but pegged their exchange rates to the dollar. This step also implied a commitment to maintain domestic price stability. If their domestic, including traded goods, prices went up in comparison to U.S. prices, their competitiveness would suffer and they would be compelled at some point to correct a balance-of-payments deficit by bringing their prices back in line with U.S. prices. Under the postwar rules, therefore, all countries incurred a basic responsibility to pursue noninflationary domestic economic policies. That commitment translated, as we see later, into relatively conservative fiscal and monetary policies throughout the Western world from 1947 to 1967.

The second commitment to lower trade barriers was not absolute (and was not even specified in the 1944 Bretton Woods agreements but developed in practice after 1950; see Chapter 4). It was a commitment to liberalize trade at the margins, that is, to take incremental steps toward lower barriers, not necessarily to wind up in a completely free market. It was also partial. It did not include agriculture, investment, or portfolio capital flows. Even in manufactured goods, it became clear and consistent only over time. With all its limitations, however, the commitment to freer trade was central to spurring competition. Postwar reconstruction began behind formidable protectionist and monopolistic barriers in most Western countries, the residue of both nationalistic economic policies in the 1930s and wartime policies. These barriers severely limited competition and impeded efficient use of resources. Although the commitment to remove such barriers was ambiguous on the question of how much free trade overall was a good thing, it was unambiguous on the conviction that *more* free trade was a good thing.

The third commitment to restrain government intervention in specific microeconomic areas such as nationalization of production or direct subsidies to industry did not necessarily imply laissez-faire policies. It implied only that countries would retain the capability through flexible labor, capital, and product markets to respond to noninflationary fiscal and monetary policies and growing external trade competition by moving resources from declining to advancing industries. Although by the late 1940s almost all Western governments had accepted some responsibility for actively managing macroeconomic policy, they differed in their propensity to intervene at the microeconomic level. France, Germany, and Japan, for example, had inherited systems from the prewar period that featured highly centralized administrative or corporatist (close ties between government, banks, and industry) industrial structures. Although, as we have noted, such command-oriented economic systems can be efficient if they are unconstrained by scarce resources or participate in competitive international markets (the latter having been the case apparently for small European states in the postwar period [see Katzenstein 1985] and for the more centrally directed development strategies of Japan and the newly industrializing economies in the same period), these systems can also be less efficient if governments indulge their greater proclivity to pursue noneconomic objectives. Government-oriented or -directed industries may be able to postpone payment of debt because they have greater access to financial resources through the broad taxing and monetary powers of governments, or they may be less inclined, for political reasons, to lay off workers and to modernize or relocate industrial plants. Thus, the more extensive government intervention is in an economy, the more that economy's behavior may be determined by factors other than competitive market signals and the more inflexible the economy may become. The policies promoted by the Marshall Plan recognized this possibility and actively discouraged extensive microeconomic intervention in Europe (less so in Japan; see Chapter 4).

The economic policy choices under the Bretton Woods system, as that system evolved from approximately 1947 to 1967, were thus highly efficient. The premise of freer trade ensured competition, especially for smaller countries; the premise of price stability ensured a stable environment for domestic investment and stable exchange rates for expanding trade; and the premise of flexible domestic economies ensured prompt adjustment to changing market conditions and comparative advantage. Financing facilitated adjustment to balance-of-payments disequilibria, but it was not so generous or unconditional that adjustment was unduly delayed.

Linking Economic Policy Choices to Performance

The fourth issue in evaluating international economic policies in a given political framework is to link policy choices in some rough manner to economic
performance. This study contends that the policy choices that operated in the
Bretton Woods system from 1947 to 1967 may be plausibly adduced to account
in good part for the unprecedented economic growth of this period and that
subsequent changes in policy choices from the mid-1960s on may also be
associated with a slowdown in economic performance in the 1970s. Since 1979,
the study argues, further policy changes have restored some of the earlier
provisions of the Bretton Woods system, particularly in the areas of monetary
policy and deregulation of some labor, capital, and product markets. But fiscal
and trade policies have continued to deteriorate, especially in the United States,
and inflexible labor and product markets continue to restrain growth and
contribute to high unemployment, especially in Europe. The partial return to
efficient policy premises, it is argued, explains both the successes and shortcomings of the most recent period of U.S. foreign economic policymaking.

Figure 1-1 offers a summary, statistical snapshot of the relationship during the
various periods of the postwar era between economic policy choices by the major
industrial countries and the economic performance of these countries. Similar data
for other countries, including developing countries, appear in the appendixes
to this book. Chapters 3–9 supply the historical details behind these numbers.
Here it suffices to note that an obvious association exists between the Bretton
Woods policy triad of conservative macroeconomic policies, restrained microeconomic policy, and incremental trade liberalization, on the one hand, and
economic performance among the principal industrial countries, on the other.

From 1947 to 1967, macroeconomic and microeconomic policies in the
United States (see IA of Figure 1-1) and the G-4 countries (France, Germany,
Japan, and the United Kingdom — see IIA) were exceptionally restrained, especially when compared to subsequent periods. As measured by budget deficits as
a percentage of GDP, fiscal policies were nearly balanced (-0.2 and -0.7,
respectively). Money growth (M_1 and M_2) was expansive in the G-4 countries
but (by later standards) unusually tight in the United States (2.5 and 5.3 percent per year, respectively), which was supplying the reserve currency for the
system and was therefore the principal guarantor of price stability. The growth
of the public sector, as measured by government expenditures as a percentage
of GDP, was actually negative in the G-4 countries (on a basis of percentage
point change per decade), whereas its growth was modest in the United States
(1.6 percentage points per decade). Trade barriers descended rapidly in this
period, particularly in Europe in the area of nontariff barriers or quantitative
restrictions (see IIA in Figure 1-1). Corresponding to these policies, the growth,
inflation, and unemployment record of the period from 1947 to 1967 was
exemplary (see IB and IIB in Figure 1-1). Growth was higher than in any
subsequent period (except for the United States, when it was higher for the
period 1983–1987), whereas inflation and unemployment were lower (except
for inflation in the G-4 countries, where it was lower still in the period from
1983 to 1987).

I. United States

A. Policies

Budget Deficits

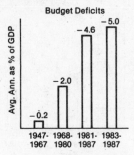

Monetary Policy
(M_1 = ☐; M_2 = ▨)

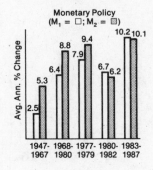

Growth of Public Sector

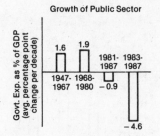

Trade Barriers

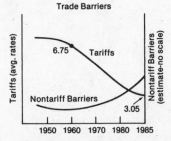

B. Performance

Real GDP

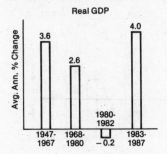

Inflation

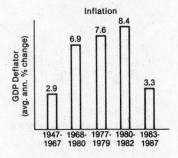

Unemployment

Figure 1-1 Economic policy and performance indicators, 1947–1987. (Sources: See Appendix tables A-1, A-2, A-3, A-9, and A-10.)

II. G-4 (France, Germany, Japan and United Kingdom)

A. Policies

B. Performance

Budget Deficits

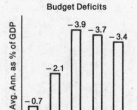

Real GDP

Monetary Policy
$(M_1 = \square; M_2 = \boxdot)$

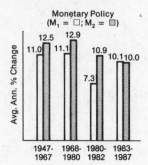

Inflation

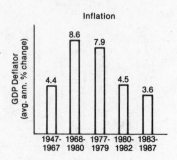

Growth of Public Sector

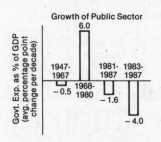

Unemployment

Trade Barriers

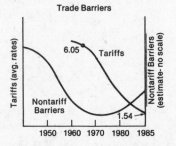

Figure 1-1 (continued).

From 1968 to 1980, fiscal and monetary discipline was lost in both the United States and G-4 countries (see IA and IIA in Figure 1-1). Budget deficits multiplied tenfold in the United States and threefold in the G-4; money growth rates roughly doubled in the United States; and the public sector exploded in the G-4 countries, from -0.5 to $+6.0$ percentage points per decade. Tariff barriers came down sharply, but nontariff barriers began to reemerge, particularly in the form of voluntary export restraints. Correspondingly, growth slowed noticeably, inflation more than doubled, and unemployment skyrocketed (see IB and IIB in Figure 1-1).

Since 1980 the picture has become more mixed. Policies reverted partially to features characteristic of the initial postwar period (see IA and IIA in Figure 1-1). Monetary policies, especially in the early 1980s, contracted sharply in both the United States and the G-4 countries, and the growth of the public sector was decisively reversed, especially after 1983. On the other hand, fiscal policies diverged further. Rising earlier, budget deficits in the G-4 countries began to descend after 1980. In the United States, however, they continued to rise, reaching the unprecedented level of 5.0 percent of GDP for the period from 1983 to 1987. Aggravated fiscal imbalances put increasing strain on monetary policy, which accelerated in both the United States and the G-4 countries after 1983, and on trade policy, which at the margin became increasingly restrictive, especially in the United States. Corresponding to these mixed policy changes was a mixed record of economic performance (see IB and IIB in Figure 1-1). Inflation declined rapidly, to postwar lows in the G-4 countries; growth accelerated, to postwar highs in the United States for noninflationary periods; but unemployment continued to rise in Europe and remained above initial postwar levels in the United States. Most important, trade imbalances led to a growing foreign indebtedness on the part of the United States, making the U.S. economy potentially vulnerable to international developments at the next turn of the business cycle.

It is both difficult and controversial to argue that the policy shifts identified in Figure 1-1 caused the related performance outcomes. The factors influencing growth are far more numerous and complex than the few policy instruments identified in Figure 1-1 (Denison 1962, Harberger 1984). Moreover, it is always possible to argue that causality runs the other way, from performance (circumstances) to policy rather than from policy to performance. Much of the literature ascribes a very large role to changes in circumstances in the early 1970s, particularly in world oil markets (e.g., Ikenberry 1988). Policy is assumed to have reacted in a more stimulative and interventionist fashion largely to offset the recessionary impact of higher oil prices. This study, however, shows clearly that the trends toward more stimulative macroeconomic policies and interventionist microeconomic policies were already in place long before world oil prices quadrupled in the winter of 1973–1974 (see Chapter 5). It relies on considerable analysis and evidence supplied by the Organization for Economic Cooperation and Development (OECD) that suggest that these policy trends may in fact have played a significant role in precipitating the large and sudden increases in oil prices (see Chapter 6).

In any event, the associations between policy shifts and performance outcomes put forward in this study are developed with considerable care and supported by substantial statistical and historical data. They allow one to say for the period from approximately 1947 to 1967 that "at least . . . the policy being followed was consistent with successful performance" (Stein 1984, p. 87). The same cannot be said for policies pursued in the period after 1967; although it is too early to evaluate definitively policies for the period since 1980, a fair judgment would acknowledge that these policies coincide with improvements in some areas, such as inflation, growth, employment, and entrepreneurship, and with new problems in other areas, such as protectionism and foreign debt (Huntington 1988–1989). The perspective in this study insists on holding policies responsible for outcomes; it does not accept the alternative of attributing economic outcomes exclusively or primarily to circumstances beyond the control of policy or of simply noting that there is no satisfactory explanation of economic growth (Gilpin 1987, p. 110).

Cocoon of Consensus Building

The argument in this book is that choices of national purpose and efficient or inefficient economic policies have more to do with international performance than power or interests. These choices are made on the basis of ideas generated in the cocoon of nongovernmental and quasi-governmental organizations surrounding the bureaucratic process of policymaking. Ideas that survive the competition in this cocoon and win the support of a majority political coalition eventually become part of the immediate policymaking process and harness the power and interests of the state to implement the substantive policy goals embodied in these ideas. The cocoon of consensus building in this sense helps to nourish the highly fragmented policy process and enables that process to coalesce and make policy with some degree of competence and coherence. Without such an enveloping cocoon, the bureaucratic process would degenerate into unconstrained conflict and chaos.

Conceived in this way, the policy process is more than interests (society), institutions (state), or the interaction between interests and institutions (state and society [Skocpol 1985, Ikenberry et al. 1988]). For too long, policy studies have viewed politics largely as a question of distribution: "who gets what, when, where and how." This perspective focuses attention exclusively on the location of autonomous actors, not on how they relate to the larger reality of community. Thus, once the autonomous actor is located, it is easily concluded that that actor expresses only its special interest and is unlikely to represent the common interest. The perspective, in short, makes it almost impossible to acknowledge, let alone investigate, common interests.

Because the distribution perspective on politics demeans the role of common interests, studies both take shortcuts and go to considerable lengths to revive the concept of common interests. Public goods theories, as I discuss in the next chapter, simply assume common interests by defining arbitrarily certain public goods, such as clean air, to be goods valued by the society as a whole (Olson 1982). Other studies look for common interests in special interests that persist over time (i.e., influence different actors sequentially) and retain a certain ordering of priorities (Krasner 1978).

By contrast, this study looks for common interests within a given place and time. It identifies these interests in the ideas of individual actors that embrace the special interests of other actors and that succeed over time through debate and actual performance in persuading these actors that their special interests can be satisfied within this larger common framework. Ideas in this sense are not just products of new knowledge but also expressions of new inspiration that derive from the unending human quest for improved political community—what Ernst Haas calls the evolution of social consciousness through the interaction of political man with both nature (i.e., circumstances) and culture (i.e., choice) (1983, pp. 24–25).

Ideas are not thereby totally divorced from interests (e.g., class) or institutions (i.e., authority). But just as power and wealth are not the sole or even primary sources of motivation in international politics, they are even less so in domestic politics, where common interests are more apparent. Common interests, it is true, can only be advanced by specific actors in specific circumstances; moreover, they usually satisfy the interests of these specific actors. But some expressions of common interests go beyond special interest. The analyst can judge which ones do so by identifying empirically after the fact which expressions of common interest proved to be most persuasive, which succeeded in encompassing more or fewer special interests, and which in fact resulted over time in benefits to all or most of the special interests involved. By these standards, the postwar experience suggests that common interests defined in terms of individual political freedom and competitive economic initiative have succeeded in both persuading and benefiting more special interests throughout the world than any other definition of common interests. On this empirical basis, the postwar period under American leadership may be said to reflect political ideas going beyond the special interests of wealth and power of any single participant.

In today's modern, competitive democratic societies, ideas that claim to represent the common interest have to meet an increasingly rigorous test to win substantial support and influence policymaking. They have to convince highly mobilized pluralistic interest groups, with greater access to information and communications facilities than ever before, that such ideas do in fact satisfy the special interests of these groups. Cultural or elite imperialism therefore is less likely today than before. What is more likely is that the ideas of policymakers may be rejected, not because politics today is more intractable, but because ideas may remain unpersuasive or ineffectively communicated—the case, one might argue, with the Reagan administration's program for spending cuts (see Chapter 8).

Like power and interests, the concept of institutions has been overworked in postwar policy studies. In 1976 Graham Allison and Peter Szanton argued in their book *Remaking Foreign Policy* that "organization matters—decisively in some instances, importantly in almost all—and that, unless substantial reforms are undertaken, the nation will continue to find the gears of foreign policy machinery failing to mesh with the problems the external world presents" (1976, p. 211). They saw three fundamental defects in the policymaking system of the mid-1970s: (1) lack of balance, due largely to the failure to include all relevant views and agencies in the decision-making process; (2) lack of competence, reflecting the failure to recruit the necessary expertise to cope with a world of complex interdependence; and (3) lack of integration of policy to ensure a coherent conception of U.S. national interests.

After more than 10 years and numerous, further organizational changes, the same criticisms could be made of foreign policymaking in the late 1980s. The Reagan administration, through its own combination of cabinet government and a relatively detached president, did reasonably well in terms of including relevant viewpoints. But it performed poorly, if one accepts the common media criticism, in the areas of competence and coherence; and in a few celebrated instances, such as the Iran–Contra controversy, it also excluded essential viewpoints, with what critics regard as disastrous consequences.

Should we assume, then, that Allison's and Szanton's prescriptions are still valid but have just not been followed? Should we blame inappropriate organizational arrangements for Iran–Contra or the failure of the Reagan administration to complete the course of economic reform it embarked upon in 1981? Should we propose the conventional organizational and bureaucratic solutions to improve U.S. foreign policymaking in the future?

The concepts in this study say we should not. Organization matters, to be sure, but it does not matter enough. Moreover, the standards of "good" organization that Allison and Szanton and other organizational experts propose would produce "good" policy only under very unique circumstances. Including all relevant viewpoints, for example, would, under most circumstances, produce either delay or negotiated trade-offs that might be considered fair but not necessarily competent or coherent. Competence might ensure logical analysis but prove to be politically unacceptable (e.g., Carter's initial energy policy). And coherence may produce an integrated policy but reduce it to generalities or exclude essential qualifications or uncertainties. As Dean Acheson once observed, "agreement can always be reached by increasing the generality of the conclusion . . . [but] when this is done, the form is preserved but only the illusion of policy is created" (1969, p. 733). Standards of comprehensiveness, competence, and coherence can ensure good policy only if one assumes that all relevant viewpoints are not antithetical, that competence is applied in a setting where political consensus already exists, and that coherence does not alienate any particular point of view.

In short, good organization and good policy assume a consensus coming out of the cocoon of consensus building that defines the broad content of policy within which the policy process takes place. If that consensus does not exist, bureaucratic competition cannot produce coherent policy. The narrower studies of the bureaucratic process popular in the 1970s themselves reflected a specific consensus at the time. The Allison and Szanton study, which grew out of the work of the Murphy Commission on the Organization of the Government for the Conduct of Foreign Policy, addressed bureaucratic issues in the mid-1970s, when the broader postwar policy consensus was disintegrating under the impact of Vietnam (see Commission 1975). They offered their proposals as a way to reconfigure the government to cope with this changing world. What is most interesting, however, is not their proposals for organizational reform but the world view that informed these proposals. This view described some characteristics of the world on which there was and probably is today a broad consensus—an emerging global economy, physical interdependence, the multiplication of actors and the multilateralization of issues. But it contained other elements that were more controversial and that were challenged in whole or in part over the next decade—diminished use of force, decline of American power, erosion of Cold War consensus, and loss of confidence in government. Moreover, some of the agreed elements, such as interdependence and a world economy, may be under attack today, as protectionists pursue policies that would in fact reduce the levels of physical and economic interdependence in the world economy.

To rescue policy studies from the iron grip of interests and institutions, this study looks for significant policy changes in the wider layers of the cocoon of consensus building that surrounds the immediate policymaking process, to include the activities of think tanks, public and private sector commissions, citizen organizations, university institutes, media organizations, and similar entities in the international or transnational arena. Table 1-6 identifies some of these layers, which suggest the areas in which broader policy ideas originate, how they acquire attention and legitimacy among larger groups, whether they penetrate sufficiently into the bureaucracy through appointments and recruitment, and how they serve to tie together bureaucratic interests and larger support groups within the society. To be sure, bureaucracies participate in the generation of new ideas. Certainly, they try to coopt broader ideas that serve their narrow interests. But bureaucracies are only the tip of the iceberg.

Table 1-6 Selective Layers of the Cocoon of Consensus Building

1. *Speech-Making Layer.* The way an administration or Congress articulates policy to the various layers of the cocoon through speeches, spokespersons, press aides, etc. (Bureaucratic models of decision making largely ignore this activity and little is known therefore about how some highly effective policymakers, such as former Secretary of State Henry Kissinger, have used this layer to considerable advantage in influencing policy.)

2. *Commission and Special Committee Layer.* The use of special presidential commissions and committees to review and help resolve foreign policy (as well as other) issues outside the immediate policy context of bureaucratic relationships. (Examples include the Murphy Commission mentioned in the text, the Greenspan Commission on Social Security in 1983, and the National Economic Commission on the budget deficit in 1988.)

3. *Policy Studies Layer.* University institutes, think tanks, professional organizations (e.g., Heritage Foundation, Brookings Institution, Council on Foreign Relations) that debate and study American foreign policy purposes and programs. This layer is much more diverse today, reflecting the growth of conservative institutions in the 1970s, and has become less effective in integrating policy ideas than it was in the 1940s (see Chapter 3).

4. *Media and Public Opinion Layer.* The media, polling organizations, and the public itself are now all important players in the American foreign policy debate.

5. *International Layer.* International study projects, conferences, organizations, commissions, and so on, that tie together the various layers of the American cocoon of consensus building with those of other countries. (Examples include Trilateral Commission, Quadrangular Forum, Annual Davos Management Seminars, etc.)

Critical policy ideas originate more often than not outside government. Whereas the State Department bureaucracy (specifically, Cordell Hull) played a helpful role in initiating the United States' liberal trade policy in the 1930s (Haggard 1988), free trade got its legitimacy from the broader economic consensus that emerged in the United States after World War II in nongovernmental organizations such as the Committee for Economic Development and the Council on Foreign Relations. This consensus endorsed not only free trade but the moderate macroeconomic and microeconomic policies that were needed to stabilize domestic prices and reallocate domestic resources as trade competition grew (see Chapters 3 and 4). What is more, when these macroeconomic and microeconomic policy ideas changed in the early 1960s, new ideas came again from outside the government. Policy advisers to Presidents Kennedy and Johnson brought these ideas into government and eventually persuaded presidents to implement them (see Chapter 5). Similarly, the various schools of economic thought that penetrated government with the Reagan administration in 1981 also developed initially outside government (see Chapter 7). The real arena of policymaking is thus much broader than the usual bureaucratic or interbranch perspective that dominates most policy studies (Pastor 1980). Yet we know relatively little about this larger arena.

The concept of cocoon in this study should not be equated simply with the larger political process in pluralist societies. The debate among policy ideas in this cocoon of consensus building is captured to some extent by the larger political competition that cycles parties into and out of power in pluralist societies in the West — but only to some extent. Table 1-7 shows the political parties that have governed in the major industrial countries since 1947. It classifies these governments in terms of center-right and center-left parties and then assesses the amount of time each type of government has ruled in the three phases of the postwar period coinciding with the evolution of economic policy choices.

Table 1-7 Center-Right Governments: Percentage of Time in Office, 1947–1987*

	1947–1967	1968–1980	1981–1987
Canada	35	8	50
France	70	42	29
Germany	100	8	71
Italy	80	54	43
Japan	96	100	100
United Kingdom	65	33	100
United States	40	66	100
G-5 countries	75	50	80
G-7 countries	70	44	70

Sources: Allen (1977), Banks (1988), Day and Degenhardt (1984).

*In the case of coalition governments, including minority coalitions, the center-right determination was made on the basis of the party that led the coalition.

From 1947 to 1967, center-right governments held power roughly 75 percent of the time in the G-5 countries—France, Germany, Japan, United Kingdom, and the United States. From 1968 to 1980, they held power only 50 percent of the time, but from 1981 to 1987 they took control again about 80 percent of the time. When we match this pattern with Figure 1-1 there is some basis to argue that center-right governments tend to pursue more conservative, growth-oriented economic policies, whereas center-left governments tend to pursue expansionist programs of full employment and microeconomic management of the domestic economy.

But the pattern breaks down in the case of individual countries. In France the conservatives lost power steadily over the 40-year period; in Japan, after a brief socialist government in 1947, they never gave it up. The most serious breakdown in the pattern occurs in the case of the United States. Here center-right governments were in power for more time in the 1970s than in the 1950s or 1960s, yet growth-oriented policies were pursued more strongly in the earlier period. As we shall see in Chapters 5 and 6, the Nixon-Ford administrations pursued policies of massive macroeconomic stimulus and pervasive microeconomic intervention, especially wage and price controls, that expansionist, center-left governments of Presidents Kennedy and Johnson only dreamed of. Something else was obviously moving policy in the 1970s besides politics. That something else was policy ideas that by the early 1970s had crossed party lines and captured a wide segment of American public opinion, rejecting containment and moderate Keynesian economic policies and advocating new policies of detente and aggressive Keynesian expansion (see Chapter 6).

In the 1970s new ideas struggled to the top once again. The Committee on the Present Danger, a private citizens' group, led a renewal of concern about American defense. On the economic front, monetarist economists advocating quantity rules for managing the money supply revived respect for monetary policy, and supply-side theorists drew attention once again to market incentives and flexibility. The Reagan administration failed to integrate these ideas and to link them in all cases with clear and consistent purposes (see Chapters 7–9), yet by the end of the 1980s some ideas, such as a new appreciation for defense and markets, appeared to have crossed party lines and may continue to influence policy well beyond the Reagan years.

In this study, therefore, ideas mix with politics to influence outcomes in a struggle that is not predetermined by deep-seated interests, institutions or politics (i.e., structural forces) but that can swing either way, depending on the competence and tenacity with which policy ideas are pressed and, most importantly, the performance that these policies actually produce. Policy is not unconnected from real-world constraints. It must contend with competition posed by international power and with scarce economic resources posed by international markets. It must ultimately perform successfully to defend national interests and to spur economic growth. This link between domestic policy choices and the international context is the subject of the next chapter of this study.

2

Power, Markets, Institutions, and Society: Linking Choice-Oriented Concepts with International Constraints

This chapter links the domestic concepts of a choice-oriented model of American foreign economic policy to performance criteria set by international constraints of power, scarce resources, bureaucratic institutions, and social coalitions. Unlike structuralist models, however, the voluntarist approach sees circumstances directly constraining choice only in extreme situations, as when a nation's economic policies ignore the constraints of scarce resources for so long that the economy suffers hyperinflation and effective bankruptcy. In less extreme situations, circumstances tolerate a much wider range of choice than structuralist models allow and are themselves influenced by competing choices that define configurations of interest and shifting balances of power.

International power, markets, institutions, and social forces all surround and influence the purposes, policies, and process (cocoon of consensus building) of American foreign economic policy. Relatively, American power has unquestionably declined over the past 40 years; international markets have become larger, more interdependent and complex (representing yearly roughly $4-5 trillion worth of goods and services — approximately the size of the U.S. domestic market — and upwards of $100 trillion in capital and currency transactions); international organizations have proliferated to include more than 200 governmental and 2,000 nongovernmental institutions; and international society for the first time in history embraces non-Western (e.g., Islamic, Confucian) as well as Western (Judeo-Christian) cultural and ethical forces. Yet these postwar international developments toward greater complexity and diffusion of power and culture were not inevitable or irreversible (as some studies imply — see Keohane and Nye 1977, Nye 1988). Over the same period of time, configurations of power, institutions, market, and society in the Eastern socialist countries were relatively static. Western countries, particularly the United States, created the circumstances of the contemporary nonsocialist world, and they retain choices to preserve and enhance or to destroy this highly integrated world system.

For example, the United States and like-minded countries joining the General Agreement on Tariffs and Trade (GATT) spawned the modern, interdependent world economy by systematically lowering trade barriers to international flows of manufactured goods. This system was no more the inevitable consequence of technological change and modernization than was the disastrous cycle of protectionist policies in the interwar period. Similarly, today's world of financial deregulation and massive capital flows is not so much the product of modern information technologies as it is the result of the policies of the member countries of the International Monetary Fund (IMF) to remove systematically restrictions on foreign exchange transactions in the 1950s and to decide subsequently, in the United States and eventually other countries, to allow markets to determine exchange rates and to seek domestic adjustment through means other than restrictions on financial transactions.

When the external situation is less permissive, as it is today, international constraints are believed to be more decisive. This assumption follows, however, only from a consideration of external factors. Dominant states, if they are pluralist rather than authoritarian, may not act any more decisively in permissive than in nonpermissive environments. After World War II, the United States, despite its preponderant power, acted haltingly to implement the Bretton Woods design, drifting in 1945–1946 and then moving more decisively in 1947–1948. Even then it took 10 years or more to implement the substantive provisions of the design—liberalization of exchange restrictions, dismantling of discriminatory quotas, and tariff reductions. Pluralist states, therefore, may act hesitantly and incrementally to shape international circumstances even when the external environment is permissive. Whether they act decisively depends on how well these states succeed in mobilizing domestic political and material support for external action.

To influence effectively external circumstances, state actions must meet certain performance tests. In the first instance, these tests are created by the military, economic, and political relationships existing in the global system. State actions must address these relationships effectively or be limited by them. For example, if a state ignores existing military power relationships long enough, it will very likely have to yield at some point to coercion by other states. Similarly, if it fails to consider scarce resource factors, it risks, at some point, economic decline and bankruptcy. And if it fails to take into account the organizational and social requirements of effective action, it may, again in the extreme, be ostracized by other states. The constraints at this level of the international environment are precisely those constraints emphasized by structuralist views of recent U.S. foreign economic policies (see the quote at the beginning of Chapter 1 from Oye, Lieber, and Rothchild).

In this study, however, structural constraints not only limit policy, but may themselves be created by policy. By failing to meet performance tests effectively, state policies allow external pressures to accumulate; and these pressures may eventually result in compelling circumstances or crises. In this sense, state policies create crises rather than just react to them. The historical analysis in this study offers several examples. From 1945 to 1947, U.S. policies exaggerated the extent of social consensus or shared political community with the Soviet Union in the postwar international system. The United States rapidly demobilized and withdrew from Europe, ignoring existing differences between radically different conceptions of political community in central Europe — democratic vs. communist. These differences festered until the United States had a choice of either accepting the potential loss of democratic institutions in Europe or reversing its policies of economic and military indifference. The Cold War was a crisis created as much by U.S. policies in 1945 as by Soviet policies in 1947. Similarly, U.S. policies in the mid-1960s failed to consider the economic costs of both domestic expansion and a foreign war and subsequently brought the U.S. economy close to the brink of bankruptcy. In that sense U.S. decisions in the 1960s created the economic malaise of the 1970s. Finally, U.S. policies in the early 1980s downplayed too severely the social and political benefits of international economic cooperation, in the process almost precipitating rejection of U.S. policies altogether and exit from the global economy by indispensable foreign partners (e.g., the heavily indebted developing countries in summer 1982 and France in early 1983; see Chapters 7 and 8).

Why do state policies fail to deal effectively with external circumstances and thereby create the structural drift that later constrains state action? There is a deeper level of interaction and performance taking place that structural studies ignore (Ashley 1984, Rosenau 1986, Kratochwil and Ruggie 1986). This level has to do not just with how states behave but with what states themselves are, why they exist, and how the state system itself changes, with the possibility that someday this system may be transformed into a world in which separate nation-states are no longer the primary social actors. This level recognizes that the state is a purposeful actor that shapes its environment and does not just react to it. As Alexander Wendt puts it, "social structures are the result of the intended and unintended consequences of human action, just as those actions presuppose or are mediated by an irreducible structural context" (1987, p. 360). State actions and structural relationships influence and determine one another reciprocally.

Although the interaction of state and structure is circular and continuous, the study of this interaction need not result in tautology or indeterminacy. Tautology arises from the possibility that state and structure, because they each determine the other, become indistinguishable from one another conceptually. These concepts are then left to be defined empirically only within a specific historical period and in that case cannot explain the transition from one historical period to another. To avoid tautology and to remove these concepts from any one historical context, appeal must be made to so-called unobserved phenomena, that is, phenomena that can be posited to exist, although unobserved, and that explain effects observed in history (Wendt 1987, pp. 350–355). Various hypothesized unobserved phenomena can then be compared in terms of which one appears to explain best the historical effects that are observed. Indeterminacy raises the question of whether the state and its environment can be differentiated from one another empirically. This can be difficult when historical events occur simultaneously. The problem can be overcome, however, by "bracketing" the investigation of state–structure interactions, that is, looking first at state actions while holding structure constant and then reversing the procedure (Wendt 1987, pp. 364–365).

The concept of national purpose in the present study operates at this deeper level of reciprocal state–structure interaction. As outlined in the previous chapter, it addresses the questions of how human beings or groups of human beings organize to apportion private and public rights and to decide public issues. It appeals to a phenomenon that transcends historical context and the state system—namely, the human search for political community that goes back to the subnational, village, town, and city communities that existed before the contemporary nation-state—and involves political communities that may arise going beyond the nation-state, such as the EC, or devolve once again to subnational levels, such as the two German states after World War II. Although this phenomenon may not always be observable in the specific statements and motivations of human beings (statesmen seldom talk in these terms), the concept of national purpose does have empirical manifestations—namely, the perceived domestic and diplomatic dimensions of global political community (see Chapter 1)—and its effects can be observed in specific historical events. But it refers to an activity that is not limited to any specific historical period.

The current state system, for example, with its concepts of sovereignty at the diplomatic level and various "civil religions" at the domestic level, is only one possible answer to the ageless questions of political community. The state is the present grouping that appears to answer these questions most satisfactorily, but it may not do so always in the future. National purpose, therefore, even though it is labeled *national*, to suggest that states *currently* operate as the primary agents of human purpose in the world community, identifies a phenomenon that cannot be reduced to the contemporary historical period alone.

A final level of interaction must be taken into account in international politics. This level goes even deeper than the structural or agent–structure interactions we have discussed and addresses some *physical* or absolute constraints that affect all phenomena in both the natural and social world. These constraints include the limits of space, time, and resources. To some extent these physical constraints are malleable by purposeful action. Resources and life expectancy can be stretched through technological change, and space can be gained through prospective habitation of other solar bodies. But these limits also ultimately constrain human behavior, especially in the sense of having adequate resources available for policy actions in the right place (space) at the right time. Such physical limits give special importance in this study to the efficiency of state economic policies. Since states are always working against an objective universe of scarce resources (even after the limits are bent through innovation, for example), a test of their performance is always possible in terms of their relative efficiency in using these resources. This test is not necessarily more important than the test at the agent–structure level of defining an acceptable political community or the one at the purely structural level of coping successfully, for example, with the military balance of power. But in a world of grinding poverty, it is at least as important.

All theorizing is itself a political exercise (Alker, Biersteker, and Inoguchi 1985). Structural studies, agent–structure theories, and the voluntarist approach in this study all advocate certain political options and exclude others. For example, as long as structuralist studies define policy options only within the irreversible constraints of structure, they advance these policy options and reject others that might change the constraints of structure. As Robert Cox notes, these approaches "appear ideologically to be a science at the service of big-power management of the international system" (1986, p. 248). This is quite ironic, since, as we have seen, many practitioners of structuralist approaches style themselves as sharp critics of assertive American policies.

On the other hand, agent–structure approaches that acknowledge a role for purposeful actors that alter structure cannot escape political bias either. For example, Cox argues on behalf of a "historicist Marxism that rejects the notion of objective laws of history and focuses upon class struggle as the heuristic model for the understanding of structural change" (1986, p. 248). But Cox fails to explain the origins and possible transformation of the concept of class, no less than structuralists fail to explain the origins and possible transformation of the state. For Cox the unobserved phenomenon is class, although he does not acknowledge it.

Similarly, John Ruggie appeals to an unobserved phenomenon in his agent–structure argument about embedded liberalism (Ruggie 1983b). Examined more fully later in this chapter, Ruggie's argument treats the emergence of the liberal postwar international economic system as a consequence of an historic shift in the social consensus among principal industrial states. This consensus accepted the application of free trade policies abroad in return for generous government intervention at home to protect social interests. The new consensus was brought about by a priority impulse of human beings to seek social solidarity. As the unobserved phenomenon in Ruggie's account, this impulse leads over long periods of time to the redefinition of property rights, from feudal to modern and now, in the mid-twentieth century, post-modern social relationships (Ruggie 1986).

This study too appeals to an unobserved phenomenon, and in that sense this study has political consequences. The unobserved phenomenon is the concept of national or, more broadly, human purpose engaged in the search for a better political community. I assume, as Aristotle wrote, that the human being is preeminently and "naturally a political animal," that "there is . . . in all persons a natural impetus to associate with each other," and that "every society is established for some good purpose; for an apparent good is the spring of all human actions" (1912, pp. 1–2). The impetus to associate is political (not social or economic) to define the spheres of private and public rights and obligations in order to achieve a more perfect union. I cannot prove that this impetus to define the good political life is the primary source of human motivation, any more than others can prove that greed (wealth), status (class), social solidarity (welfare), or aggression (power) is. All we can do is analyze historical events that are the effects of these unobserved phenomena and compare accounts. In the historical chapters of this study, I have reinterpreted postwar events from the perspective of national purpose and found that it explains more than studies that focus primarily on power, institutions, markets, or social forces. That result, as well as a desire to believe better of human beings than that they are motivated primarily by base instincts, leaves me quite comfortable with the perspective adopted in this study.

In the rest of this chapter I lay out the dynamic characteristics of my choice-oriented model of American foreign economic policy and then suggest how it draws upon but also significantly differs from various structuralist models and the agent–structure approach that underlies the argument about embedded liberalism.

A Competitive Model of International Economic Relations

The choice-oriented model of U.S. foreign economic policy focuses on the competition among major national actors in the world economy to define the political framework of economic exchanges and to choose the content and mechanisms of international economic policy relations. It takes as its starting point the continuous debate in the cocoon of consensus building, foremost at the national level, that formulates alternative concepts of national purpose and economic policy that national societies seek to pursue in international relations. This debate reflects the ageless quest of human purpose to define an ever more perfect political community. At the present stage of history, the nation-state reflects the strongest manifestation of political community. Thus, the choice-oriented approach looks primarily to national power and authority to establish the terms of international political community and to influence the development of international markets and institutions. It considers this approach both more pragmatic, because national institutions are stronger than international institutions (and if national institutions cannot bring about appropriate actions, international institutions may be even less likely to do so), and more moral, because national institutions are directly accountable, in most countries in the West at least, to popularly elected parliaments, whereas international institutions are not. At the same time, the choice-oriented perspective recognizes that the national political community is itself continuously changing and that this community is continuously influencing and being influenced by the shared political community among nations. The debate in the cocoon of consensus building therefore is about both what the national and international communities are *and* what they may be becoming.

The choice-oriented approach does not ignore structural constraints or power, interdependence, and interest group politics but regards these constraints as malleable under the influence of various domestic choices of national purpose and economic policy. Not all domestic choices of national purpose and economic policy succeed equally in influencing structural factors. As noted earlier, domestic choices must meet certain tests of performance. Specifically, national purpose must accurately assess the feasible mix between domestic and diplomatic components of international political reality.

If national purpose focuses too exclusively on domestic components, emphasizing the uniqueness of any one country's domestic political community, it may lead that country either to withdraw from international community (as the United States did before World War II) or to overextend its national presence in the world community (as the United States did in Vietnam). In either case, international power realities will eventually reassert themselves and "invite" a correction. (Structuralists would say compel, but that is too strong a word for this voluntarist perspective.) If, on the other hand, national purpose focuses too exclusively on the diplomatic components of international political community, emphasizing process over substantive policy, it may lead to cooperation that assumes political agreement where none exists (as U.S. policies did in 1945 vis-à-vis the Soviet Union) or to cooperation that advances antithetical political goals (as the United States may have done through cooperation with Germany in the 1930s or with a domestically unreformed Soviet Union in the detente period of the 1970s). In either case, social differences will eventually intervene to invite a correction. Although there are no precise quantitative measures to judge the mesh between national purpose and international realities at any given time, the concept of national purpose is not unhinged from its historical environment. It can both change the existing mix of domestic and diplomatic realities, if it is accurate, and be changed (eventually) by those realities, if it is off the mark.

Similarly, economic policies must meet a certain test of reality. Not every combination of domestic, trade, exchange rate, and financial policies is equally efficient. Because resources are scarce economically (and that means physically at any given price), some economic policies will prove to be more efficient than others. Although it may be difficult to know *ex ante* which policies are more efficient (except in a theoretical sense), it is possible to establish *ex post* which policies seem to correlate best with more efficient performance.

National purposes compete to establish the terms of shared international community. This shared community must exceed a certain threshold to support an international economic system. There has to be a reason for economic activity other than that activity itself. Markets do not define their own content. Traditional studies speak about this requirement in terms of an international political framework, by which they usually mean the balance of power (Carr 1964). But this study goes beyond the balance of power to define the international political community in terms of how states choose to identify private rights that are protected from government (public) intervention and to organize public institutions to decide everything else that is not in the private domain.

The degree of similarity in national conceptions of the good political life define the boundaries of shared international political community. If that similarity is not sufficiently perceived, as the United States failed to perceive it in relations with free societies in Europe in 1945, or if it is insufficiently present, as it has been in East–West conflicts over human rights and democratic institutions, economic activity will be limited. Again, it is difficult to know quantitatively where this threshold of sufficient similarity lies. And because national political communities and shared international communities change continuously, economic activity can always break out or die down in any particular area at any particular time. But economic communities do not emerge in a vacuum; they can be predicted roughly where values and institutions converge, and they grow more or less depending on the pace at and extent to which such convergence occurs.

Once a sufficient level of shared political community exists, participants collaborate to decide the content of economic policies they will pursue. This decision leads either to the creation of international markets or to the direct bargaining of all economic exchanges in international institutions or to some combination of the two. National policies create open markets, and national policies close them. Moreover, national policy choices have different consequences for the growth and stability of open markets. Postwar evidence suggests that inflationary policies, extensive government intervention in individual labor, product, and capital markets, and trade restrictions correlate with slower growth and greater instability in open markets. Although it is not possible to establish direct cause-and-effect relationships for these interactions, it is plausible to conclude that policy choices may have had at least as much to do with postwar growth and stability as exogenous structural circumstances.

The emergence of initial agreement on political community and the content of economic policies does not end the matter (as embedded liberalism assumes). From a choice-oriented perspective, the consensus on the purposes and policies of economic cooperation must be continuously nurtured. Participants influence this consensus through both institutions and markets. These mechanisms can reinforce one another, as Marshall Plan institutions did to promote market-oriented growth in the late 1940s and 1950s, or they can work at cross-purposes to one another, as U.S. market actions to realign exchange rates did in 1971 when these actions disrupted the Bretton Woods institutions.

The exercise of influence through markets or institutions is not equally available to all actors in the contemporary world economy, nor is it unconstrained for those actors to whom it is available. In a perfectly competitive setting, as we know from the theory of markets, no single actor has sufficient power to influence other actors. In the present international setting, however, several actors have such power; and they now increasingly compete and provide greater opportunities for other, smaller actors to gain economic benefits and political influence. The United States, Japan, and the European Community (assuming the latter acts in a unified fashion) possess sufficient weight in the world economy to project powerful national influences onto the world market and to restructure the incentives and options facing other countries. Theoretically, countries with trade or capital flows that represent a small percentage of GNP but a large percentage of world trade or capital markets have the greatest economic power (Bergsten 1975, Chapter 2).

Such power, however, even for these actors, is not unconstrained. If the United States or the EC acts in ways that alienate less powerful actors, the latter can always exit the market. If enough participants withdraw, a market ceases to exist or becomes thinner and more imperfect. The more powerful members of the system lose not only economic benefits from the closing or thinning of the market but also an indirect means to exercise influence over other countries. Thus, a prerequisite for influencing international economic events through markets is to maintain the minimum degree of international consensus necessary to sustain an open system.

Maintaining a minimum social consensus is also necessary to support direct bargaining in international conferences and institutions. No bargaining process, let alone the agreements that emerge from such a process, can be sustained without some degree of consensus and organizational coordination. Whether through institutions or markets, therefore, the avenues for asserting national power and policy are inextricably linked with national purposes and hence the shared political community a nation advances.

In this sense, while the choice-oriented model focuses on the purposes, policies, and power of major national actors in the world economy, it does not ignore the interests of small nations or fail to impose standards of international accountability on the major actors. It accepts power where it exists but requires that that power be used to shape a shared political community, which implies eventually a more equitable distribution of power and of participation in the world economy.

The choice-oriented perspective stresses the growing role of markets in contemporary international economic policy coordination, whereas structuralist approaches stress coordination through institutions and direct bargaining (see later discussion of markets). The choice-oriented perspective recognizes that, as the number of participants in international institutions increases, it becomes more and more difficult to reach agreements through direct bargaining processes. The vicissitudes of even small groups, such as the G-5 (France, Germany, Japan, the United Kingdom, and the United States), reflect this difficulty. The ability to take unilateral actions in the international marketplace, therefore, may help to break deadlocks in multilateral negotiations. In the early 1980s, for example, French and American views on the acceptable level of inflation were so far apart that the prospects for agreement on the content of economic policies in summit or other international negotiations were essentially nonexistent (see Chapter 7). The existence of highly interdependent markets offered another means to narrow these differences. Market actions, of course, must be supplemented ultimately by institutional initiatives or they run the risk of threatening the shared political community without which markets cannot exist. United States actions in the early 1980s came close to destroying this community.

The various concepts of the choice-oriented alternative are pulled together graphically in Table 2-1. The solid lines represent the indirect influences of national policies on one another through the national and international marketplace. The broken lines suggest the direct bargaining influences between national policies within international institutions. The solid lines identify the avenues of influence most emphasized by the choice-oriented perspective.

Table 2-1 The Choice-Oriented Model of International Economic Relations

Country A		Other Countries
National purpose and policy		National purpose and policy
Politics and power		Politics and power
National economy		National economy

International conferences, regimes, and institutions

Consensus and performance

World trade and capital markets

Balance of power

Balance of power

Note: All lines suggest two-way influences.

59

Structuralist Models of International Economic Relations

Structuralist models of international relations in the world economy focus on only selective aspects of the factors and relationships identified in Table 2-1. A first group of structuralist approaches focuses primarily on material capabilities and balance-of-power relationships (Waltz 1979). A second group focuses principally on interactions and processes (interdependence) within a given structure of power. This group emphasizes regimes, institutions, and markets that deflect power across issue areas, constrain it, or shift it over time (Keohane and Nye 1977, Keohane 1984, Krasner 1983, Gilpin 1981). A third group emphasizes political or social forces, either deep-seated, state–society relationships (Katzenstein 1977) or special-interest groups (Olson 1982).

None of these approaches allows for purposeful action significant enough to bring about a basic transformation of structures in the state system. One explanation in the second category does reach down to the level of agent-structure interaction and seek to explain the origins of the state system, as well as its possible modification today under conditions of "intense" interdependence. The embedded liberalism approach by John Ruggie traces the transition from the medieval to the modern, state-centered world in terms of a basic change in the social definition of property and foresees the possible transformation of the state system today in terms of another such change from laissez-faire liberalism to embedded liberalism (Ruggie 1983b, 1986). According to this argument, interdependence and process level interactions, if they become intense enough, can expedite the transformation of basic social purposes and, in that sense, produce system-level changes.

Each of these structuralist models has important limitations. The power and interdependence models cannot adequately explain the purpose or timing of economic policy actions, and the embedded liberalism thesis sees system transformation coming about as a result of unintended consequences, not purposeful action as in the choice-oriented approach. The embedded liberalism explanation also covers such long periods of time that it overlooks many changes in social purpose and economic policy that occur in between, and it does not appear to apply adequate performance tests to judge which social purposes and economic policies cope effectively with structural and physical circumstances and which do not (except to indulge a kind of historicism that concludes that what survives had to survive).

Power Models

Power models predict economic policy coordination when a dominant or hegemonic power exists that has a large stake in the international system and is willing to assume the costs of supporting a liberal economic order, including opening its market to foreign imports, providing capital for investment in other countries, and allowing its currency to serve as the reserve monetary unit of the system (Kindleberger 1973). According to this explanation, America created and led an open world economic order after World War II as long as its international power was disproportionate. As competitors emerged, however, the United States became less willing to absorb the costs and less able to extract the benefits of liberal policies (Keohane 1980).

Most analysts adopting this perspective have been puzzled by why an open economic system seems to have persisted beyond the date of America's relative decline as a dominant power (or, for that matter, beyond the date of Britain's relative decline in the early twentieth century; see Krasner 1976). If the system persists only because of lag effects, then analysts ponder whether it must inevitably degenerate into hegemonic war that eventually produces a new dominant power (Gilpin 1981) or whether collective action can replace the hegemon's role and reconstruct the world economic system along liberal or other lines (Keohane 1984).

Not only do liberal economic regimes seem to outlive the relative decline of dominant power, but dominant power does not always give rise to liberal regimes. After World War II, for example, the Soviet Union was the dominant power in Eastern Europe. Even in 1981 its GNP represented a larger share of total GNP in the socialist bloc than the U.S. share represented among Western nations (70 percent compared to 40 percent). Similarly, Germany and Japan were clearly the dominant powers in Europe and Asia before the war. Yet none of these countries established liberal trade policies. About the only thing dominant power seems to predict is that dominant powers will seek access to subordinate states (Hirschman 1980). In this sense, hegemony predicts the form but not the content of an economic regime (Ruggie 1983b, p. 198). Dominant powers may impose command-oriented or market-oriented policies. The resulting hegemonic order may be exploitative or benevolent (Snidal 1985).

More recently, analysts have argued that only dominant powers with high productivity will prefer liberal regimes, because the "free functioning of the international market is assumed to concentrate wealth in nations of high productivity" (Lake 1984, p. 149). In effect, dominant power is redefined to include not only dominant existing power but also a dominant capability (i.e., high productivity) to create future wealth and power. But the association between relative productivity and liberal policies follows only if one assumes that free markets concentrate wealth disproportionately in high productivity nations. As we noted in the previous chapter, there is some reason to doubt this assumption. Relatively free markets not only concentrate wealth but spread it (Myrdal 1971). In the short run the concentration effects may predominate, but over the longer run the "rate of growth in the core tends to slow and the location of economic activities tends to diffuse to new growth centers in the periphery" (Gilpin 1987, p. 95). This was certainly the case for the distribution of wealth in the postwar world economy (see Table 1-2).

The question, then, is why a dominant power would prefer liberal policies to begin with. It could conceivably gain as much in the short run from exploitative policies (Conybeare 1984), and its dominant position at the outset, in effect a monopolistic position, is not a factor conducive to the competitive and efficient operation of international markets. From purely economic considerations, therefore, liberal policies are not an obvious choice for dominant powers. Perhaps dominant powers choose them for noneconomic reasons. In this study, I argue that the United States chose an efficiency-based economic system in 1947 primarily for security reasons, as the best way to harness Western resources to defend Western freedom (see Chapter 4). Others find similar motivation in British behavior in the mid-nineteenth century (see Lawson 1983 on British Middle East policy). But if security requirements provide the explanation, why did the Soviets not opt in 1945 for a liberal economic system with Eastern Europe, or why did Germany and Japan not adopt such a policy with their allies before the war? Certainly from their perspectives, these countries faced security threats. Security policies, however, are not just a function of external imperatives; they also reflect how a country defines itself and hence what it is that is being secured. It is doubtful, for example, that a command-oriented domestic economic system would decide to protect itself by pursuing market-oriented international policies. Ultimately, the content of security and economic policies is also a reflection of domestic policies and institutions (Katzenstein 1977, Comisso and Tyson 1986).

The hegemonic thesis also suffers because it does not tell us how much power suffices to establish a liberal international economic system or when that power has declined sufficiently to make the liberal order unsustainable. The measurements of power in this respect are very weak (Russett 1985, p. 213). Moreover, what measures exist tell us little about why America used its power after World War II and not before. Table 2-2 shows two measures of U.S. economic power during the twentieth century. On the basis of these two measures it is difficult to understand why the United States acted to create an international economic system in 1948 and not in 1928. The U.S. share of world industrial production in 1928 was 39.3 percent by one measure (Bairoch) and 42.0 percent by another (Rostow). Although no numbers are available for the period immediately after the war, the U.S. share in 1953 was only 44.7 percent. Even if this share was higher still right after the war, are we to conclude that 40 percent or so is the magic threshold that triggers initiatives by dominant powers? Or, because the U.S. share of world trade went up from 14 percent in 1928 to 16 percent in 1948 (see third row in Table 2-2), are we to arrive at 15 percent or so as the magic threshold for hegemony in terms of world trade? The data alone are clearly unrevealing.

Even more interesting, the data in Tables 2-3 and 2-4, as well as in Table 2-2, offer no strong evidence that American power has declined all that significantly since World War II. From 1950 to 1970, when exchange rates were relatively constant and conversions of GNP to dollars involved few statistical distortions, the U.S. share of total OECD GNP declined only modestly, from 59.2 to 47.6 percent (first column in Table 2-3). After 1970, floating exchange rates introduced significant potential distortions into the conversions, but OECD calculations based on purchasing power parity exchange rates designed to minimize these distortions (second column in Table 2-3) still show that the American share declined by less than one percentage point from 1970 to 1985. Finally, Table 2-4 shows that America's share of OECD and total world trade has also declined only modestly over the 35 years since 1950—from 33.63 to 31.22

Table 2-2 U.S. Economic Power in the Twentieth Century

	1913	1928	1938	1948	1953	1958	1963	1968	1971	1973	1978	1980	1985
U.S. share of world industrial production:													
Bairoch	32.0	39.3	31.4		44.7		35.1			33.0		31.5	
Rostow	36.0	42.0	32.0				32.0	34.0		33.0			
U.S. share of world trade:													
Rostow	11	14	10	16		14	11		13		12.5		14.5

Source: First row: Biaroch (1982), p. 304; second row: Rostow (1978), p. 52; third row: Rostow (1978), pp. 72–73 for 1913–1971; data for 1978 and 1985 are calculated from UNCTAD (1987), pp. 2–3.

Table 2-3 U.S. Economy as Share of Total OECD Economy, 1950–1985

	U.S. Share of GNP* (at average exchange rates for given year)	U.S. Share of GNP† (at purchasing power parity exchange rates)
1950	59.2	
1955	56.8	
1960	54.0	
1965	50.1	
1970	47.6	40.47
1975	38.5	39.00
1980	35.3	38.88
1985	45.8‡	39.74

Sources: First column: IMF *International Financial Statistics: Yearbook*, various issues; second column: OECD (1987a), p. 14.

*GNP of other countries besides the United States is converted to dollars on basis of the average dollar exchange rate for the currency of that country in each given year.

†This series does not go back before 1970.

‡Some GNP figures for 1985 were not available. For France, Luxembourg, Finland, and Portugal, GDP was used instead of GNP; for Spain, GNP and average exchange rates are 1984 figures; and for Turkey, GDP was used instead of GNP and was taken, along with the average exchange rate, from 1984.

percent for world trade and from 52.55 to 43.50 percent for OECD trade. The composition of U.S. trade has shifted sharply from exports to imports, which some analysts may wish to interpret as a sign of decline. But a growing U.S. dependence on world imports also reflects a growing world dependence on U.S. markets. When the role of the U.S. dollar is added to all of this, a role that has grown despite predictions in the late 1960s that a pure dollar standard (i.e., the

Table 2-4 U.S. Trade as Percentage of Total World and OECD Trade, 1950–1985

	Percentage of World Trade			Percentage of OECD Trade		
	Exports	Imports	Total	Exports	Imports	Total
1950	17.61	16.02	33.63	28.25	24.30	52.55
1955	17.88	13.59	31.47	26.59	20.21	46.80
1960	17.43	13.20	30.63	24.55	19.16	43.71
1965	16.17	12.94	29.11	21.90	17.68	39.58
1970	15.10	14.21	29.31	19.66	18.81	38.47
1975	13.32	12.81	26.13	19.03	17.97	37.00
1980	11.73	13.47	25.20	17.81	18.77	36.58
1985	11.97	19.25	31.22	16.94	26.56	43.50

Source: IMF, *International Financial Statistics: Yearbook*, various issues.

dollar without gold backing) was unsustainable (see Chapter 6), the evidence of continuing U.S. economic power in world markets is impressive. Although hegemonic theories may still be right and American power may have to fall further to test the model, how studies of the decline of American power have been able to make so much out of so little is puzzling to say the least.

Game Theory

If dominant power predicts the possibility but not the content of international economic regimes, in the absence of dominant power, not even the possibility of such regimes can be predicted. Thus, to explain the possibility of cooperation under conditions of relatively equal distribution of power, structuralist studies turn to game theoretic, regime, institutional, and market approaches. All these approaches stress interactions among states or processes within a given structure of goals (payoffs), norms, power relationships, and incentives. Thus, none of these approaches can explain how goals or purposes change and hence how institutions and structures may be transformed (Axelrod and Keohane 1986, pp. 248–254).

Game theory seeks to explain the possibility of cooperation even in the absence of regimes or institutions. In its purest form, it assumes conflicting goals among the participants and no communications between them (in effect, the absence of any regime). Payoffs are specified and rank-ordered for various strategies of cooperation and defection. For example, in the Prisoner's Dilemma game, mutual defection results in both prisoners getting a heavier prison sentence than they would get if both cooperated. But they decline to cooperate because if one does and the other does not, the cooperating prisoner receives a still heavier penalty (Axelrod 1984).

Now the question becomes how to induce cooperation and achieve more optimal outcomes. One way is to alter the rank ordering of the payoffs. If the benefits of mutual cooperation can be raised significantly compared to those of unilateral defection, the participants will eventually prefer mutual cooperation over unilateral defection (Oye 1986). Game theory itself, however, offers little insight into how these payoffs are changed. Payoff conditions are given. They are part of the structure within which interactions occur. Thus, game theory in its purest form is a series of static situations, which cannot explain how participants get into a particular situation or move from one situation into another. What is more, few international situations seem to fit the static conditions of the various games (Gowa 1986, Keohane 1986). Some of the more interesting situations in international relations occur where self-interest and common interest converge (harmony), as in market-based economic activities, or where no perceived common interests exist at all (deadlock), as in war-threatening crisis situations.

To rescue itself, game theory has to turn to the development of regimes. Other ways to induce cooperation are to permit repeated plays and to reduce the numbers of participants. The possibility of repeated plays permits the participants to begin to anticipate each other's moves and thus to develop a primitive form of communications through reciprocal responses (the so-called shadow of the future). Reducing the number of players facilitates such communications by permitting easier recognition and retaliation against defectors (Oye 1986). Thus, regimes emerge defined "as sets of implicit or explicit principles, norms, rules, and decision-making procedures around which actors' expectations converge" (Krasner 1983, p. 2). Such regimes, it is argued, help to increase the likelihood and intensity of communications and interactions among players by establishing legal liability (e.g., property rights), increasing information, and lowering transaction costs (Keohane 1984, Chapters 5 and 6).

Regimes mediate between structure and behavior and therefore may account for why behavior, such as the open-market policies of industrial countries in the postwar period, persists beyond the decline of initial structures, such as the dominance of American power. According to some theorists, regimes may also account for actual changes in the norms and structures themselves. This happens when the interactions within regimes become so intense that these interactions influence directly the basic payoffs of the regimes, bringing about, for example, a change in the definition of property rights (Krasner 1983, pp. 362–364; Ruggie 1986, pp. 148–149). Such changes occur unintentionally, however, not as a consequence of purposeful actions, and hence are likely to emerge only in issue areas where the stakes of participating actors are relatively low (e.g., in nonsecurity issues; Krasner 1983, pp. 7–8).

Moreover, this possibility of change in basic structures depends upon the relatively benign view of institutions that regime theorists employ. Regimes are regarded as soft or "decentralized institutions," which are "relatively efficient . . . compared with the alternative of having a myriad of unrelated agreements . . . " (Keohane 1984, pp. 97–98). Although such institutions may lower transactions costs and make more information available, accelerating the growth of interactions and hence the impact of these interactions on structures, other institutions, especially more centralized ones, may do just the opposite. Institutions order relationships hierarchically as well as horizontally, and whether they facilitate communications and interactions depends entirely on whether they are open, responsive, and cost-effective (Williamson 1975, Lindblom 1977). Institutions or regimes, in short, may produce externalities no less than markets (Haggard and Simmons 1987, p. 508). We have no theory of bureaucratic incentives that would enable us to say whether institutional intervention is superior to unregulated market exchanges or not (Buchanan 1985, pp. 109–116). Because institutions and regimes frequently reflect the interests of strong powers, they may actually work to perpetuate existing structures rather than to change these structures through intensifying interactions (Kratochwil and Ruggie 1986, Nau 1979, Rochester 1986).

Markets

Game theoretic approaches operate in situations in which conflict is assumed. The classical concept of markets, by contrast, orders situations in which interests are assumed to be in harmony. To achieve optimal outcomes in markets, direct coordination among the participants is unnecessary. Each participant can act without any knowledge or anticipation of the other participant's response and still achieve optimal outcomes for both. Indeed, in the market model, attempts to communicate or collude with other participants are not only unnecessary but undesirable. They distort otherwise mutually beneficial outcomes. Participants act only on the basis of anonymous information in the marketplace, which reflects the sum total of all the participants' actions rather than the action of any specific participant.

Nevertheless, market models share certain characteristics with power and game theory approaches. Markets assume rational behavior based on cost-benefit analysis within a given set of external circumstances. The difference is that, unlike power models, market approaches assume an egalitarian distribution of power (i.e., no one actor can influence the system), and unlike game theoretic approaches, they assume an indifference on the part of the actors to relative as opposed to absolute gains. The latter condition removes the pairing of payoffs that characterizes game theoretical approaches (i.e., what one actor does being influenced by what the other actor does). Each actor competes against an undifferentiated marketplace that rewards efficiency and penalizes inefficiency solely on the basis of individual effort. The marketplace works on the principle of opportunity costs: Actors make independent assessments of the use of scarce resources, and competition ensures the survival of only the most optimal use of these resources (Rhoads 1985).

Markets maximize efficiency, whereas hegemonic and game theoretic approaches seek to maximize direct cooperation or coordination. None of these approaches specifies the *content* of what is being maximized or coordinated. Generally speaking, such studies consider that "more cooperation is better than less," even while acknowledging that "this does not mean that all endeavors to promote international cooperation will yield good results" (Axelrod and Keohane 1986, pp. 253–254). Markets too do not specify the content of actions or results. Although it is often assumed that markets operate on motives of material greed, they operate in fact only on the basis of *self*-interest. That is, they specify the location of interests—namely, the individual actor—but not necessarily the content of those interests. As Herbert Stein writes,

> the classical argument about the superiority of the free market system only maintained that such a system would satisfy the wants of the population efficiently. . . . It did not judge the merits of those wants, but took them as given. The free market argument did not exclude the legitimacy of a collective decision to affect the wants the economy served, either by education, regulation or government expenditure [1984, p. 93].

The content of market interests, in other words, is a function of consumer choice or sovereignty. Consumers may choose to maximize wealth, power, morality, public goods (such as education, environment, etc.), or any other asset. Indeed, analysts have applied market models to study wealth (Smith 1937), bureaucratic power (Buchanan and Tullock 1962), and other phenomena. The one criterion in all these models is not greed but competition in the assessment of alternative uses of a scarce commodity.

Like regimes or institutions, markets do not transcend social context. But the fact that markets cannot be abstracted from their social context does not mean that they should be buried in that context. Some analysts argue that "markets are themselves institutions . . . whose character is historically determined" (Hall 1986, p. 35). By the same logic, regimes and institutions would also have to be equated rigidly with existing power structures. There would then be little room for maneuver or change either through markets or regimes and institutions.

A better approach would seem to be to regard markets and regimes as alternative mechanisms for bargaining among state actors to decide the purposes and policy of international economic activities. Regimes and direct bargaining in international institutions may be more useful mechanisms when conflicts exist that may shatter markets and when regimes and institutions are relatively open, responsive, and cost effective. Markets and indirect bargaining through competition in the marketplace may be more useful mechanisms when regimes and institutions are not working efficiently and when unilateral actions may be presumed to promote the common good (i.e., when a harmony of interests exists). What is surprising is that regime approaches to international political economy all but ignore market processes (e.g., Keohane explicitly limits his inquiry to "coordination achieved through bargaining" and excludes "mutual adjustment [that] take[s] place without direct communications between the participants;" 1984, quote from p. 76; see also pp. 52–53). This choice reflects a clear bias in the political economy literature against markets, matched only by a comparable bias in the free market literature against government institutions.

Yet the case can be made that markets are more important today than institutions. In the 1950s, institutions—sometimes quite centrally organized institutions, particularly at the regional level (e.g., EC)—were instrumental in building markets. With fewer actors and vested interests being present, institutions in this period were also relatively efficient. Today, however, institutions are more bureaucratic and contested, and interdependent markets may provide an alternative arena for flexible interaction and indirect leverage. While regimes are said to be necessary to handle the increased complexity of contemporary international relations, markets are uniquely suited to handle even more complexity. As long as they are sufficiently competitive, markets process far more information than institutions, particularly hierarchically organized institutions, which may be prone to information overload. In addition, markets provide anonymity for participants. Parameters are set by competitive processes in the marketplace, not by dominant actors at the bargaining table. This is a feature that can be particularly valuable in a world community, where national sensitivities are high and nations resist yielding to other nations in publicly visible international conferences and institutions. For example, how much more difficult might it have been than it already was to persuade Japan to restructure its domestic economy in the late 1980s if the participants had had to negotiate the appreciation of the yen in full international view rather than have it appear as a consequence in large part of anonymous market forces?

Markets have no inherent tendencies of their own to expand, innovate, or consume, as may be implied by some historical patterns (Gilpin 1987, Chapter 3). They do so primarily as a consequence of policy. Thus, markets, like regimes and institutions, are constantly being shaped by social forces. Viewing markets and institutions more neutrally as alternative arenas of interaction and bargaining does not detract from the role of these social forces but actually helps to identify them more clearly. For the content of interaction in both bargaining (game theoretic) and nonbargaining (market) situations is ultimately a question of neither structure (power) nor process (interdependence). It is a question of domestic politics and social interests.

Domestic Politics and Social Interests

The analysis of social forces has concentrated at two widely separate levels. One level taps what we might call the deep structure of society. Marxists see this deep structure in class struggle. Other analysts see it in the patterns of state-society networks and institutions in various countries (Katzenstein 1977). What is common to these approaches is that change comes about not by design or intention but by a process of contradictions, stresses, and conflicts in the system (Katzenstein 1977, p. 333). Although objectives of purposeful actors, such as labor and business groups, play a role in this process, these objectives do not change except over very long periods of time (indeed, for Katzenstein, domestic structures persist for centuries), and change in the system itself is more a result of unintended consequences than of directed initiatives. Thus, deep structure implies that the sources of change and continuity are deeply embedded in society and are not subject to purposeful influence in any immediate time frame.

The other level at which the analysis of social forces concentrates is bureaucratic or interest group politics. Although this literature goes back to the seminal postwar works in political science (Truman 1951), its most recent theoretical elaboration is by an economist, Mancur Olson (1965, 1982). In *The Logic of Collective Action* Olson argued that groups behaving on the basis of rational self-interest would not cooperate voluntarily to produce public or collective goods unless government intervened to organize side benefits. Public goods he defined as goods that are indivisible, so that one person's consumption of the good cannot subtract from another's, and they are nonexclusive, so that one person cannot exclude another from consuming the good. Examples include such things as clean air, public transportation, and national security. Such public goods may be necessary to support systems that produce largely private goods. Markets, for example, require clearly defined property rights and dispute settlement mechanisms to function effectively. Olson inquired: "How, then, are such goods produced?"

Market activities alone cannot produce these goods, because gains from public goods cannot be privatized. Public policy cooperation is required. But now a paradox emerges. From the standpoint of rational self-interest, individuals will not have an interest in contributing to public goods because the gains generated will have to be shared with others and would have to at least equal the number of participants in the group times the cost incurred by any one participant before any one individual could be assured of recouping his or her costs. The larger the number of participants, therefore, the less likely collective goods will be supplied. Hence, a small number of participants or even a dominant participant is the best situation for the supply of collective goods. Participants then identify more readily with the interests of the system and contribute their fair share to the creation of the public good.

But what about situations in which interest groups proliferate, the more common situation in pluralist democracies? Then each individual has a tendency to take a free ride. Not being able to achieve sufficient private gain to offset the cost of public goods, individuals or numerous small interest groups concentrate on redistributing the public goods that already exist. This holds particularly for economic goods. Because any one individual or group would have to reduce the output of the system as a whole by a factor equal to the total number of participants, before that individual or group could expect to lose more than it might gain, the individual or group would have an interest in redistributing resources rather than creating them. Thus, in groups of large numbers, the logic of collective action works both to discourage the creation of public goods and to encourage the redistribution of existing goods.

On the basis of this logic, Olson subsequently explained *The Rise and Decline of Nations* (1982). In young societies a small number of groups tends to dominate and produce the collective goods needed to grow; in more stable societies a large number of groups emerges and creates coalitions that redistribute existing resources. Democratic societies have the greatest difficulty resisting the formation and dominance over time of these distributional coalitions.

The only way to offset the logic of collective action is for governments to intervene and either provide selective incentives to groups, such as subsidies, to encourage them to contribute to the production of public goods or coerce these groups to contribute through mechanisms such as compulsory taxation. But now Olson's logic reaches an impasse. Governments too, we may assume, are dependent on the logic of collective action. They can take decisions to create public goods and provide selective incentives only in situations in which there is a small number of actors within government. But in these situations as long as we are talking about governments that are pluralistic and democratic (i.e., there is a rough correspondence between decision-making groups in society and decision-making groups in government), the number of groups in the society as a whole will also be small; and in this set of circumstances, the theory tells us that government intervention is not necessary to produce public goods. Thus, democratic governments either become superfluous to the production of public goods in social situations where a small number of groups exists or become exogenous factors in social situations where a large number of groups exists — exogenous in the sense that we cannot explain, in the terms of the theory itself, how governments in that situation reach decisions to define public goods because their internal decision-making process is subject to the same free rider problem that afflicts decision making in the society as a whole.

The problem for collective goods theory, like game theory, arises from its focus only on situations of conflict. Olson notes that it is not easy to find special-interest groups that represent their own interests and those of the society as a whole (1982, p. 46). He denies the possibility of classical market situations in which special and collective interests converge.[1] Yet, curiously and arbitrarily, he allows for this possibility in the case of special-interest groups (e.g., government agencies) operating within government. Suddenly, these groups, although presumably reflecting the fragmentation of special interests in the society at large, acquire the capability within government to represent common interests and define public goods. Because the theory cannot explain how this comes about (and if it were possible in government, it would also be possible in the society at large, and the theory would no longer have a free rider problem to resolve), it in effect takes public goods to be self-evident.

But public goods are not self-evident. They have to be valued by the public. Air, for example, is the classic public good. No one's consumption detracts from anyone else's, and no one can be excluded from it. But *clean* air is not a priori a public good. Neither is public transportation, parks, a liberal economic regime for the protection of property rights, or any other politically interesting public good. Public goods have to be defined, and in the process of defining them, participants organize, debate, and eventually agree or fail to agree to designate certain goods as public. Collective goods logic tells us nothing about this prior political process. As an interest group theory, therefore, it skips right over the largest and most interesting group activity—that is, the representation and conflict of special-interest groups within government to define public goods. It deals in the end only with policy implementation (how to supply public goods), not with policy formation (how to define these goods in the first place).

The logic of collective action, therefore, leaves us where we started in this examination of structuralist alternatives to a more choice-oriented, competitive model of international economic policy relations. We are back to situations that start from a premise of conflict that offer no internally consistent mechanism for explaining how conflict can be resolved or structures transformed short of war (power models), the intervention of utopian or at least benevolent regimes and institutions (game theory), or the operation of competitive (markets), cooperative (public goods), or long-term social (deep structure) processes of interaction devoid of explicit values and intended results.

Embedded Liberalism

There is no escape from seeking a model to explain postwar international economic events that goes beyond power, process, and domestic social structures and interests. Many studies recognize this fact (see, for example, Gilpin

---

[1] As a consequence, the logic of collective action works in a manner exactly opposite to that of markets: More competition (i.e., a larger number of groups) breeds less public goods or welfare, whereas in markets it breeds more; and fewer or more encompassing organizations breed more welfare, whereas in markets such oligopolies breed less welfare.

1987). But few, if any, elaborate a theory of purposeful action by states, especially by the United States, that emphasizes the influence of policy in altering structural constraints significantly. The hegemonic, game theoretic, market, and interest group approaches all take their cue primarily from circumstances. They lack content except as theorists impute to them a capability to define what is good (e.g., governments in the logic of collective action) that cannot be explained by circumstances. If hegemony, liberal ideology, and common interests are all prerequisites of liberal market policies, as some studies suggest (Gilpin 1987, p. 73), what accounts for the emergence of liberal ideologies or common interests? These studies tell us more about the origins and decline of hegemonic power than they do about the origins and decline of ideology or common interests. Why is it that some actors develop common interests in the welfare of others (empathy) and other actors retain only egotistic interests in their own welfare (Keohane 1984, p. 125)? The postwar period, as I noted in the previous chapter, has been most marked by the prevalence of liberal ideologies and the spread of common democratic institutions and values. Are these developments accidents of history, or should their rise and decline be explained by our conceptual models, no less than the rise and decline of American power?

One prominent attempt to explain postwar events in terms that go beyond structural constraints is John Ruggie's concept of embedded liberalism (1983b). This attempt reaches down to the agent–structure level of interaction and seeks to account for the emergence of the postwar liberal international economic system in terms of a fusion of structural power and social purpose. According to this argument, the United States and its allies constructed the Bretton Woods system after World War II on the basis of a social compromise between advocates of international liberalism, who sought to overcome the economic nationalism of the 1930s, and advocates of domestic intervention, who sought to avoid the social disruption at home of previous international economic systems based on gold and free trade. The compromise called upon governments to reduce their intervention in the international economy—refrain from exchange rates changes and dismantle foreign exchange and trade restrictions—but to increase their intervention in the domestic economy to maintain full employment and to facilitate domestic adjustment to international competition. It represented a synthesis, as Ruggie explains, of the two great social movements of capitalism and socialism—one of which built unrestrained free markets in the nineteenth century and the other of which introduced Keynesian social techniques to manage markets in the twentieth century (see also Maier 1977 and Hogan 1987, who trace this synthesis back to the associationalism of business and labor interests already in the 1920s, and Gourevitch 1986).

The embedded liberalism argument provides a sweeping interpretation of the postwar evolution of domestic social purpose and international economic structures. It suggests a transformation of the contemporary international system from the modern to the "postmodern" world not unlike the transformation four centuries ago from medieval to modern society (Ruggie 1986, pp. 146–148). Because of its sweep, however, the embedded liberalism argument goes beyond practical means for testing its validity. How do we know that we are in the postmodern world? Ruggie argues that embedded liberalism reflects *The Great Transformation* (1944) from market rationality to social regulation described by Karl Polanyi. But Friedrick Hayek, writing about the same time in *The Road to Serfdom* (1944) foresaw a movement away from social regulation back to free markets, and since the late 1970s, the world seems to have been moving more in this direction than toward Keynesianism, producing another great historical divide — all within two generations.

The concept of embedded liberalism is too elastic to be tested empirically (Haggard and Simmons 1987, pp. 494, 510). It encompasses practically any compromise between economic efficiency and social regulation. Ruggie argues, for example, that this compromise persists even though the guiding rules of the postwar system for trade policy have shifted increasingly toward protectionism. But as Ruggie himself acknowledges elsewhere (Kratochwil and Ruggie 1986, p. 769), the commitments to free trade and domestic intervention are ultimately in conflict with one another. Policymakers cannot give both commitments equal weight and expect to liberalize trade dramatically, as they did in the 1950s and 1960s; nor can they balance these commitments precisely and bring about the slide toward protectionism that has occurred in the 1970s and 1980s. The different outcomes in these two periods had to be to some extent a consequence of differing policy emphasis — in the first period toward freer trade and in the second toward social protection.

Embedded liberalism needs to be tied to some standard of performance to judge its shifting content. Otherwise, it becomes anything the theorist wants it to be. This is a danger, of course, with all agent–structure concepts that seek to transcend structural context. Embedded liberalism, like historical Marxism, becomes an inevitable unfolding over time of a social process (whether reflecting intensifying interdependence or class struggle) that is largely predetermined. This danger can be avoided by testing the purpose and policy content of the social consensus from time to time against changes in outcomes or performance. This assumes, in contrast to structuralist approaches, that intended or purposeful action bears some relationship to actual outcomes and is not overwhelmed by feedback and unintended consequences (i.e., historical accident).

Embedded liberalism may well explain the shift in social purposes and economic policies after World War II that gave governments an active role in the management of aggregate demand in domestic economies. In that sense, the postwar consensus repudiated absolute laissez-faire in national and international markets. But embedded liberalism goes too far when it suggests that this shift sanctioned any degree of government stimulus in macroeconomic policy and gave governments further carte blanche to intervene heavily in microeconomic areas. By failing to specify the more limited, conservative character of the postwar economic consensus (see Chapters 3 and 4), embedded liberalism assumes that the expansion of this consensus in the mid-1960s to include more aggressive macroeconomic and microeconomic policies was merely the continuation of the same postwar compromise. It cannot explain, therefore, at least in terms of policy, the very different performance outcomes of the two postwar periods (pre- and post-1967; see Figure 1-1) and leaves us to conclude either that policy has little to do with outcomes (and only policy process matters) or that the content of policy is so elastic that it can account for any outcomes.

Purpose, Policy, and Performance

The conceptual model in this study therefore insists on a link between the purpose and content of policy actions and actual performance outcomes. This link is forged on a continuing basis within a broader context of deeper social structures, power relationships, and process mechanisms, but it is not determined by this context. Indeed, over time, consistent policy actions can change significant features of the broader context; that is, they can alter basic structures and circumstances. The next part of this study (Chapters 3 and 4) tells the story of how the policy actions of the United States and other industrial countries, after World War II, eventually altered prevailing conditions of inflation, interventionism, and bilaterally managed trade and, in place of these conditions, created the Bretton Woods world of stable prices, flexible markets, and multilaterally freer trade.

II

Making Bretton Woods

3

The Bretton Woods Agreements: Lacking Sufficient Purpose and Policy

Contrary to popular myth, the postwar international economic system did not begin with the agreements signed in Bretton Woods, New Hampshire, in July 1944. Those agreements were effectively stillborn. The institutions they created did not play a significant role in postwar economic relations until well into the 1950s.

Why was this so? American power was never greater than it was in 1944–1945, and American politics was more coherent than it would be in 1946–1947, when the Republicans took control of Congress while the Democrats remained in the White House. Yet American power and public opinion were not harnessed to launch postwar economic reconstruction. The reason is that American purposes and policies failed to direct and mobilize American power and public opinion. America projected a utopian vision of world community based on law and commerce that muddled important differences of political community between democratic and totalitarian societies in central Europe. On the basis of this vision, American security policy demobilized and withdrew U.S. military forces from abroad, despite America's unilateral possession of nuclear weapons; and American economic policies sought to promote open trade and commerce across massive and ultimately unbridgeable gaps in domestic economic institutions and policies between the centralized, planned economy of the Soviet Union, on the one hand, and the competitive, market-oriented economy of the United States, on the other. Except in the earliest days of 1942–1943, when the outcome of the war was still in doubt, external circumstances exerted few constraints on American power in this period. The reasons for American behavior were internal and had to do with America's utopian definition of its political self-image, its unwillingness to assert relatively competitive and efficient economic policies over more interventionist policy preferences abroad, particularly in Great Britain, and its inability to inspire sufficient support in Congress and the larger nongovernmental arena of consensus building to commit the necessary resources for postwar reconstruction.

The interpretation in this chapter thus challenges some of the more common explanations of U.S. policy in this period that stress American power (Kindleberger 1981), the desire of American policymakers to impose New Deal interventionist policies on the rest of the world (Block 1977), or the benign influence in this period of growth and productivity on American politics (Maier 1977).

If American power was central in this period, it was neither American military power, which was withdrawn, nor American economic power, which would have prescribed more aggressive funding to finance U.S. exports than the meager resources Congress provided the IMF in 1945. It may have been American political or ideological power, that is, the projection of America's liberal self-image onto the rest of the world (Russett 1985), but if that was the case, America's power in 1945 proved to be insufficient. The universal–utopian view of the world that the United States advanced during wartime negotiations (see Table 1-3) died at the United Nations in 1945, despite considerable American sensitivity to Soviet feelings after the war (Gaddis 1987, Pollard 1985).

American economic policy choices in this period were also far from the "New Deal for the world economy" that some students of this period have claimed. The White Plan, designed by U.S. Treasury officials in 1942, was, in fact, closer to the premises of the discredited gold standard than New Deal interventionism. To be sure, the White Plan was substantially compromised by 1944 to accommodate British, Soviet, and congressional concerns, making the final product at Bretton Woods look more interventionist. But these compromises also robbed the Bretton Woods agreements of their economic effectiveness and account for the general irrelevance of these agreements during the initial postwar years. The British loan experience of 1946–1947, discussed later in this chapter, demonstrated that highly interventionist domestic policies were generally incompatible with liberal external policies (e.g., reduced foreign exchange and import restrictions). In 1947–1948 the Marshall Plan went back to the more efficient economic policies of the original U.S. wartime proposals (see Chapter 4).

Finally, it stretches the evidence of this period to argue that the domestic economic policy consensus that eventually emerged was the product of a grand synthesis, "the politics of productivity," in which the prospects of growth and increasing productivity somehow overcame otherwise irreconcilable class differences between business and labor (Maier 1977). The domestic consensus was, in fact, the product of intense debate and struggle in which some economic policy options were accepted and others were rejected. The final consensus rejected most of the New Deal options supported by labor groups. It embraced, as Robert Collins writes, "the right wing of the Keynesian spectrum . . . which opted for an active monetary policy and a passive fiscal policy, for automatic [fiscal] stabilizers over discretionary [fiscal] management, for reductions in taxation and increases in private spending, for a modicum of unemployment over a modicum of inflation, and for economic stability over the redistribution of income and the reallocation of resources" (1981, pp. 16–17). Indeed, conservative Keynesianism, which dominated U.S. postwar economic thinking for two decades, accepted only one tenet of interventionist dogma, the notion that fiscal policy should be set such that deficits increased and acted as automatic stabilizers in times of recession but slight surpluses resulted in times of high (not full) employment to be used to retire the national debt. Otherwise, the postwar consensus rejected discretionary fine tuning of fiscal policy, affirmed policies of monetary restraint to maintain price stability, advocated tax reductions to promote private incentives, and saw microeconomic policy as a means to promote rather than suppress competition (e.g., traditional antitrust legislation).

It was this domestic consensus on market-oriented policies that permitted the external liberalization of trade and ultimately ushered in the era of postwar productivity and prosperity. This consensus did not try to accommodate every policy option, however incompatible, as the Bretton Woods agreements had tried to do at the international level. But it did represent a set of ideas that offered benefits to every group. In that sense it produced the politics of productivity, rather than derived from it, as individual groups (labor, business, etc.) were won over to the consensus by the promise of mutual gain.

World Community Through Commerce and Law

The dominant definition of America's national purpose prior to World War II was isolationism. Isolationism negated the diplomatic dimension of international politics and defined America's role in the world as essentially one of noninvolvement (see Table 1-3). World War II changed all that. It made America self-conscious on the global scene. A searing national experience, it weakened isolationist forces on both the left (radical New Dealers) and the right (mid-Western Republicans). A majority in the United States came to realize that the country would have to put its imprint on the international system or it would never feel safe again.

As soon as war broke out in Europe, America began to plan for the postwar world. Discussions began outside the government in various working groups of the Council on Foreign Relations already in 1940. At the State Department, Secretary Cordell Hull ordered planning to begin in early 1940 and pushed Roosevelt, in the Atlantic Charter discussions of August 1941, to obtain British commitments to nondiscriminatory postwar trading arrangements even before the United States entered the war (Eckes 1975, pp. 34–35). Four days after Pearl Harbor, Secretary of the Treasury Henry Morgenthau, Jr., asked his assistant secretary, Harry Dexter White, to draft a blueprint for the postwar monetary system (Blum 1967, pp. 228–229). At this point, planning was highly abstract. No one knew whether the war would be won, let alone what kind of world would be left in its wake.

The unreality of it all facilitated the design of an integrated and ideal vision of postwar order. Security (United Nations) and economic (Bretton Woods) proposals went forward hand in hand, "as interdependent as the blades of a pair of scissors," as Morgenthau would later describe it (House of Representatives 1945, pp. 4–5). Together, the U.S. proposals projected a comprehensive vision of shared postwar political community based on liberal American values of political freedom (law) and economic competition (commerce). "Economic security," as Robert Pollard points out, meant that "American interests would be best served by an open and integrated economic system, as opposed to a large peacetime military establishment" (1985, p.13). Although going beyond isolationism, the legal–commercial view was not inconsistent with America's self-definition under isolationism. The latter had rejected American involvement in the international system because the system relied on traditional military and balance-of-power politics. Now the United States would remake the system in terms of its own domestic prescription for world political community based exclusively on legal and economic relations.

Hull and Morgenthau had much in common. They shared longevity in office (from 1933 to 1944 and 1945, respectively), the common experiences of the 1930s, and above all a set of common ideas. Both men regarded economic conflict as the primary source of war (Senate 1945, Blum 1967). Hull, who had been developing his views on commerce and peace since World War I and had contributed point 3 to Wilson's 14 points calling for an international trade conference after World War I, believed that open commerce could not only avoid jealousy among nations but also increase output, employment, and standards of living, bringing about "the firm establishment of a basis of friendship and confidence on which permanent peace could be built" (Hull 1948, vol. I, pp. 81, 174). For Hull, who initiated the postwar freer trade policy with his Reciprocal Trade Agreements Act of 1934, the point of it all was not wealth but the "constant pursuit of human liberties" (1948, vol. I, p. 175). Morgenthau too saw commerce as a way to banish power politics. But he also worried that "power economics may be just as dangerous" (House of Representatives 1945, p. 6) and initiated efforts already in the Tripartite Agreement of 1936 (which included initially France, Great Britain, and the United States and by 1945 some nine other countries) to promote financial cooperation to prevent the "economic scuffling of the 1930s [which] developed the gangsters who finally discarded their economic blackjacks and brass knuckles in favor of the tanks and bombs that bathed Europe, and most of the world, in blood" (Senate 1945, p. 8).

The legal–commercial view of the world saw a "reciprocal identity between national and international interests" (Blum 1967, p. 228). International politics did not concern the management of fundamental domestic interests, because these interests could be assumed to be shared. It concerned only how diplomatic relationships were structured to solve common functional problems of commerce, money, food, health, and other needs. Whereas isolationism had denied the diplomatic dimension of international politics, the legal–commercial view denied the domestic dimension (see Table 1-3). This view, as John Gaddis observes, tried "to construct a new world economic order without first resolving the deep political differences which divided the United States and the Soviet Union" (1972, p. 23). Although the view was not shared by all American officials in this period (see, for example, Dean Acheson's biting characterization of U.S. planning in this period as "singularly sterile . . . platonic planning of a Utopia"; 1969, pp. 64, 88), it was clearly the dominant thinking behind early U.S. economic proposals that stressed technical, neoclassical solutions to postwar economic relations.

The White Plan: No New Deal

The earliest Treasury plan for the postwar world economy — known as the White Plan, after its principal drafter, Harry Dexter White — was a textbook case of highly competitive trade and disciplined domestic economic policies (see draft of this plan dated 1942 in Horsefield 1969, vol. III, pp. 37–83). The plan placed a premium on stable domestic and international prices, especially prices of national currencies or exchange rates. As the notes accompanying the plan made clear, "wide swings in price levels are one of the most destructive elements in domestic as well as international trade." Stabilizing exchange rates would reduce the "costs of conducting foreign trade" and offer "more chance of avoiding the disrupting effects of flights of capital and of inflation" (Horsefield 1969, vol. III, pp. 46, 47). Thus, the plan obligated members not to alter exchange rates without the consent of the other members, to reduce foreign exchange restrictions, and to eliminate discriminating exchange and trade prac-ticies. It also called for lowering trade barriers to increase exposure to competition. The emphasis on stable prices and lower trade barriers, coupled with the relatively modest levels of external financing the plan offered to help members cope with temporary deficits in their balance of payments (see below), ensured that member states would adjust to external imbalances largely by means of domestic macroeconomic and microeconomic policies. In that sense, the first Treasury plan came astonishingly close to the discredited gold standard of the 1920s.

The purpose of the original Treasury plan was not laissez-faire. It was, in fact, to prevent the kind of unconstrained laissez-faire competition among nations that had led to economic warfare in the 1930s. But the disciplines that were proposed to prevent such competition applied to governments as well as to private actors. Both Morgenthau and Hull wanted governments subjected to a set of rules that would lend openness and predictability to market forces and prevent the kind of government manipulation of market forces that Morgenthau identified with the "gangsters" of the 1930s. Thus, the White Plan repudiated the national capitalism of the 1930s, the command-oriented policies of Hjalmar Schacht, president of the Reichsbank in Nazi Germany. It called for government intervention to stabilize exchange rates but created a process of international consultation on other key national economic policies that effectively limited government intervention or required governments to promote the flexible reallocation of resources through domestic policies, if they did intervene, rather than to restrict trade.

The U.N. Stabilization Fund called for in the White Plan of April 1942 (forerunner of the IMF) fixed exchange rates within a narrow margin that countries could change only with the approval of 80 percent of the membership of the Fund. Although all other voting would be on a weighted basis, White recommended that for voting on exchange rate changes, each country have only one vote, to avoid giving richer nations too much authority over weaker ones. This meant the United States could not veto such a decision with respect to the dollar, an incredible surrender of national sovereignty on the part of the dominant economic power. White acknowledged this concession but explained that changing exchange rates affected other countries and "unless nations are willing to sacrifice some of their power to take unilateral action in matters of international economic relations, there is very little hope of any significant international cooperation" (Horsefield 1969, vol. III, p. 66).

The White Plan did not stop there. It went on to require member countries (1) to abandon all restrictions and controls over foreign exchange transactions within 1 year after becoming members (Section III-1 of original outline; the very first draft set a deadline of only 6 months; Gardner 1980, p. 89), (2) not to adopt any domestic monetary measure that in the opinion of the majority of the Fund members could lead to serious disequilibrium in the balance of payments if 80 percent of the members disapproved of that measure (Section III-5), (3) to eliminate discrimination in currency and trade relations (Section III-4), and (4) to embark on a gradual reduction of existing trade barriers (Section III-6).

White explained that these requirements were not based on the nineteenth-century economic creed that all interference in international trade, capital, and gold movements was harmful. The plan actually called for strict government controls on the flow of private capital, designed particularly to staunch capital flight and speculation. Moreover, in other areas, the intent of the White Plan was not "to prohibit instruments of control" but "to develop those measures . . . as will be most effective in obtaining the objectives of worldwide sustained prosperity" (Horsefield 1969, vol. III, p 64). Exchange controls and restrictions might be used, for example, but "only when they [are] clearly justified by the circumstances, and only to the extent necessary to carry out a purpose contributing to general prosperity" (p. 64). Reduction of trade restrictions would focus only on "unnecessary barriers to international trade" (p. 48).

The right of the Fund to disapprove domestic policy measures, White conceded, was the most controversial yet important power of the Fund because "there will be instances in which the case is clear enough and the consequences important enough to justify the exercise of that power" (p. 68). The requirements that a majority of the Fund members consider the domestic policy measure to be significant enough to cause serious balance-of-payments problems and that 80 percent of the members be needed to disapprove it were designed to ensure that the Fund did not use this power arbitrarily. Nevertheless, even with these qualifications, the early White Plan made unmistakably clear that domestic policy measures might ultimately conflict with the requirements of international price (exchange rate) stability and that, if that were the case, domestic policy measures would have to be changed.

The efficient premises of the White Plan become even more apparent when one looks at the provisions for finance. The plan offered only small and temporary balance-of-payments financing, $5 billion altogether, "large enough to take care of fluctuations in the country's balance of payments that occur within a year or two"(Horsefield 1969, vol. III, p. 50). Within a year or two, in short, the borrowing country would be expected to put its domestic economic house back in order. Fund resources would come from quota subscriptions by each member paid partly in gold and the rest in local currencies, which legislatures would have to approve. The plan stated emphatically that the "purpose of the Fund could not be to supply an unlimited amount of foreign exchange to any country which might not wish to adopt the proper measures to correct a prolonged disequilibrium in its transactions with foreign countries" (p. 50). Thus, a country could draw up to 100 percent of its original quota, but for amounts beyond that it would require approval of 80 percent of the Fund's membership and would have to agree "to adopt and carry out measures designed to correct the disequilibrium in the country's balance of payments, which the Fund recommended . . . " (p. 42). Given the limits on finance, the emphasis clearly fell on early adjustment actions, and given the constraints on trade and foreign exchange policies, the brunt of the adjustment fell on domestic policies (see Table 1-4).

White did propose, along with the Stabilization Fund, a World Bank for postwar reconstruction that would have had substantially larger resources ($10 billion). The Bank also had powers to raise additional resources on private capital markets and to engage in anticyclical stabilization measures. But the World Bank plan was never pursued seriously by Treasury officials in the early deliberations; and when it was decided in 1942 to give priority to the Stabilization Fund (Oliver 1975, p. 138), Treasury officials did not raise the amount to be provided by the Fund, even though other executive agencies warned that the planned resources would be sorely inadequate (Gardner 1980, p. 75). White himself considered the direct loans of the Bank to be far less significant than the guarantees for private loans that the Bank would provide (Blum 1967, p. 256).

Reviewing the details of the original White Plan makes it hard to go along with some interpretations of U.S. policy in this period. For example, Fred Block argues that White's "concern was to create an international monetary order in which countries would be able to pursue systematically full employment policies without the danger of exhausting their international reserves" and that, although the Fund and the Bank acquired a "great deal of influence over domestic economic policies . . . this influence was to be used to coerce member nations to pursue policies consistent with global full employment" (1977, pp. 44–45). Richard Gardner claims that the "primary aim of the Treasury planners was not to restore a regime of private enterprise but to create a climate of world expansion consistent with the social and economic objectives of the New Deal" (Gardner 1980, p. 76).

These views seem to go well beyond the evidence of this period. "Generally," as Robert Pollard writes, "U.S. wartime planners favored private enterprise over state ownership of the means of production and commerce through private channels over state trading" (1985, p. 8). What is more, recent archival research by G. William Domhoff and others shows that White and especially

Morgenthau were neither the radical Keynesians or New Dealers they are some-times presumed to be nor at odds during this period with the capitalist interna-tionalist elite in the State Department (Domhoff 1990, Eckes 1975). According to this research, White's earliest ideas for the postwar monetary system were discussed already in fall 1941 by members of the Economic and Financial Group of the Council on Foreign Relations' War–Peace Study Groups. The Economic and Financial Group was chaired by Professor Alvin Hansen of Harvard, a fully committed Keynesian, and Professor Jacob Viner of the Uni-versity of Chicago, a more traditional market economist. Viner apparently had substantial influence in drafting the White Plan, and according to the Morgen-thau diaries (transcriptions of his telephone and office conversations), Morgenthau and Viner held considerably more conservative economic views than White (Domhoff 1990, Chapter 6). White, like Hansen, favored greater autonomy for member states to pursue expansionist domestic policies, but as Alfred Eckes notes, "there is no evidence . . . that White found autarky an acceptable second-best alternative to multilateralism" (1975, p. 46). Moreover, White's biographer, David Rees, describes White as a "moderate Keynesian" and notes that "liberal commercial policies, involving a lowering of trade barri-ers, were clearly a corollary of the new postwar world of the White Plan" (1973, pp. 66, 139). Because, as we have noted, liberal trade policies, along with fixed exchange rates and limited financing, imposed certain constraints on expan-sionary domestic economic policies, White and other U.S. officials un-doubtedly had mixed views (Pollard 1985, p. 7). But the White Plan of 1942 clearly resolved these conflicts in favor of freer trade and more conservative domestic economic policies.

On the one hand, therefore, the White Plan did represent a break with the unconstrained laissez-faire of the 1930s. It sanctioned government intervention in exchange rates and capital flows. On the other hand, however, it carefully disciplined this intervention, and it sought to place both the authority to inter-vene and the restrictions on intervention under international controls. It re-quired members "to part with a considerable measure of economic sovereignty" (Gardner 1980, p. 74) and even gave the Fund, in the words of Richard Cooper, the "character of an economic ministry in a world government" (1968, p. 33). By setting steep requirements to change exchange rates (majority vote with no country having veto power), to borrow (approval by 80 percent of Fund mem-bership to draw beyond relatively small initial quotas), and to restrict trade (although permitting "necessary" trade barriers), the plan ensured that domes-tic policies would have to absorb the largest burden of adjustment. It was not a new gold standard, but it differed from the old gold standard more because it shed the automaticity of that standard and relied on the judgment and good faith of national and international policymakers than because it abandoned the policy priorities of the gold standard. (Friedrich Hayek noted this fact when he compared the White Plan to the subsequent, much looser Bretton Woods agreements; House of Representatives 1945, p. 849.)

Why, then, it might be asked, did the Treasury Department ever propose such a politically severe economic system? Two answers seem most plausible:

because the system followed from the then dominant Wilsonian definition of America's national purpose that stressed technical, legal, and commercial solutions to world problems, and because early postwar planning took place in a highly abstract context in which political realities did not significantly intrude. As those political realities came to the fore in 1943–1944 in the negotiations with the British, the Soviet Union, and the U.S. Congress, the White Plan had to be significantly amended.

The Keynes Plan: More Domestic Expansion

The British proposal developed in 1941–1942 by John Maynard Keynes, already well known for his interventionist economic policies, called for more generous international financing than the White Plan (see copy of February 1942 draft; Horsefield 1969, vol. III, pp. 3–19). It envisioned a Clearing Union with overdraft rights denominated in a new international reserve currency known as bancors. These overdraft rights would be established on the basis of a country's share of international trade during a representative period before the war. No limits were set on these rights, although Keynes foresaw discussions between creditors and the Clearing Union once debtors reached a certain level of indebtedness and he later imposed symmetrical interest charges on both creditor and debtor balances. The plan also allowed for greater exchange rate flexibility, up to 5 percent a year without approval of the Fund and an interim transition period of 5 years or longer during which exchange restrictions could be maintained. In all cases, including exchange rate changes beyond 5 percent, Britain and the United States would control the majority of votes.

Further, the British proposal allowed for discriminatory trade measures and, in sharp contrast to the White Plan, did not mention explicitly the reduction of trade barriers, although it hoped that with the ample financing provided by the proposal, members would "feel sufficiently free from anxiety to contemplate the ultimate removal of the more directly dislocating forms of protection and discrimination and . . . the prohibition of some of the worst of [these forms] from the outset" (Horsefield 1969, vol. III, p. 8). As Roy Harrod reports, Keynes was hopelessly skeptical about liberal trade and envisioned (quite erroneously as it turned out in the 1950s and 1960s) a future of state trading, international cartels, and quotas (1971, p. 568).

The British plan thus allowed for more liberal financing and more adjustment through trade restrictions and exchange rate changes, leaving greater room for inflationary domestic expansion. After the disastrous domestic consequences of the attempt to return to the gold standard in the 1920s, and with Keynes hoping to apply his expansionist economic program in Great Britain after the war, British authorities tilted toward internal expansion over external stability.

Even then, however, it would be wrong to conclude that British planning sought to preserve unlimited national flexibility at the expense of international stability. Earlier drafts of the Keynes Plan (there were three such drafts before the February 1942 draft printed in the Horsefield volume) imposed more discipline on adjustment policies. For example, drafts developed in the fall of 1941 gave the Clearing Union authority to require a country to depreciate or appreciate its currency (not just authority to approve a requested change), if the country exceeded a certain level of credit or debit on its balances with the Union. The Union also had the right to transfer at the end of each year surplus accounts of creditor countries to a reserve fund to be used by supranational bodies to provide relief and preserve the peace (Van Dormael 1978, pp. 34–35). The February 1942 draft dropped these provisions. Keynes explained that these provisions had been criticized for being too dependent on rules and not allowing sufficient discretion for the administrative authority. The fourth draft, he said, was now deliberately biased toward discretion because "it may be better not to attempt to settle too much beforehand" (Horsefield 1969, vol. III, p. 6). Nevertheless, Keynes too initially foresaw significant limitations on national sovereignty, albeit with less emphasis on domestic price stability, liberalized trade, and rigid exchange rates.

The real difference between the early British and American plans was not so much the degree of internal discretion versus external stability but the question of which countries should bear primary responsibility for external stability, the creditor or debtor countries. U.S. planners sought to place important responsibilities on debtor countries, willing to accept severe limitations on the internal policies of these countries because the United States never expected to be in the position of a debtor. The British, on the other hand, expected to be a debtor country for some time after the war and hoped therefore to saddle creditor countries with significant adjustment responsibilities so these countries could not hoard international reserves, as Keynes believed the United States did in the interwar period, and force deflationary policies on the rest of the world economy.

Creditors, Keynes hoped, would either have to lend their surplus reserves, which was accomplished automatically by the overdraft provisions in the Clearing Union, increase their imports, or accept discrimination against their exports. The United States eventually accepted the latter possibility in the so-called scarce currency clause of the IMF. This clause allowed a currency that was being hoarded by a surplus country to be declared scarce and thereby effectively rationed to reduce deficit countries' imports of that country's exports. In this way the surplus country would not be allowed to continue to hoard export revenues but would have either to lend more or import more if it wished to continue to export.

Keynes, in short, wanted to see external stability biased toward higher levels of domestic expansion even at the risk of higher inflation, whereas the original White Plan preferred external stability based on more moderate levels of domestic expansion to ensure stable prices. Fortunately for the early postwar world economy, the system as it was applied after 1947 was biased toward low inflation. The world learned much later in the 1970s that more employment could not necessarily be bought at higher levels of inflation (see Chapter 6).

Bretton Woods: Compromise Without Purpose

The compromise that emerged at Bretton Woods involved a trade-off between financing and flexibility. The British accepted much less financing than they initially wanted ($8.8 billion in total IMF resources compared to the $26 billion provided for in the Clearing Union), and the Americans accepted less discipline and international jurisdiction over adjustment policies (exchange rate, domestic and trade policies) than they originally desired (at least as laid out in the provisions of the White Plan of 1942). The final IMF agreement permitted a one-time 10 percent change in a member's exchange rate without Fund approval but subsequent changes only with Fund approval and then only to correct so-called fundamental disequilibria in the balance of payments. What qualified as "fundamental" was left ambiguous, opening up a loophole that Fund critics sharply attacked. Moreover, in approving or disapproving an exchange rate change, the Fund was prohibited from taking into account any domestic social or political policies. This meant, as one critic of the Bretton Woods agreements explained, that

> these provisions would seem to open the door wide to any member state . . . to debase its monetary unit, that is, to reduce its gold par. . . . [For] it is precisely by reason of expensive domestic social and political policies that nations usually spend beyond their means, incur heavy budget deficits, borrow excessively, inflate their currencies and finally recognize the inflation as a fait accompli by formally debasing their monetary unit [House of Representatives 1945, p. 833].

The United States did manage to insert back into the agreements at Bretton Woods the authority for the Fund to communicate its views to member states on any matter arising under the agreement. By a two-thirds vote, the Fund could issue a public report "regarding [a given country's] monetary or economic conditions and developments which directly tend to produce a serious disequilibrium in the international balance of payments" (Article XII, Section 8). This provision made it clear that domestic policies were not completely immune to Fund criticism (Gardner 1980, pp. 115–116). Even Keynes, in spelling out his original plan, recognized that the international authority had the right to be somewhat, just not "too grandmotherly" (Gardner 1980, p. 88).

The compromise, as U.S. officials later explained, led to a "stable, if *moderately* flexible" system of exchange rates (White 1945, p. 199, emphasis added). White compared the flexibility of the exchange rate system to the sway of the Empire State building (House of Representatives 1945, p. 235). As long as this flexibility was built into the design of the system, it would not destroy it. The idea was to prevent manipulations or arbitrary changes in exchange rates, not to dictate specific domestic social or economic policies underlying a country's basic institutions. "The architects at Bretton Woods," as Richard Gardner concludes, "apparently hoped that, with the aid of Fund resources, deficit and surplus countries could be relied on to restore a balance within a relatively short time by *reasonable domestic policies* and by occasional changes in exchange rates to correct 'fundamental disequilibrium'" (Gardner 1980, p. xi, emphasis added).

The Fund agreement permitted a transition period of indefinite length during which exchange restrictions could be maintained. After 5 years, maintaining such restrictions required Fund approval, but no predetermined termination date was called for. Because it also permitted restrictions on capital transactions, there was considerable doubt whether the Fund would be able to distinguish between capital and current account restrictions and reduce trade restrictions. Moreover, the agreement said nothing explicit about the need to end discrimination and liberalize trade barriers (a major concession to Keynes' Plan over the White Plan). The scarce currency clause actually sanctioned restrictions against the exports of a creditor country. On the other hand, this clause also carried the potential for liberalizing import policies in the creditor country. If the creditor country wanted to avoid discrimination against its exports yet was unwilling to lend more resources to the Fund, it could agree to accept more imports. As Senator Fulbright pointed out for the United States, the "operation of this fund successfully would be inconsistent with the maintenance of a high tariff exclusive policy on the part of this country" (Senate 1945, p. 111).

This commitment to freer trade lagged well behind the other commitments at Bretton Woods. The United States persistently attacked discrimination, that is, applying trade barriers differentially among trading partners, but it was ambivalent or at best "selective, reciprocal and moderate" with respect to the commitment to lower the trade barriers themselves (Gardner 1980, p. 22). From the London Conference of 1933 to the Atlantic Charter discussions in August 1941 and the Article 7, Lend Lease disputes throughout the war, Cordell Hull strove mightily to break up the British imperial preference system, which gave British exports preferential access to the markets of the Commonwealth countries. The British, however, conceded little until 1949 on either discrimination or quantitative restrictions (see discussion in Chapter 4). Meanwhile, initial progress to reduce tariff barriers was also slow. By 1945 the United States had negotiated 32 reciprocal trade agreements with 27 countries granting tariff concessions on 64 percent of all dutiable imports and reducing rates by an average of 44 percent (Destler 1986, p. 10). Trade with these countries was growing faster than trade with other countries (Hull 1948, vol. I, p. 746); the case for freer trade was being made by the results. But attitudes changed slowly. The closest vote to renew the Reciprocal Trade Agreements Act came in 1940 (42–37 in the Senate and 218–168 in the House). The margin in 1945 was better (54–21 in the Senate and 239–153 in the House) but still not overwhelming (Congressional Record 1940, 1945).

The Bretton Woods agreements did not establish an international reserve currency, as much as Keynes had hoped it would. White was sympathetic, having added and then taken out of his plan again a currency unit known as unitas. But White recognized, as he indicated in his early drafts of the Bretton Woods accords, that the "adoption of a common currency by several countries is possible only if they surrender separate sovereignty in monetary and credit policies in favor of sovereignty exercised by one over all of them, or by an international organization" (Horsefield 1969, vol. III, p. 80). White knew Congress would never accept the surrender of monetary authority and creation of a "new-fangled international currency" (Van Dormael 1978, p. 110), and Keynes hoped to avoid either a gold or a dollar standard, although he recognized that the United States held the trump cards on this issue.

The key to the emergence of an international reserve currency, as John H. Williams, then vice-president of the New York Federal Reserve Bank testified, was confidence in the issuing country or authority (Senate 1945, p. 337). That confidence depended upon political factors, and although the dollar was the only usable international currency after the war, the United States in 1944–1945 had no strong political reasons to supply the dollars. In 1945 it was withdrawing from world responsibilities and advocating a concept of shared political community with the Soviet Union that suggested no urgency to begin postwar economic reconstruction. The Bretton Woods agreements established only the shell or process of international economic cooperation, while their principal signatories, the United States and the United Kingdom, turned to deal with their individual postwar priorities at home or to address their international problems through more limited bilateral arrangements (e.g., British loan—see later section). In addition, compromises at Bretton Woods had introduced so much ambiguity and even contradiction into the content of economic policies (e.g., the contradiction between inflationary domestic policies and stable exchange rates or between discriminatory trade restrictions and the free convertibility of currencies) that the agreements did not offer particularly effective guidance for organizing postwar reconstruction in the face of severely limited resources and productive capacities.

The Bretton Woods conference, later mythologized as the ideal way to resolve conflicting policy differences and consummate a grand design for the world economy (see Chapters 8 and 9), actually provided very little economic or political impetus to the postwar world. It did create some institutions that established the habit of consultation and symbolized a commitment to act in good faith. These were factors that were not without effect on later decisions to cooperate more substantively (see Chapter 4). But good faith then as now would prove to be insufficient without stronger political incentives and more consistent economic policies.

Failure of Bretton Woods

The Bretton Woods agreements passed Congress in 1945 by substantial margins, but Congress provided only $2.75 billion as the U.S. contribution to the IMF. In addition, in August 1945, Congress cut off bilateral aid under lend lease programs through which the United States had financed wartime military sales to the allies. The United States created a grand design for postwar peace and reconstruction and then failed to fund it.

Preponderant power was not enough to activate U.S. policy. As John Gaddis writes, "American omnipotence turned out to be an illusion because Washington policy-makers failed to devise strategies for applying their newly gained power effectively in practical diplomacy" (1972, p. 356). Economic motives were also not enough. Recent scholarship has sharply questioned the revisionist accounts of postwar U.S. foreign economic policy that stressed the U.S. need for export markets (for revisionist accounts, see Kolko and Kolko 1972 and Williams 1962; for critiques of these accounts, see Pollard 1985 and Gaddis 1987). If export motivations had been primary, U.S. policymakers would have certainly funded the Bretton Woods agreements more generously or accelerated bilateral assistance, since war-ravaged countries could buy U.S. products only if they could borrow U.S. dollars.

As it was, U.S. merchandise exports as a share of GNP were smaller by half in 1946 — 5.5 percent — than they had been in 1916 — 12 percent (ERP 1988, pp. 248, 364; Frieden 1988, p. 70). Sizable amounts (over 30 percent) of some agricultural products were exported (cotton, tobacco, and rice), and people close to these industries, such as Assistant Secretary of State William Clayton, a former executive from the cotton industry, stressed export opportunities when testifying on behalf of the Bretton Woods agreements (House of Representatives 1945, p. 275). But manufacturing exports were much less significant and studies of the origins of the free trade movement in the United States show that the initiative came primarily "not from the private sector, nor from the White House, but from champions of freer trade in the State Department" (Haggard 1988, p. 118). Neither then nor now has the State Department been considered the most likely bastion of American capitalism.

American ideology was a factor in motivating U.S. policy in this period (Russett 1985, pp. 228–231; Keohane 1984, pp. 136–137; see also Gilpin 1977, p. 73; and Krasner 1978, pp. 337–342). But this was not an undifferentiated American liberalism, let alone a virulent anti-Communism, as the reference to ideology usually implies. It was a specific definition of America's national purpose, the legal–commercial view, which sought extensive cooperation with the Soviet Union. This view was quite different from the containment view that emerged in 1947–1948 or subsequent definitions of American purpose that followed in the 1960s. National purpose, as used in this study, is not a constant that derives from an unchanging political culture or is unrelated to real and changing structural and physical factors in the external environment (see Chapter 2). Accordingly, it has little in common with what other studies refer to as ideology or general American liberalism.

The Bretton Woods design in 1944–1945 reflected a specific, voluntarist formulation of American purpose that did not address well either the internal or external realities at the time and therefore did not significantly affect those realities. At home, U.S. officials mobilized public support for the Bretton Woods agreements, but the awareness of these agreements among the general public, particularly in comparison to Marshall Plan proposals later in fall–winter 1947 (see Chapter 4), was relatively thin. One Gallup poll showed only 23 percent of the respondents could even relate Bretton Woods to world affairs (Eckes 1975, p. 196). In the end the Bretton Woods ideas did not succeed in persuading a broad range of special interests in the United States or elsewhere that their interests would be served under these arrangements. Maybe that was because there was no perceived threat to American security in 1945 or because no scare tactics (i.e., anti-Communism) were used by Bretton Woods proponents. But maybe it was also because the Bretton Woods ideas simply did not correspond sufficiently to reality and were therefore not credible to a large enough number of American citizens. As it turned out, the Bretton Woods ideas vastly overstated the extent of common domestic political interest between the United States and the Soviet Union and stretched the diplomatic processes of international economic cooperation between these countries so thin that economic policy became devoid of coherent content.

John Williams, vice-president of the Federal Reserve Bank in New York and a critic of the Bretton Woods agreement at the time, noted the fact that free trade was ultimately incompatible with managed economies. "In a world comprising a fully managed economy like that of Russia, a centrally planned economy in England, if anything like the Beveridge model should be adopted, and some kind of modified free enterprise system in [the United States]," he warned, "there will be much room for honest doubt as to whether a system of multilateral trade and free exchange is any longer workable" (Senate 1945, p. 323). Williams and other critics were right to note that the content of the Bretton Woods agreements had been compromised beyond any economic effectiveness. But they were wrong to believe, as they did, that the multilateral process of the Bretton Woods arrangements could be replaced by bilateral arrangements. The British loan experience would prove that multilateral cooperation was necessary even for more limited bilateral arrangements to succeed.

The British Loan Experience

After cutting off lend lease in August, the United States extended a loan to Britain in December 1945, the price some analysts say was necessary to ensure British acceptance of the Bretton Woods agreements (which passed Parliament in late December; see Gardner 1980, p. 224). In the debate over this loan, which lasted 6 months, Congress began to recognize the large inconsistencies between the policy provisions of the Bretton Woods agreements and the domestic requirements of effective international economic cooperation. It saw the need for Britain to exercise more domestic policy discipline and to put an end to discriminatory and state trading practices (Gardner 1980, p. 198). In the end, however, it approved the loan without such conditions, primarily, as Richard Gardner notes, because of deteriorating relations between the United States and the Soviet Union (1980, pp. 249–252; see also Pollard 1985, pp. 66–73).

The British loan thus became a test case of the loose and muddled economic policy provisions of the Bretton Woods agreements. Not unexpectedly, it failed. When the British resumed full convertibility 1 year later, the pound came under immediate and intense pressure; and within 1 month, Britain suspended convertibility once again.

The British loan experience, however, provided some useful policy lessons. First, Congress, U.S. officials, and officials of other countries came to understand better that domestic policies were ultimately more important for restoring exchange rate stability than the amount of international finance that might be made available. The British loan had provided a total of $3.75 billion, $1 billion more than the entire U.S. contribution to the Fund. Yet the loan was not enough, and there was a widespread feeling that the failure was due chiefly to British domestic policies of full employment, easy money, and extensive social services (Gardner 1980, p. 311). As Margaret de Vries and others noted, officials in the Fund and elsewhere became more aware during this period of the underlying importance of national policies (de Vries 1986, p. 32; Van der Wee 1986, p. 437). To the extent that it operated in this period, the Fund tightened its requirements. More important, Marshall Plan aid was subsequently tied much closer to domestic policy performance (see Chapter 4).

Second, the British loan suggested the overriding importance of security objectives or a convincing concept of national purpose to persuade Congress and the American people to make sizable resources available for postwar recovery. Given national differences, universal approaches and organizations were too vague. Experience with functional organizations such as the U.N. Relief and Rehabilitation Agency (UNRRA) had demonstrated that international organizations lacked adequate direction and accountability and that the United States failed to get sufficient acknowledgment for the resources provided (Castle 1957). As Seymour Harris wrote in 1948, "UNRRA had taught American authorities that dollars might be used by an international organization to support economies at political loggerheads with the United States" (Harris 1948, p. 24). The subsequent European Recovery Program, he notes, reflected the "determination to tie appropriations to control of use by the United States" (1948, p. 25).

Third, the British loan demonstrated that bilateral arrangements, which critics of Bretton Woods had championed (such as John Williams; see earlier), were also insufficient to deal effectively with postwar economic reconstruction. The British loan failed in part because the British paid for their imports in pounds, which their suppliers could immediately convert to dollars, whereas the British received payments for exports in currencies that were often inconvertible. Thus, unless all countries moved toward convertibility simultaneously, no one country could do so alone for very long. This experience was an important factor affecting the design of the subsequent European Payments Union (see Chapter 4).

Domestic Politics of Postwar Liberalism

Bureaucratic and domestic politics offer poor explanations of why the United States acted indecisively from 1942 to 1945. During this period relations between the State and Treasury departments, although sometimes fractious, were relatively harmonious as long as Hull and Morgenthau were in office. By contrast, between 1945 and 1947, when the United States finally did mobilize to act, three secretaries of state and two secretaries of treasury occupied these posts. Similarly, the New Deal coalition in Congress peaked in 1943, and by 1947 the Democrats lost the Congress altogether to the Republicans. Thus, the United States did not act effectively from 1942 to 1945, when domestic bureaucratic and congressional relations were relatively coherent, but did act effectively in 1947, when domestic political relations were more divided. Something other than politics had to account for this difference. From the perspective of this study, that something else was the substantive debate going on in the larger nongovernmental arena, or what we called the cocoon of consensus building surrounding the immediate policymaking process (see Chapter 1).

After 1943 a more pragmatic, internationally minded group of business, labor, and professional leaders began to emerge as an influential group outside the official policymaking process. The large investment houses and capital-intensive multinationals, along with allies in the trade unions, agricultural community, academic circles, and private foundations, led this group. The war boosted their influence. In 1945–1946, however, this group still had greater influence outside the channels of official policy than in the policymaking process. The U.S. Congress was still not strongly internationally minded. As we have seen, the universal, legal–commercial approach promoted at the United Nations involved only a limited and happily inexpensive type of international intervention on the part of the United States. As the flip side of isolationism—that is, going from the view that the American system was unique to the view that the American system was embraced by the entire world—this approach did not demand strong efforts to shape foreign perceptions or policies. Thus, the new group of internationalists exercised influence largely through nongovernmental organizations, such as the Council on Foreign Relations and the Committee for Economic Development (for more on the consensus in security policy see Chapter 4).

By 1945–1946 a consensus on domestic economic policy began to gel in the nongovernmental sector. Although Congress and the administration continued to flirt with the inconsistent economic policies of the Bretton Woods and British loan agreements, business views began to converge around three critical issues—a conservative version of Keynesian macroeconomic theory, a labor-management partnership that tied economic gains for workers to increased productivity, and a firmer commitment to multilateralism and intergovernmental organization of the world economy that blended government action with private enterprise and free markets (Hogan 1987, pp. 14–18). The key to this consensus was the recognition that domestic as well as international prosperity required some discipline over macroeconomic policy and some restraint on microeconomic policy.

It is necessary to elaborate the different positions advanced in this debate over domestic economic policy and to identify the policy options that prevailed and those that were rejected, because subsequent arguments about embedded liberalism and the domestic politics of productivity have failed to make adequate distinctions among various fiscal, monetary, and microeconomic policy options in this period. The most careful studies of the period are provided by

Robert Collins and Herbert Stein (1981 and 1984, respectively; for an account more narrowly focused on the CED, see Schriftgiesser 1960). Stein was a staff economist at the Committee for Economic Development, which Collins identifies as the lead organization in the development of the postwar business–labor consensus (1981, p. 129). According to Stein, four schools of economic thought were represented in the debate in 1945–1946: reformers and planners, strict and exclusive Keynesians, conservative macroeconomists, and conventional conservatives (for the following discussion, see 1984, pp. 71–75). The reformers and planners embraced Keynesian macroeconomic prescriptions for full employment but did not think that that was enough and therefore also advocated more detailed government intervention to achieve high wages, promote industrial development, and expand welfare benefits such as housing and health.

The second group of strict and exclusive Keynesians was concerned primarily with the achievement of full employment through the use of fiscal policy and was not that much concerned with inflation, microeconomic intervention, or extensive regulations, although this group was not averse to price controls if necessary to control inflation. (I refer to this group later as "deep" Keynesian once it embraced more enthusiastically the idea of microeconomic intervention to control prices and to restructure industry.) The strict Keynesians considered monetary policy ineffective or effective only in limiting, but not stimulating, demand.

The third school of conservative macroeconomists agreed to the management of aggregate demand through macroeconomic policy but sought high, not necessarily full, employment and were more skeptical about the fine tuning of fiscal policy to achieve these goals. They worried more about inflation and saw a greater role for monetary policy. Indeed, once fiscal policy was set to achieve high employment, they considered monetary policy the chief instrument of macroeconomic stabilization. They opposed any significant further intervention in microeconomic affairs through subsidies, regulations, or price controls, although they supported light public works such as road repairs (see Yntema et al. 1946).

Finally, the fourth school of conventional conservatives rejected any Keynesian manipulation of the economy and favored lower taxes for corporations and less power for unions. (Collins identifies the same four groups, although he divides the reformers and planners into two groups—the compensatory spenders, who viewed capitalism as capable of only short periods of stability, and the interventionists or stagnationists, who considered capitalism incapable of regeneration; 1981, pp. 9–10.)

The consensus that emerged at the time clearly rejected the policy options advocated by the reformers and planners and by the conventional conservatives. The interesting and decisive debate, as Stein records, was between the strict and exclusive Keynesians, on the one hand, and the conservative macroeconomists, on the other (for following discussion, see 1984, pp. 76–87). The two views clashed in the legislative debate over the Employment Act of 1946. The key issues were fourfold:

- Did full employment imply a target so ambitious that it could be achieved only by harsh government controls or by inflation?
- How much emphasis should be given to the manipulation of the budget, especially the deficit, to achieve full employment?
- To achieve full employment, could the government direct or suppress the private economy?
- Should the legislation give implementing authority to the Keynesian economists in the Budget Bureau or to an outside, nongovernmental council?

"On all of these issues," as Stein reports, "compromises were reached in a conservative direction." After a major national debate, Congress removed from the bill the term "full employment" and substituted the goal of "maximum employment, production and purchasing power." It "rejected an overly ambitious, inflationary definition of [that] goal, rejected exclusive reliance on deficit financing as the means [to achieve it] and reaffirmed its devotion to the private enterprise system." It also established the Council of Economic Advisors, which although within the government, created a potential counterweight to the Budget Bureau, which was thought to be biased toward more ambitious Keynesian policies. When the Republicans took control of both houses of Congress in 1947, the conservative economic coalition went on to "domesticate" exclusive Keynesianism by proposing a fiscal policy standard of balanced or slightly surplus budgets at times of high employment (rather than fiscal deficits of any size to achieve full employment), by using monetary policy to supplement demand and contain inflation, and by declaring a "truce" against further microeconomic intervention. As Stein points out, "explicitly or implicitly, this became the standard approach to fiscal policy in the Truman and Eisenhower years and into the Kennedy years" (Stein 1984, p. 80).

Several points about the conservative policy standard require emphasizing. First, the CED, which was in the vanguard of the consensus process, was the most "liberal" of the business groups. The Chamber of Commerce and National Association of Manufacturers were more reluctant supporters of even the most conservative elements of Keynesianism (Collins 1981, Chapter 4). As Collins notes, the consensus embraced the "right wing" of Keynesianism, the high-employment, not the full-employment wing. The key formulation on fiscal policy was that tax rates should be set "to balance the budget and provide a *surplus for debt retirement* at an agreed *high level of employment* and national income" (Collins 1981, p. 133; emphasis added). In addition, "tax rates, having been set to yield a surplus at high employment, would then be left alone barring 'some major change in national policy or condition of national life'" (Collins 1981, p. 133). The conservative macroeconomists, in short, were concerned about balancing budgets at high employment (though they accepted budget deficits in a recession), and they were opposed to fiscal fine tuning.

Even more important, the conservative macroeconomists saw a key role for monetary policy, which, as Collins notes, "is far more conservative in its implications than the fiscal alternatives" (1981, p. 9). In the late 1940s the Federal Reserve Board was still "serving as an adjunct of Treasury finance," pegging the interest rates on Treasury notes and bonds to absorb the large wartime debts of the federal government (Melton 1985, p. 20). In 1948 President Truman appointed Thomas McCabe, a CED alumnus and proponent of the conservative macroeconomic policy consensus, to chair the Federal Reserve Board, and McCabe "became an important figure in the Fed's attempt to free itself from the domination of the Treasury Department" (Collins 1981, p. 148). Eventually, in 1951, the Fed gained its independence in the celebrated "March Accord" between the Treasury and the Federal Reserve. The conservative macroeconomists were not monetarists, emphasizing a quantity rule for monetary creation. They expected the Fed to use discretion, but they also expected the Fed to give priority to price stability and not the financing of fiscal deficits.

Finally, the conservative economic policy consensus emphatically eschewed microeconomic intervention. The aggressive interventionists or stagnationists, as Collins calls them, were at the far left of the Keynesian spectrum. The postwar consensus "defeated . . . those left-wing liberals whose gloomy predictions of economic stagnation had led them to demand more centralized planning, greater government direction of the economy, and permanent programs of public works" (Hogan 1987, p. 14). "Acceptance of a Keynesian role for the federal government and of the wisdom of occasional deficit spending," writes Collins, "did not . . . signify adoption of the liberal, stagnationist position" (1981, p. 86). During this period, for example, the U.S. government turned back a "New Deal offensive" to nationalize control of U.S. petroleum reserves in Saudi Arabia and, when that failed, to negotiate a bilateral government petroleum cartel with Great Britain (Keohane 1984, pp. 151–159; Krasner 1978). The ideas of the early New Deal and centralized wartime administration did not triumph in the aftermath of the war; they were roundly rejected. Long-cycle evolutionary explanations of postwar U.S. economic policy miss these critical policy distinctions when they equate postwar policies with the New Deal.

What accounts for the emergence of this conservative economic policy consensus, "despite what seemed at the end of the war a wide range of opinion about national economic policy" (Stein 1984, p. 74)? According to the argument in this book, the consensus emerged from a competition among ideas initially carried on outside the governmental arena (e.g., Council on Foreign Relations and CED) and then focusing in on specific congressional legislation in the Employment Act of 1946. This competition was not predetermined, and although some elements of the resulting consensus may have had their roots in earlier developments such as business–labor associations under Herbert Hoover or extensive government intervention in the early New Deal days and during the war, it is misleading to conclude that the postwar consensus was a linear descendant of interventionist planning. Charles Maier has argued, for example, that the wartime experience and favorable postwar economic conditions promoted a "politics of productivity" in which business and labor groups were able to break the stalemate that the New Deal had encountered in the late 1930s and forge a "new consensus on interventionist planning" to boost productive efficiency (Maier 1977, pp. 613–614). These groups, according to Maier, approached "the transition to a society of abundance as a problem of engineering, not of politics." For them the "mission of planning became one of expanding aggregate economic performance and eliminating poverty by enriching everyone, not one of redressing the balance among economic classes or political parties" (Maier 1977, p. 615).

There are several reasons to be skeptical of this argument. First, its logic seems out of sequence. Normally, the pursuit of efficiency, whether it emphasizes market competition or the planning, engineering approach that Maier seems to have in mind, assumes a preexisting consensus. Such a consensus is necessary to motivate efficient economic policies (see discussion of national purpose in Chapter 1) and to encourage leading classes or hegemonic powers to absorb the costs of an open market system (Kindleberger 1973). This consensus is also necessary, in a planned economy approach, to establish and sustain the legitimacy of centralized institutions and planning, without which such an approach cannot be productive. Now, where did this consensus suddenly come from in the postwar period? Maier argues, on the one hand, that an evolving "associationalism" (i.e., cooperation) among business, labor, and other groups provided the basis for this consensus; on the other hand, he argues that the productive process itself is "apolitical" and provides the society with a means to overcome class conflicts that arise from scarcity. The reasoning is circular. Why would labor and business groups, if they were sharply divided by class conflicts in the first place, find it easier to agree on productive policies than on less productive ones, especially when it was believed by many at the time that efficient policies might actually exacerbate distributional and hence class differences? Moreover, if a New Deal style of government intervention and planning would ensure more equitable distribution of benefits, what would ensure that this intervention was also efficient and consistent with productivity and growth? It is at least possible that such intervention would weaken efficiency in favor of equity. And if government intervention was not efficient, then growth would slow and class conflict would not be eased but might even intensify.

Second, the argument vastly overstates the degree to which the postwar consensus embraced interventionism. Maier acknowledges that government planning "was accepted only in a restricted sense" and applied more to regional (e.g., TVA) than national affairs (1977, p. 614). But he tends to view Hoover's associationalism, Roosevelt's highly interventionist National Recovery Administration, centralized wartime planning, and the postwar politics of productivity as one long evolutionary cycle of increasing government intervention and domestic social transformation. This argument passes over all the policy and performance differences during this cycle, especially in the postwar period, which this study identifies and seeks to explain. By most accounts, for example, the politics of productivity, which greases social conflict with material wealth, should have increased inflation in the first decade or two after World War II (for an explanation of why this is so, see Zysman 1983, pp. 138–144). Yet just the opposite occurred as we observed in Chapter 1 (Figure 1-1). Inflation in the G-5 countries (France, Germany, Japan, United Kingdom and United States), indeed throughout the industrial world, was exceedingly modest from 1947 to 1967, especially in comparison to the decade after 1968.

The postwar consensus in the United States on relatively conservative, efficient economic policies did not emerge in a domestic or international political vacuum or mysteriously materialize out of unresolved class conflicts. It arose out of an intense struggle among competing ideas. (See Collins 1981, Chapter 6, for an excellent account of how the CED and other organizations participated in this interplay of ideas to establish the ascendance of moderate Keynesianism.) Eventually that struggle was decided by human energy and commitments, by policy choices that took into account relevant circumstances (e.g., opportunities for growth) but also sought to alter these circumstances (e.g., reverse the wartime propensity toward intervention, in areas such as price controls). The consensus was also not dictated by the Cold War; it was largely in place by 1946, before the onset of the Cold War between the Soviet Union and the United States.

To be sure, the Cold War reinforced the consensus. Indeed what may have sealed the fate of the deep planners and exclusive Keynesians was the gradual tightening of external conditions in 1946–1947. Conditions had deteriorated under ineffective U.S. policies in 1945 and now a higher premium was put on more efficient policies to revive Europe economically and politically. Without the growing recognition in the winter of 1946–1947 that a struggle was taking place in central Europe over the terms of political community and that something more efficient than the policies of the British loan agreement was needed to help reconstruct Europe and thereby assert and defend Western freedoms, it is at least arguable that American national purpose and economic policies would have continued to drift. The Cold War now intruded with greater force and led to both a redefinition of national purpose under the perspective of containment and an internationalization of conservative economic policy standards through the programs of the Marshall Plan.

4

The Marshall Plan:
Purpose and Policy for Prosperity

The policy of containment to defend the Western world from Soviet expansionism acknowledged the limits of shared political community in the postwar world. It recognized that the United States and the Soviet Union championed different conceptions of political society in central Europe and that the United States could not assume that its Wilsonian values of individual freedom and economic competition would be preserved, let alone advanced, in the postwar world without a commitment of American power. Initially, however, the United States hoped to exert this power primarily through political and economic means, holding to its historic rejection of military balances. In the Marshall Plan, containment retained the emphasis on law and commerce but now applied these principles to central Europe and the Western world, rather than to global relations as a whole. By the values it advocated, the Marshall Plan could not include the Soviet Union, any more than the universal Wilsonian view could. And when the Soviet Union decided that the Marshall Plan could not include the countries of Eastern Europe as well, the iron curtain fell. Eventually, after the Soviet Union threatened to cut off Western access to Berlin in 1948 and supported North Korea's invasion of South Korea in 1950, containment became militarized.

Containment brought a practical clarity to America's national purpose in 1947, just as the universal–utopian view had provided a theoretical clarity in 1942. It supplied the intensity but not necessarily the content of the international economic community that emerged. Containment made urgent the need to reconstruct Western Europe and to do so as efficiently as possible, given the severe constraints on economic resources. Not surprisingly, containment and its economic flagship, the Marshall Plan, retrieved some of the efficient economic policies that had influenced Treasury Department planning in 1942 and that returned in 1946 in the form of the policy triad of price stability, freer trade, and flexibility of labor, capital, and product markets that marked the conservative Keynesian consensus on economic and social policies in the United States. Marshall Plan institutions projected this consensus onto the international stage. As we examine in this chapter, national policies in the major Western European countries and, to a lesser extent, in Japan moved after 1947 incrementally toward more conservative macroeconomic policies, more moderate microeconomic intervention, and more liberal trade.

But the content of the postwar economic system was not self-evident. The choice of economic and social policies cannot be understood primarily in terms of the long-cycle evolution of social ideologies (emphasized, for example, by the embedded liberalism argument) or a domestic political process of growth (i.e., the politics of productivity). It reflects a continuous struggle between different conceptions of national purpose and different choices of economic policies that went on within each of the Western countries as well as among them. The struggle could have gone either way. The first postwar German government took power in 1949 by a one-vote margin. Politics in the United States was unusually turbulent. For the only time in the postwar period, both houses of Congress changed hands in the United States twice within 2 years (to Republican in 1946 and back to Democratic in 1948). Moreover, bureaucracy was fragmented. When the United States moved in the winter of 1946–1947 to initiate the Marshall Plan, it had no unified policy process. The National Security Council was not created until the summer of 1947. Yet American policy in this period integrated security and economic concerns better than in any subsequent postwar period.

The consensus that emerged in 1947–1948 was not a direct product of social history (i.e., culmination of the New Deal), growth politics, or coherent bureaucracy; it was a result of an intense discussion of alternative ideas debated in large part outside the official policymaking process in what we have called the cocoon of consensus building. In that cocoon, such organizations as the Council on Foreign Relations and the Committee for Economic Development worked to shape the consensus on containment and efficient economic policies that defined the structures of international politics for the next two decades.

International institutions played a powerful role in this period, not the global Bretton Woods institutions whose purposes and policies remained muddled, but the regional institutions of the Marshall Plan inspired by containment and the economic policy triad of price stability, flexible markets, and freer trade. These institutions—the Economic Cooperation Administration (ECA), the Organization for European Economic Cooperation (OEEC), the European Payments Union (EPU), and the institutions of the European Communities (Coal and Steel Community, Common Market, and Atomic Energy Community) "blended government action . . . with private enterprise" to build the massive, highly interdependent world markets that emerged in the 1950s and 1960s (Hogan 1987, p. 18). Institutions played a more dominant role in this period than they would in later periods (see Chapter 6), because markets barely existed in 1947 and direct bargaining among governments was the only means to remove the obstacles to create international markets. But government institutions never became too dominant. Many of them in this period actually disappeared once their job was done (the ECA, EPU, and in its original form, the

OEEC). Similarly, market policies never became too dogmatic. The ambitious long-term goal of steering Western economies toward more competitive and noninflationary markets was combined with pragmatic accommodation of short-term political and economic requirements.

This period holds powerful lessons for students of international economic cooperation, both for those who believe that institutions are the sole answer to international economic cooperation and for those who believe that unfettered markets, on their own, build stable, prosperous communities. International institutions did not obstruct or substitute for international markets, but played a central role in creating them. At the same time, market forces did not race ahead so rapidly that they threatened the sense of social and political community among the participating countries. Throughout, the United States played a key role. It imbued regional Western institutions with a clear sense of purpose and policy direction. It did not shy away from asserting firmly its owns preferences, particularly the policy triad of low inflation, flexible markets, and freer trade. Yet it did so in a way that was not heavy-handed (Hogan 1987, p. 443) or even in opposition to European values with which American leaders in this period shared a broad affinity (Isaacson and Thomas 1986, pp. 32–33). Containment and the Marshall Plan took political and economic structures into account even as they transformed these structures from the divided politics and markets of Europe in 1947 to the more homogeneous democracies and economic relations of the Western world in the 1960s.

The Shaping of the Postwar Policy Triad: 1947–1958

Containment: Limits of Community

Containment dispelled the notion that peace depended primarily on how international relations were structured rather than on the fundamental values and interests of the participants. It resurrected the domestic dimension of international politics and reduced the scope of the diplomatic dimension from the global focus of Wilsonianism to the more regional initiatives of the European and Atlantic area (see Chapter 1). According to the containment perspective, the Soviet Union behaved the way it did not just because it was "externally insecure," the popular explanation in 1945–1946, but because it was "internally totalitarian" (Gaddis 1987, pp. 34–40). The Soviet Union and the United States therefore could not cooperate extensively, given the gulf between their domestic purposes, policies, and institutions (for an elaboration of this point, see Chapter 10). Each country pursued foreign policies whose substantive content was closely linked to the policy choices and institutional arrangements of its own domestic system. If these choices differed — the one valuing individual freedom and markets, the other valuing totalitarian society and centralized planning — their international policies also differed.

Thus, containment, unlike Wilsonian universalism, did not assume an identity of national and international interests. It saw immediate and potential threats to national interests. In 1947 U.S. officials no longer feared international economic rivalries leading to war, as in the 1930s, but rather the loss of domestic political freedoms and democratic institutions in Western Europe. The threat was internal political instability, not military or even economic rivalries. And at the root of this instability was the economic chaos and drift that the U.N. and Bretton Woods agreements of 1944–1945 had done very little to prevent. Diplomacy now had to address these domestic economic conditions to ward off political outcomes that might make diplomacy as ineffective among Western countries as it had become between the United States and the Soviet Union.

Marshall Planners were anxious to avoid the impression that America was reacting in traditional military terms. The Truman Doctrine, which had been announced in March 1947, had promised American aid to forces fighting Communism. But this Doctrine, a key State Department memorandum said at the time, did not imply "that the United States' approach to world problems is a defensive reaction to communist pressures and that the effort to restore sound economic conditions is only a by-product of this reaction and not something we would be interested in doing if there were no communist menace." Nor did it imply, the memo stated, "a blank check to give economic and military aid to any area in the world where the communists show signs of being successful" (Price 1955, p. 23). America's diplomacy toward Europe was intended "not to combat communism, but the economic maladjustment which makes European society vulnerable to exploitation by any and all totalitarian movements and which Russian communism is now exploiting" (Price 1955, p. 22). The United States sought to emphasize positive goals, especially the virtues of economic progress; and it preferred long-term, self-reliant solutions to Europe's problems through market-oriented policies rather than short-term palliatives through aid (Pogue 1987, pp. 209–210).

General Marshall highlighted these points in his famous June 5th announcement of the plan:

> Our policy is not directed against any country or doctrine but against hunger, poverty, desperation and chaos. . . . Any assistance that this Government may render in the future should provide a cure rather than a mere palliative. . . . Before the United States can proceed much further in its efforts . . . there must be some agreement among the countries of Europe as to the requirements of the situation and the part those countries themselves will take. . . . The initiative, I think, must come from Europe [U.S. Department of State 1947].

Marshall knew, as he subsequently acknowledged, that "any proposal for more funds . . . would be ruthlessly repulsed" by Congress (Price 1955, p. 25); and Acheson believed that Marshall added the appeal for Europe to take the initiative primarily to ensure Congress that any U.S. aid would be closely tied to efficient self-help policies in Europe (Acheson 1969, pp. 233–234; for other inside views on the background of Marshall's initiative, see Kindleberger 1987, pp. 25–33). Thus, the tendency of American liberalism in this period to see the world in political and economic rather than military terms and to look for diplomatic solutions with least cost accounted in good part for the relative emphasis of Marshall Plan programs on domestic economic adjustment policies rather than international aid.

Why was Europe receptive to a plan that offered so little economic aid and promised such harsh adjustment measures? The answer lies in the common perception of political danger and the process of economic consultation and cooperation that the United States and Europe had begun at Bretton Woods in 1944. Majority elements of European society shared American liberal values. Once the United States expressed these values pragmatically, focusing on the specific economic and political conditions in Western Europe in 1946–1947 rather than idealistic aspirations for the entire world (the earlier universal-utopian view), Europe responded. Europe's response, according to some studies, might have been the same even without the American initiative (Milward 1984). Although this view probably goes too far, it does suggest that America did not impose its values on Europe after 1947 (a conclusion also reached by Pollard 1985, pp. 156–161). The United States and its European partners shared similar concepts of domestic political community and had also established by 1947 the practice of cooperation on postwar economic policy. In this sense, the Bretton Woods institutions, although playing a small role in defining the substance of economic policies after 1944, played an important role in ensuring the process of cooperation in 1947.

Without the experience of common danger and economic cooperation, the Marshall Plan might well have fallen on fallow ground. As it was, the Europeans seized upon the idea without even inquiring what the United States had in mind (Acheson 1969, p. 234). Bevin of Great Britain met in early July with Bidault of France and Molotov of the Soviet Union. Although the latter dropped out immediately, Bevin and Bidault met again in Paris within a month and convened a group of 16 European nations that eventually worked out a 4-year plan of economic assistance and cooperation and presented it to the U.S. government in September.

The origins of the Marshall Plan suggest the groping that characterized U.S. policy in 1947. In the case of the Truman Doctrine announced in March and the Marshall Plan in June, the United States knew it had to act but continued to define the threat in "soft" economic terms—to the point that Europeans included the Soviet Union in initial reactions to the U.S. Plan—and proceeded gingerly with a Congress that was now controlled by more conservative Republicans. Government and private leaders collaborated through organizations such as the Council on Foreign Relations, whose journal published in April 1947 the influential containment article by George Kennan (under the pseudonym X), Marshall's close aide in the State Department. But more work would have to be done over the next year to lay the groundwork in Congress and the larger nongovernmental arena for a significant recovery program.

Eventually, belligerent Soviet moves in Berlin and then Korea reinforced the tentative steps of the initial Marshall Plan, increasing in particular the urgency to deal with the German question. United States policy shifted gradually toward political and military responses and even compromised for a decade or more its economic desire for free commerce among the Western allies by accepting discrimination against U.S. exports in both Europe and Japan in return for allied support to position American forces in Europe and Asia (Gilpin 1972, pp. 63–67). Economic liberalization pressed ahead in Europe, however.

And rearmament efforts, as Hogan reports, posed only "temporary reversals" to the process of reducing foreign exchange restrictions, reestablishing the convertibility of currencies, and creating a common market in Western Europe (Hogan 1987, pp. 430–431). The battle to square regional integration with global liberalization and redress discrimination against American exports would be fought later, largely with success in Europe (Dillon and Kennedy Rounds) but with much more mixed outcomes in Japan and other Asian countries.

The Marshall Plan: Conservative Domestic Policies

The Marshall Plan projected onto the international economy the conservative domestic economic policy consensus that emerged in the United States in the employment legislation of 1946 (see Chapter 3). It did so, in large part, by coopting both the people and the organizational ideas associated with the U.S. consensus.

From the beginning, Marshall Plan discussions emphasized the need to combat inflation, revive domestic investment and production, and establish cooperation in trade, transport, and other key areas of international economic relations (in effect, the policy triad that characterized the subsequent Bretton Woods system). Collaborating with American advisors, the 16 European nations meeting in Paris in July and August of 1947 identified four main areas of action: restoring internal financial stability; launching a strong production effort; setting up a continuing organization to spur cooperation in production, resources, trade, transport, and the movement of persons; and resolving the dollar deficit (Price 1955, p. 37).

The President's Committee on Foreign Aid under the chairmanship of W. Averell Harriman, which Truman had set up in summer 1947 to alleviate congressional doubts about the Marshall aid program, immediately praised the European report for emphasizing that "European production can expand only as currencies and exchange rates are stabilized, as budgets are balanced, and as trade barriers are reduced." The Harriman report also called for "private capital to go to work in Europe." This meant, it noted, "the reduction of endless restrictions and regulations and a freer movement of men and money" (President's Committee 1947, pp. 5, 97). In congressional testimony, both advocates and critics of the Marshall Plan "shared the conviction that recovery policy should revive industry, liberalize commercial and currency arrangements and encourage integration" (Hogan 1987, p. 108). The legislation for the European Recovery Program, which passed Congress in April 1948, not only incorporated these objectives but mandated specific programs in each recipient country to achieve them (Wexler 1983, p. 5).

The people and institutions called upon to run the Marshall Program reflected the market-oriented policy and organizational convictions of the conservative domestic economic policy consensus in the United States. Roughly half of the 19 members of the Harriman Committee, which according to one study "exerted the single most cohesive influence" on the Marshall Plan legislation (Wexler 1983, p. 27), were also members of the Committee for Economic Development (CED), the Commerce Department's Business Advisory Council (BAC), and the National Planning Association (NPA)—all organizations instrumental in developing the U.S. consensus on conservative economic policies (see Chapter 3). Among the Harriman Committee members was Paul G. Hoffman, president of the Studebaker Corporation and the former first chairman of the CED, whom Truman "dragooned" to head the Economic Cooperation Administration (ECA), the U.S. agency to implement the Marshall Plan (announcing the appointment publicly before Hoffman had accepted it; Price 1955, pp. 68, 72). Harriman himself went on to head the Marshall Plan's Office of Special Representative in Europe, putting both a Democrat (Harriman) and a Republican (Hoffman) in the highest positions of the American program. Altogether, the Marshall Plan became, as one observer notes, "something like a coordinated campaign mounted by an interlocking directorate of public and private figures" to implement the U.S. domestic policy consensus (Hogan 1987, pp. 88–101, quote from p. 98).

The ECA set up to run the Marshall Plan was an independent agency staffed with managerial talent largely drawn from the private sector and linked to industrial and labor groups through a network of advisory committees. This arrangement had been recommended by a Brookings Institution study requested by Senator Vandenberg, who was suspicious of the aid and planning mentality in the State Department (Vandenberg 1952, p. 388; Arkes 1972, pp. 84–85). Congress wanted strong business management of the program and relaxed administrative provisions to allow higher salaries to recruit the best brains from the private sector.

The ECA model influenced the structure of regional organizations in Europe. As Hogan notes, "Marshall Planners . . . urged participating countries to replicate this administrative system." The result was a "series of partnerships that blended public and private power, much like the Marshall Planners tried to fuse free-market forces and institutional coordinators to clear the obstacles of a single market in Western Europe." "This strategy," Hogan continues, "inspired [the] plan for a European payments union through which a supervisory board of experts was to use both administrative controls and market incentives to adjust national monetary and fiscal policies in the interest of European stabilization and integration." It also "lay behind [the] efforts to make the Organization for European Economic Cooperation into a corporate body with a competent professional staff and limited authority to coordinate national recovery programs" (Hogan 1987, pp. 19–20; on the interweaving of markets and institutions, see also Gordon 1984, pp. 53–59).

The policies and organizations of the Marshall Plan were thus fully consistent. Market policies were implemented largely by private sector leaders operating through public institutions with close accountability to Congress and the parliaments of other participating nations. Congress did not necessarily refrain from micromanagement of many aspects of the Marshall Plan policy. Although Truman originally requested a 4-year authorization, Congress authorized Marshall Plan funds only on a year-by-year basis, reducing the amounts steadily over the 4-year period. When Congress mandated a 25 percent cut in aid in the second year, however, ECA administrators used this cut to increase incentives for adjustment in recipient countries (Price 1955, p. 128). Private sector leaders understood the opportunity that less finance offered to implement more imaginative adjustment policies.

In 1948–1949 U.S. policymakers in the ECA "laid greater emphasis than did the OEEC on a rapid upswing in production" (Price 1955, p. 92). To accomplish this they acted to restrain demand, dampen inflation (and thereby increase savings and investment), and persuade labor to modernize production facilities. Monetary reform soaked up excess liquidity and restored financial confidence. Marshall Plan aid, which generated so-called counterpart funds when recipients of imports from the United States paid for these goods in local currencies, was used to reconstruct infrastructure—electric power, water systems, roads, and so on—and to influence macroeconomic policies. Unlike the separation of program and project assistance that later characterized IMF and World Bank programs, Marshall Planners used counterpart funds to supplement capital formation and reduce government borrowing (e.g., in France), to offset the deflationary impact of macroeconomic policies on housing and employment (e.g., in Italy), and to stabilize currencies and encourage more liberal trade policies (e.g., in Great Britain; see later discussion in this chapter).

Marshall Plan programs also launched a broad effort to educate management and labor in Europe on the secrets of American productivity. These programs created bilateral productivity councils, encouraged hundreds of visits to American plants, and advocated a more collaborative pattern of labor–management relations in European industry. The objective was to break up the archaic work rules of European labor unions and to encourage voluntary wage restraint, thereby creating greater flexibility in the movement of labor and the adoption of new production techniques. Marshall Planners pressed for the close collaboration of labor unions in both the ECA and the OEEC, creating special liaisons with labor that survive to this day in the State Department and the OECD (the successor organization to the OEEC).

By the spring of 1949, the focus on restraining inflation and reviving production was paying off. Inflation was being brought under control in most of Western Europe, and industrial production was 18 percent higher in the OEEC than it had been in 1938 (Price 1955, p. 100; Hogan 1987, p. 189). Progressively, the focus shifted to the third leg of the postwar recovery triad—the liberalization of trade. As Hogan reports, "the recovery of prewar production levels and the growing signs of financial stability . . . led inevitably to . . . pressure for the devaluation of currencies, the liberalization of trade, and the formation of a European payments union" (1987, p. 191).

The Marshall Plan: Trade Liberalization

The consensus on trade liberalization lagged behind the domestic policy consensus. Priority went to reviving domestic production, but from the beginning U.S. policymakers converged on the "twin concepts of production and integration . . . [in which] gains in production were to be accompanied by greater European efforts to stabilize finances and multilateralize intra-European trade" (Hogan 1987, p. 87). The impulse toward integration picked up momentum after the Berlin crisis of 1948 as the need to meet the requirements of both rearmament and recovery put still further emphasis on the efficient use of scarce resources among the allied countries.

The Bretton Woods organization to promote liberal trade, the International Trade Organization (ITO), was never established (Wilcox 1949, Brown 1950). In 1945 Congress had insisted on an exemption for agriculture. In 1948 it inserted the peril point provision in the Reciprocal Trade Agreements legislation, requiring the Tariff Commission to estimate the point below which tariff reductions would place American industries in "peril." In 1949 Congress failed to ratify the ITO. With the ITO died the commitments in Chapter II of the ITO charter to full employment and in Chapter III to microeconomic industrial strategies. These commitments were out of line with the spreading postwar policy consensus that restrained macroeconomic and microeconomic intervention. Nevertheless, the part of the ITO dealing with manufactured trade survived. The General Agreement on Tariffs and Trade (GATT) conducted its first multilateral trade negotiations in 1946–1947, and some 23 negotiating countries signed 123 agreements lowering tariffs, in the case of the United States by 18.9 percent (Evans 1971, p. 11; Van der Wee 1986, p. 349).

Events in the winter of 1948–1949 tightened the case for freer trade. In Britain the Berlin crisis moved the Labour Party away from the hope that Britain could align with Western Europe as a third force between the United States and the Soviet Union, and encouraged the thought that Britain could align with the United States but exercise a "third role" between the United States and the Continent, drawing on its influence in the Commonwealth. France was alarmed at this prospect of British defection from the Continent and began not only to contemplate economic union with Italy and the Benelux countries (known as Fritalux or Finebel) but also to consider including a reconstructed West Germany. Thus, by summer 1949, the Western countries were poised for decisions that would align British and American interests more closely and throw these interests, albeit in Britain's case grudgingly, behind French–German integration on the Continent (Hogan 1987, Chapter 3).

In the summer of 1949 the British economy was booming while the American economy was sliding into recession. The United States, concerned about further U.S. surpluses in Atlantic trade, pressed for devaluation of the pound and more rapid liberalization of trade in Europe. Britain, seeking to retain a role distinct from that of the Continent, sought American support to underwrite its sterling obligations and maintain its exports to Commonwealth countries. Eventually, compromises were reached that committed Britain to greater trade liberalization, the United States to a special relationship with Britain (and transition support for the pound), and France to economic integration with Germany. Out of these decisions came the devaluation of the pound sterling and broad realignment of other European currencies in September 1949, the acceleration of commitments in the OEEC in November 1949 to

dismantle quantitative trade restrictions, the plans in early 1950 for the creation of a multilateral European payments union that went into effect on July 1,
1950, and the French proposal in May 1950 for a common market in coal and
steel with Germany and other continental countries.

The outbreak of the Korean War in June 1950 reinforced the long-term
rationale for integration (i.e., to finance the rearmament effort) but also
caused temporary shortages of raw materials and higher inflation, which
slowed the liberalization process (see the following discussion of Germany).
The pace of integration varied now at different levels. Among the six countries
that became involved in European economic integration, liberalization moved
rapidly to the elimination of quotas and the lowering of internal tariffs. Within
the OEEC and EPU, liberalization proceeded steadily toward the elimination
of quantitative restrictions but lagged in the area of tariff reductions. Great
Britain continued to be ambivalent about tariff reductions with the Continent
as well as with the United States. Elements in the United States, particularly in
the Treasury Department, worried that economic integration in Europe might
proceed too rapidly and eventually detract from global integration, creating a
permanent discriminatory bloc in Europe.

The progress of trade liberalization in this period is a fascinating story.
Slowly but steadily, regional trade liberalization proceeded and eventually
merged with global liberalization in the GATT. What accounted for this outcome? Two immediate factors seem to stand out. First, the principles of trade
liberalization were never surrendered to the exceptions. As Gardner Patterson
notes, "the general commitment in the Fund Articles of Agreement and in the
General Agreement (GATT) not to discriminate, despite all the exceptions, was
in fact taken seriously" (Patterson 1966, p. 63). The proponents of nondiscrimination were not dogmatic, but as Patterson points out, "they believed that the
'general drift' of the conclusions of classical doctrine for policy was more
convincing than anything else they had heard" (1966, p. 108). They recognized
that "each stage in the process of liberalization would require painful adjustments and guarantee opposition from a variety of vested interests" (Hogan
1987, p. 325). But they proceeded pragmatically, removing discrimination first
by facilitating bilateral and then multilateral payment clearance within the
OEEC and EPU, initially with special provisions to protect the pound, and
then by removing all quantitative trade and foreign exchange restrictions, leading to full convertibility for all currencies, including the pound.

Second, the financing of trade liberalization in this period was never far
removed from the domestic economic policy adjustments needed to permit
lowering of trade restrictions. The EPU accomplished the link between finance
and adjustment by gradually increasing the amount of credit that a deficit
country had to repay in dollars or gold. Because dollars and gold were scarce,
this had the effect of increasing the pressure on deficit countries to reduce their
current account deficits through other policy means, principally macroeconomic and structural (i.e., microeconomic) policy reforms. Reducing current
account deficits, in turn, had the effect of reducing deficit countries' demand
for the exports of creditor countries within the EPU. Meanwhile, capacity to

export was being enhanced by the removal of quota restrictions within the OEEC. Creditor countries therefore had more capacity to export to dollar areas, gradually reducing the EPU's overall dollar gap and eliminating the reason for the discriminatory trading bloc in the first place. The EPU and OEEC arrangements worked, in short, to reduce both intra-European trade imbalances and Europe's external imbalances with the United States (on the operation of the EPU, see in particular Triffin 1957).

Both security and economic considerations facilitated liberalization in this period. The United States had initially security reasons to encourage economic liberalization in Europe in the 1950s and then economic reasons, as the strongest creditor country, to end discrimination between Europe and the United States through the Dillon and Kennedy Rounds of multilateral trade negotiation in the 1960s (on the latter point, see Patterson 1966, p. 107). Both incentives worked more strongly in Europe than in Japan. By 1949 the need to surmount internal political turmoil and the Berlin crisis had pushed industrial production in Europe above prewar levels. By contrast, industrial production in Japan in 1950, where the security threat did not emerge as sharply until the revolution in China in October 1949 and the invasion of South Korea in June 1950, was only one third that of prewar levels (Yamamura 1967, p. 39). The Korean War changed all that; but without other major U.S. allies in Asia and a reluctance in Europe to accept imports from Japan, U.S. trade ties with Japan took a bilateral rather than a multilateral direction. The United States welcomed Japanese exports to reorient Japanese trade firmly toward the West (Nau 1981a) but accepted discrimination against U.S. exports in Japan as much less of a threat than trade diversion by the European Common Market. The legacy of persisting protection against U.S. exports in Japan today is one of the results.

Within these larger considerations, liberalization worked because it was premised on efficient domestic adjustment policies. Otherwise, postwar economies in Europe and Japan, surging in 1947 with pent-up demand from a decade or more of wartime deprivation, would have never been able to absorb increasing international competition. Adjustment policies put a premium on savings and investment, rather than consumption, and thus permitted the necessary reallocation of labor and capital resources to more efficient uses.

Shaping the Domestic Consensus

What distinguishes the period from 1947 to 1951 from later periods in U.S. foreign economic policymaking is not the absence of multiple private players and a meddling Congress in the domestic policy process but the active and successful cultivation of support for the Marshall Plan in the larger arena of nongovernmental and quasi-governmental organizations that we have called the cocoon of consensus building. In addition to the Harriman Committee, the administration appointed private sector committees under the Secretary of the Interior, Julius A. Krug, and the chairman of the newly established Council of Economic Advisors, Edwin G. Nourse, to address concerns about the domestic economic burdens of the Marshall Plan. The House of Representatives dispatched a congressional delegation to Europe (known as the Herter Committee, after its vice-chairman, Christian A. Herter) to educate reluctant congressmen about the grim realities of that part of the world. Dean Acheson, whom Harriman credits with the initial idea of the Harriman Committee (Isaacson and Thomas 1986, p. 425), returned to private life in July 1947 and immediately created the Citizen's Committee for the Marshall Plan, which pulled together more than 300 prominent private citizens from all walks of life to launch a grassroots campaign to promote the Marshall Plan. Altogether, as Acheson describes it, the "efforts of hundreds, perhaps thousands, of . . . speakers and other workers . . . reached the minds, or at least the attention, of innumerable others" and eventually shaped the support needed to launch the Marshall Plan (Acheson 1969; see also Pollard 1985, pp. 145–153; for an excellent summary of the activities going on in this period, see Van Hoozer 1988).

The payoff for this extraordinary effort in the larger nongovernmental and quasi-governmental arena in 1947–1948 was absolutely astonishing. In early October 1947, only 49 percent of the American public had even heard of the Marshall Plan and 60 percent felt that the Europeans were not working hard enough for their own recovery (*Public Opinion Quarterly* 1947–1948, p. 675; Wexler 1983, p. 26). By February 1948, 71 percent of the American public had heard of the Plan and 50 percent agreed that it was necessary to help Europe. By contrast, only 8 percent saw its purpose as curbing Communism (*Public Opinion Quarterly* Summer 1948, p. 366). This result, it should be noted, was registered before the communist coup in Czechoslovakia. Thus, it seems hard to argue that the original containment and Marshall Plan responses were an ideological response to Communism or an inevitable response to crisis; they constituted much more a voluntarist expression of basic American values mobilized by a responsible and energetic leadership at the time in both the government and the private sector. Examining these data, Charles Mee concludes that "in the autumn of 1947 the Americans were by and large still an astonishingly generous people . . . prepared to support the Marshall Plan and to support it for largely humanitarian reasons" (1984, p. 241).

The Spread of the Postwar Policy Triad: 1947–1958

No doubt, in the climate that greeted conservative economic policies in the United States in the early 1980s, the Marshall Planners of the late 1940s would have been accused of exporting "Americanomics." United States officials self-confidently advocated a set of market-oriented policy preferences that they believed were essential to economic growth, but they did so through intensive international cooperation (in contrast to the early 1980s, see Chapters 7 and 8) and in a spirit best reflected in the Harriman Committee report. "While this Committee firmly believes," the report asserted, "that the American system of free enterprise is the best method of obtaining high productivity, it does not believe that any foreign aid program should be used as a means of requiring other countries to adopt [this system]" (Price 1955, p. 44).

American influence neither revolutionized economic policies and institutions in Europe and Japan nor left them unaffected. (U.S. influence was undoubtedly more revolutionary in connection with political changes in Germany and Japan.) The Western countries all started with different economic institutions and policy traditions (see Table 1-4). The United States and Great Britain had more pluralistic relationships between public and private actors, Japan and France more statist or centralized relationships, and Germany and the small European states more corporatist relationships involving close business, banking and labor associations (Katzenstein 1977, 1985). The Marshall Plan did not change these deep-seated relationships or structures. On the other hand, these structures did not determine the content of postwar policy in each country. Domestic structural differences tell us how countries line up with respect to one another along common comparative criteria, but they do not tell us how policy in individual countries is moving at the margin or indeed whether the policies of all of the countries involved may or may not be moving in a common direction toward one end of the spectrum or the other.

The argument here is that the policies of all of the Western countries shifted relatively in the period from 1947 to 1958 toward more conservative fiscal and monetary policies, more flexible and competitive domestic labor markets, and more liberal trade policies. This shift was perhaps least significant in Japan, but it was evident there as well, especially when compared to prewar Japan. Western policies shifted again in the 1960s toward less conservative or efficient standards. And in the 1980s there was a shift back toward more conservative standards. These policy shifts are marginal, but when they are widespread and sustained, they can affect performance. They occur under the influence of competing economic policy ideas and may move in either conservative and market-oriented or liberal and interventionist directions. This possibility of policy shifts that relate significantly to performance outcomes distinguishes the argument in this book from that of embedded liberalism, which tends to see policy moving in one direction only (i.e., government intervention to achieve social protection) and from that of deep-seated structuralist arguments, which regard marginal policy shifts as largely irrelevant.

Marshall Plan policies accepted the need for demand management of the economy but rejected the more ambitious social goals of full employment and broad-scale welfare programs. These policies also emphasized productivity and flexibility of labor markets but rejected the deep supply-side tax cuts or other broad-scale structural reforms, such as privatization, that later characterized the U.S. and other Western countries' policies in the 1980s. As we see in the subsequent discussion, Marshall Plan officials pulled Germany back from what were perceived as excessively free market–oriented policies, just as they pulled Britain, France, and Italy back from more radical Keynesian policies. The Marshall Plan consensus walked a fine line between market-oriented policies and government responsibilities; it embraced neither the exclusive, let alone deep, Keynesian approach that became prominent only in the late 1960s and 1970s (see Chapters 5 and 6) nor the radical, supply-side reforms that became popular in the 1980s and that some supply-siders traced back to German policies in the 1950s (see Chapters 7 and 8; Wanniski 1978, p. 194; Stockman 1986, p. 39).

At the margins, therefore, it is reasonable to argue that Marshall Plan programs tilted the balance of both policy and resources in Europe and, to a lesser extent, Japan. As Stephen A. Schuker concludes, Marshall Plan programs provided the "crucial margin" that made European self-help and self-transformation after the war possible (1981). As we now discuss, these programs clearly shifted the emphasis in each country toward the efficient policy triad of price stability, flexible markets, and freer trade.

Great Britain

Great Britain emerged from the war and appeared to be heading toward a paroxysm of nationalization and planning. The Labour Party took control in spring of 1945 and quickly nationalized the coal, steel, electricity, and transport industries as well as the Bank of England. Lord Beveridge promulgated his plan for *Full Employment in a Free Society* (1945) which established ambitious goals for full employment (no more than 3 percent unemployment) and a broad-scale expansion of social services (education, housing, etc.). British policy seemed well on its way to deep planning and unrestrained Keynesian budget deficits.

However, as Stephen Blank points out, the "debates over nationalization . . . were largely symbolic" (1977, p. 685). Once the fever of wartime mobilization and direct controls subsided, British policy settled on more indirect, global instruments of economic management. From 1947 until the very end of the 1950s, "government policymakers as well as business and labor leaders sought . . . to minimize government's role in the entire range of decisions affecting investment, the determination of wages and the conditions of work, and industrial relations . . . [and] attempted to limit the government's economic policy instruments to global mechanisms that would minimize administrative discretion" (1977, p. 685). As John Zysman explains, "the period of most direct government controls was 1945–47; in the period from 1947 until the conservatives took power, the controls were much more limited" (1983, p. 183).

None of this happened automatically or inevitably. Throughout the period, there was a constant struggle going on both between the Labour and Conservative parties and within the Conservative Party. In the prewar period, even the Conservatives had adopted interventionist formulas. The shift back to neoliberal strategies, according to John Zysman, "took place between 1947 and 1949." If Jean Monnet and his group of indicative central planners (see later discussion of France) had been the model for the British conservatives in 1947, Ludwig Erhard and his free market advocates (see later discussion of Germany) had become the model by 1949. Thus, when the conservatives returned to power in 1951, "they were more committed to liberal [i.e., market-oriented] doctrine than anytime since the 1920s" (Zysman 1983, p. 189).

Zysman credits American policy, through the British loan negotiations and subsequent Marshall Plan funds, with helping to bring about this postwar shift in both the Labour and Conservative parties. "Indeed," within the Labour

Party, he points out, "the American connection may have helped to consolidate the Morrisonians' victory over the left." Lord Herbert Morrison, a minister in the Labour government, advocated less direct control of the economy and of public enterprises. His influence helped to achieve a "victory for the Keynesian theory of Labour party's managers over the socialist ideals of the Labour left." And "that victory," Zysman believes, "smoothed the way for the Conservatives to establish a consensus in favor of demand management policies," rejecting detailed planning and intervention (1983, pp. 183, 185).

By 1951, therefore, both parties "were Keynesian in their approach to macro-policy." Their differences "were at the micro-level of sectoral and industrial policy." The Conservatives "emphasized policies that reinforced the liberal arm's-length relation between business and the state . . . abolishing the remaining administrative controls (particularly food rationing), rejecting planning that had become associated with the use of these direct controls, and denationalizing the steel industry" (Zysman 1983, p. 187). What is more, as Stephen Blank points out, "the Conservatives after 1951 tried to rely almost exclusively on global forms of economic regulation and emphasized, in particular, monetary policy as the master control of the level of economic activity" (1977, p. 686). The British state, to be sure, remained far more interventionist than the United States, and British macroeconomic policies continued to be, arguably, more expansionist than those in the United States. But the consensus that emerged on domestic economic policy in Britain in the late 1940s and 1950s clearly moved noticeably in the direction of all the planks of the conservative Keynesian consensus in the United States — no comprehensive planning, less direct intervention, and fiscal policy tempered by a concern for inflation and a significant role for monetary policy.

On trade and exchange rate policy, the British were much slower to move toward the postwar consensus. Britain's desire to continue to play a global role gave priority to the pound sterling. While this concern for the value and role of the pound was one of the factors favoring more conservative domestic economic policies, it was also a factor militating against the liberalization of trade and exchange restrictions. Exchange and trade restrictions protected the pound, particularly in the face of expansionist domestic policies. The blocked sterling balances were an additional problem. These were sterling earnings of the Commonwealth countries for exports to Britain during the war that could only be used by these countries to purchase imports within the sterling area, particularly from Britain. If these balances were unblocked, as Britain learned in the failed attempt to restore convertibility in 1947, Commonwealth countries could redeem these earnings in other currencies, particularly dollars, and import from other countries despite trade preferences with the British. This not only endangered the pound immediately but also threatened British exports over the longer run. The British therefore clung tenaciously to exchange restrictions and trade preferences. These were the very practices, of course, that were anathema to American postwar trade policymakers.

The turning point came, as we noted earlier, in late 1949. The British saw the recession in the United States as confirmation of its postwar fears that creditor countries, especially the United States, would not avoid the deflationary policies that had crippled the open trading system in the 1930s. At the same time, the Berlin crisis had convinced them that they could not go it alone as a third force between the Soviet Union and the United States. Thus, the British prepared to compromise. If they could get the Americans to provide special support for the pound in the sterling area, they would agree to devalue and support multilateralism and nondiscrimination. For their part, the Americans agreed both to provide additional Marshall Plan aid and to protect the pound by setting a limit on the amount that deficit countries could convert to dollars and by permitting surplus countries to accept sterling in payment along with dollars and gold (Hogan 1987, pp. 320–321). Once again Marshall Plan priorities may have tipped the balance in reversing the decades-long policy of British trade discrimination.

The compromise did not solve all the problems. The British continued to renege on their support for integration in Europe. But after 1949 the British began to dismantle their quota restrictions through the OEEC and their exchange restrictions through the EPU, accepting as inevitable the eventual return to full convertibility for the pound. The British still saw trade liberalization, as Stafford Cripps, the Labour government's Chancellor of the Exchequer, told his American counterparts, as a "fifty-year programme," but they no longer opposed it in principle in favor of bilateral preference arrangements (Hogan 1987, p. 291).

The tightening of British domestic policy in the 1950s went hand in hand with this program of trade and exchange liberalization. There was no other way to return the pound to free convertibility. Whether the "cross of the pound" was the cause of relatively slow British growth in the 1950s, compared to its continental partners, is another issue. Some argue that the British tried through this period to defend an overvalued pound (Blank 1977). But Britain devalued in 1949 along with other countries, some of which became so competitive they were accused of maintaining undervalued currencies (e.g., Germany). So conservative macroeconomic policies were not necessarily more of a burden for Britain than for other countries. The difference may lie in the flexibility of the underlying domestic economies in each country to adjust to macroeconomic stimuli. Germany, as we will show, initiated major efforts in this period to make the German economy more flexible. The British started with a highly regulated economy, and the Conservative governments in the 1950s, although not increasing microeconomic intervention, were also reluctant to roll it back (Ashford 1981, pp. 99–105). Structural rigidities, therefore, were probably an important factor contributing to the relatively sluggish performance of the British economy in the 1950s.

France

France emerged from the war determined to overcome the traditional small business and agrarian interests that had kept the French economy backward and relatively immobile. At the same time, it was also wary of the totalitarian economic methods that had marked the Nazi and Vichy regimes. As in Great

Britain, early steps emphasized planning and state intervention. The government that took power in 1944, which included the communist and socialist parties, established a new Ministry of the National Economy and charged it, under Pierre Méndes-France, with the overall direction of the economy. It was to prepare and execute a national plan and guide and control the actions of other ministries charged with industrial production, food supply, reconstruction, and other agencies in the economic domain (Cohen 1977, pp. 37–38). This was a clear step toward the comprehensive planning of the far left. In addition, from 1944 to 1946 the French government nationalized the coal mines, electric and gas industries, Air France, Renault, Bank of France, four of the largest deposit banks, and 32 insurance companies.

However, the Méndes-France experiment failed quickly. Méndes-France himself had already left the government by April of 1945. In his place emerged Jean Monnet and a group of technocrats who advocated a much looser form of voluntary national planning known as indicative planning. This group was bent above all on modernizing French industry through a "soft" form of targeting investment and technological innovation in a small group of industrial sectors. This approach was neither excessively rigid nor dependent on close control and supervision of other government ministries. In fact, the *Commissariat du Plan* headed by Monnet enjoyed an independent status within the French government. Eventually it allied with the *Trésor* in the Finance Ministry and concentrated on the allocation of credit to implement its plans (Zysman 1983, pp. 106–108). At the same time, it did not concern itself with the broad range of other government economic policies, leaving macroeconomic policy largely in the hands of the traditional ministries responsible for those policies.

The Marshall Plan appears to have had significant effect, both in consolidating the influence of the Monnet modernizers and in pressing the traditional French ministries toward more conservative macroeconomic policies. As Zysman points out, the "Marshall Plan helped solidify the link between the Plan and the *Trésor*" and was apparently an instrumental factor in causing the largest communist trade union, CGT, to withdraw from the French plan, taking with it the emphasis on comprehensive planning. At this point, as Stephen Cohen notes, the "Monnet planners' most powerful support in their struggles to maintain their investment programmes came from foreigners, from the administrators of the Marshall Plan" (Cohen 1977, p. 101).

The support of the Marshall planners was vital not only as reinforcement for the investment and growth ideology of the modernizers but also to help them cope with disastrous macroeconomic conditions in postwar France. From 1945 to 1948, inflation in France increased at a rate of more than 50 percent a year (Lieberman 1977, p. 7). In this climate no industries were going to be persuaded to undertake costly long-term investments. The Marshall Plan exerted the necessary pressure to convince Finance Ministry officials to rein in inflation. Counterpart funds were offered to finance Monnet Plan investments and thereby to relieve further financing demands on the already overextended French government budget. But in return, Marshall Plan officials "asked the French to control private credits, restrain wages and prices, reduce government borrowing, and put state-owned enterprises on a self-supporting basis" (Hogan 1987, p. 155).

The result, as François Caron reports, was that the year 1948 became a "decisive turning-point in the history of inflation in France." In that year, René Mayer, the Minister of Finance, introduced a plan that succeeded in bringing inflation under control. According to Caron, it

> used taxation to reduce liquidity; and above all sought to do away with the rigidity of the price fixing system. It continued the policy of abolishing subsidies which began in the spring [of 1947] and allowed market laws to operate; this encouraged prices to become more stable and shortages began to disappear [Caron 1979, p. 275].

As a result, the mean annual rise in retail prices in France declined from 33 percent from 1939 to 1948 to 12 percent from 1948 to 1951. And from 1950 to 1957, the consumer price index in France grew at an annual rate of only 5.3 percent, during the last four years of 1954–1957 at an average rate of less than 1.5 percent (Lieberman 1977, p. 7; Zysman 1983, p. 321). Simultaneously, output grew from 1954 to 1957 at an average rate of more than 5 percent per year. By French standards, or any standards for that matter, growth in France for a decade after 1947 reflected exemplary noninflationary macroeconomic policies.

After 1947, French policy also moved progressively toward greater trade liberalization and eventually toward integration with Germany and other continental countries in the European Coal and Steel Community and Common Market. Initially, after the war, France sought direct controls over Germany, its principal economic rival, to protect the output targets of the French plan. But as soon as it became clear that American plans included the industrial reconstruction of Germany, France began to discuss plans for economic union — with Great Britain, Italy, and other continental countries (Milward 1984, Chapter VII).

Although the French may not have voluntarily chosen freer trade as their preferred method of modernization, the spur to competition from trade was not unwelcome to the modernizers. As John Zysman explains, the modernizers sought "to force competition within the domestic market and to undermine the restrictive self-regulating arrangements of traditional French business" (Zysman 1983, p. 137). They had already attempted in 1947 to win approval in Parliament for an Italian–French customs union. Although they failed, they got another chance in the winter of 1949–1950. The Schuman Plan to establish a Coal and Steel Community, which was presented in May 1950, and the Monnet Plan to create the Common Market, which followed, were in all respects the brainchildren of the French modernizers. The modernizers accepted agricultural protection within the Common Market as the price to be paid to traditional interests in France to secure industrial integration. But overall, as Zysman elaborates, the Coal and Steel and Common Market treaties

> burst the insulation of the French economy and forced firms to compete, first in Europe and later in world markets. The presence of outside competitors made it harder for French firms to arrange markets, that is, to set prices or

production by agreement. Whereas firms had once sought protection and state certification of private arrangements to control domestic markets, they now had to seek assistance for competitive adjustment. The modernizers had created market forces that could push the economy in the direction they favored [1983, p. 137].

Again, the French economy, like the British economy, retained its unique character. *Dirigisme*, or centralized direction of policy, in France is more deeply embedded than in Great Britain or the United States. Some controls have never been removed in France, even when they have not been used (e.g., price controls). But in comparison with prewar France and with France after 1958 under de Gaulle, the French economy from 1947 to 1958 moved steadily in the direction of greater market competition. The modernizers were determined to use market forces, not obstruct them. The cold shower of industrial competition in the Common Market probably did more than anything else to spur France's superior growth performance in this period over that of Great Britain, which stayed away from tariff reductions on the Continent even as it removed discriminatory quotas. On the other hand, French growth still lagged behind even more sensational performances in other countries, such as Germany and Japan, suggesting that government intervention in France was not always entirely on the side of efficient economic policies.

Germany

In Germany, the challenge to a conservative Keynesian economic policy consensus after the war came from the right, not the left. In 1945, the occupying powers introduced policies to democratize the German economy and break up the concentrated banking and industrial groups that had fueled the Nazi war machine. Along with reparations, these policies crippled industrial recovery, and commodities, mostly imported under emergency programs, replaced paper money in what was essentially a barter economy. The growing external threat, as we have already noted, eventually prompted the currency reform in June 1948, which in turn precipitated the Berlin blockade and the eventual integration of the unified western zones of Germany into Western Europe.

The democratization policies of the allies and the strong reaction to Nazi economic policies strengthened the influence of conventional free market advocates in postwar Germany, supported intellectually by the University of Freiburg's School of *soziale Markwirtschaft* (a rough equivalent of the Chicago School of free market economics in the United States). Led politically by Germany's first postwar Economic Minister, Ludwig Erhard, these advocates rejected the deep planning of the more radical left, the full employment fiscal policies of the exclusive Keynesians, and even the discretionary fiscal and monetary program of the conservative Keynesians. The only government intervention this group accepted was intervention "to prevent the formation of monopolies and to insure competition." Where this approach acknowledged a role for anticyclical stabilizers in the federal budget, it insisted that these "stabilizing measures be as automatic as possible to avoid arbitrary intervention" (Wallich 1955, p. 115).

In practice, the policies of the conservative coalition that assumed control of the national government in September 1949 were never that pure. The German social market consensus and corporatist traditions called for concessions to intervention in macroeconomic policy, industrial relations and social policy (in each of these areas, see Katzenstein 1987, pp. 87, 132, and 176). Nevertheless, more so than other Western countries in this period and more so than German policy in subsequent periods (a perspective that seems to be confirmed by Katzenstein, who notes that the social welfare agenda expanded again only in the late 1950s; 1987, pp. 180ff.), German economic policy focused in the early to mid-1950s on production and supply-side (e.g., tax incentives) factors rather than on the aggregate demand considerations that influenced economic thinking in other countries (which, as Wallich pointed out at the time, was "hard to understand for American economists;" 1955, p. 110).

The hallmark of German policies in the 1950s was tight money and tax rate reductions. After the currency reform in June 1948, German officials adamantly insisted on maintaining a sound currency, which, as Henry Wallich notes, "meant giving priority to stable money and international balance over domestic expansion and full employment." Fiscal policy, for the most part, played a supportive role to monetary policy. It did not operate on an anticyclical basis. After the Occupation authorities raised taxes to exorbitant levels in 1946, German authorities systematically lowered both the rates and income thresholds for taxes. These tax cuts were viewed as structural (i.e., microeconomic) not cyclical (or macroeconomic) measures. As Wallich reports, "according to the [German] Finance Ministry's view the important effect of a deficit-creating tax cut is not the increase in purchasing power of consumers . . . [but] the greater incentive given to producers who now can keep a larger fraction of their income" (1955, p. 110). German fiscal policy after the war looked much more like the supply-side tax policies of the 1980s than the conservative Keynesian policies prevailing in the 1950s.

Domestic pressures to pursue more expansionary policies surfaced in Germany on several occasions—during the slowdown and trade deficit in late 1949 and during the relatively stagnant economic situation from 1951 to 1952. Marshall Plan officials actually encouraged more moderate expansion in 1949, and as Michael Hogan notes, the "Americans poured a steady stream of criticism on the laissez-faire strategies of the government in Bonn" (Hogan 1987, p. 356). U.S. military authorities in Germany, which had always pressed for more intervention in German economic reconstruction, no doubt advocated more aggressive policies (Hogan 1987, p. 197; see also Katzenstein 1987, p. 87). Nevertheless, "there is no evidence," as Henry Wallich reports, "that other United States government agencies lent support to aggressive expansionary plans" (1955, p. 84).

Instead U.S. policy in this period stressed trade liberalization (Hogan 1987, p. 357). Whereas memories of two periods of hyperinflation drove German officials toward conservative macroeconomic policies, there were no comparable memories driving them toward free trade. German trade policies of the 1930s, which had stressed bilaterally negotiated trade agreements and had been quite successful economically, had embraced a completely opposite approach.

The original impulses toward trade liberalization therefore came from the ECA (Wallich 1955, p. 372). The German export boom and subsequent identification of German growth with export-led growth and freer trade came about primarily because of German macroeconomic policies. These policies discouraged consumption and imports while encouraging investment and exports. Given the constraints on domestic consumption and the high rates of capital investment, the result was necessarily substantial surpluses for export.

When the external balance deteriorated, as it did in late 1949, German authorities resisted picking up the slack of declining exports by stimulating domestic demand and diverting resources out of exports into import-competing sectors. They prepared an expansion program but were slow to implement it. Eventually, export demand from the Korean War bailed them out. By late 1950, external demand was so great that Germany's economy had become overheated. Prices rose and Germany's trade balance deteriorated once again. This time German authorities chose to suspend liberalization commitments under the OEEC. For 1 year Germany reverted to administrative controls on imports. In early 1952, however, the liberalization process resumed, nurtured by the EPU and the emerging European integration process. Without these regional arrangements, the Germans "may have relied less on anti-inflationary means to correct the deficit and more on a raising of import barriers," so strong was their reluctance in this period to adopt Keynesian practices of demand expansion (Patterson 1966, p. 92).

Japan

As in Germany, early postwar policies in Japan concentrated on breaking up the highly integrated industrial and banking structures and democratizing Japanese agricultural and economic life more generally. The challenge in Japan was even greater than that in Germany. Japan had no tradition or philosophy of arm's length competitive markets or free trade whatsoever. Industrial modernization in Japan had been controlled since the Meiji Restoration by a handful of large family-owned corporations known as *zaibatsu*, which organized the largest industrial, banking, and trading companies vertically into mammoth conglomerates. Early postwar plans abolished the *zaibatsu*, breaking up the conglomerates into smaller firms, forbidding the use of the companies' traditional names and trademarks, selling off the stock to Japanese citizens, and preventing (through a rigid antimonopoly law) any reintegration of these groups in the future. In addition, Japanese agriculture was modernized. Land was sold off in small plots to Japanese farmers, over two thirds of whom were still tenant farmers in 1945 (Yamamura 1967). Altogether, the reforms were radical and far-reaching. Had they been sustained, Japan would look quite different today and would probably be far more accessible to Western imports and capital than it has been.

The reforms, however, were not carried through. Initially, they stymied recovery, as similar policies did in Germany. Then in 1948, when authorities relaxed the effort to break up Japanese industry and made more funds available for investment through the Reconstruction Finance Bank, inflation flared. A stabilization panic ensued, and the American Occupation authorities imposed the so-called Dodge line, which called for orthodox fiscal and monetary policies to fight inflation. These policies subsequently strangled Japanese recovery. From 1945 to 1949, therefore, U.S. policy vacillated and little was accomplished toward Japanese recovery.

Even then, the reforms clearly pushed Japan away from the command, totalitarian economy it operated before the war. In that sense, the movement of Japan after the war, although it remained by far the most interventionist of the Western economies, was still in the direction of less interventionist policies.

In late 1949 and 1950, the communist take-over in China and the outbreak of the Korean War brought about a reappraisal of U.S. policy toward Japan. Almost 3 years after external threats in Europe had spurred Marshall Plan efforts to rebuild Germany, external threats in Asia now shifted American policy away from democratizing the Japanese economy and society to rebuilding it to support U.S. security operations in Korea. Policies to break up the *zaibatsu* were all but abandoned, and Japanese bureaucracies began to exert a growing influence on reconstruction. To provide the investment capital needed to meet burgeoning American orders, Japanese finance officials devised the two-tier structure of bank lending that became the basis of Japanese monetary policy during the postwar period (Johnson 1982, pp. 200 ff.). At one level, the central bank or Bank of Japan made loans to the commercial or so-called city banks, which in turn made loans to industry. At another level, the government created new institutions, such as the Japan Development Bank, that made direct loans to industry. As laws to decentralize Japanese industry eased, enterprises developed close working relationships again with particular banks, somewhat like the old *zaibatsu* system, but now called *keiretsu*. This system was not unlike the German pattern of interlocking directorates between bank and industry officials. In the case of Japan, however, the government exercised much greater control over these relationships, eventually using them not only to manage the money supply but also to steer the development of Japanese industry in specific sectors. Eventually, bank lending in postwar Japan came to account for as much as 80 percent of all capital in Japanese industry, compared to less than 30 percent in prewar Japan.

The Japanese "high-speed" growth model did not emerge all at once, and its origins suggest that it had roots in the same conservative economic policy consensus that dominated in the United States and Europe (Pempel 1982, p. 52). But this was now conservative Keynesian policies applied Japanese style. It involved fiscal and monetary conservatism exercised through a highly concentrated or corporatist banking and industrial community and focused on controlled competition in the home market and "cutthroat" competition in export markets (Borrus and Zysman 1986). The Japanese development model would not have worked without fiscal policies that encouraged enormous savings, monetary and exchange rate policies that encouraged stability, and open markets abroad, particularly in the United States, that welcomed Japanese exports. The Japanese experiment involved different instruments but not different policy criteria from those pursued in other Western countries in the 1950s. The one premise of the Bretton Woods policy triad Japan did not accept initially was opening its own market to boost consumer imports and welfare.

The Japanese model, therefore, was not expansionist or closed in the sense advocated by exclusive Keynesians or comprehensive planners (see Chapter 3). As Chalmers Johnson describes, monetary and fiscal policy in the 1950s was quite restrained. Ichimada Naoto, who ran the Bank of Japan from 1946 to 1954, "came to stand for a deflationary, balanced-budget fiscal policy and for a mildly expansionist monetary policy" (Johnson 1982, p. 202). If anything, Ichimada's approach reflected greater conservatism than the postwar Keynesian consensus in Europe, relying more, as in the case of Germany, on monetary policy and less on fiscal policy as a tool of cyclical demand management. The debate in Japan was about the ease or restraint of money policy, not about the size of fiscal deficits or unemployment. Some officials, such as Ikeda Hayato, Finance Minister from 1949 to 1952 and then Minister of the Ministry of International Trade and Industry (MITI) and other agencies before becoming Prime Minister at the end of the decade, advocated a concept of "positive finance" in which the government expanded the money supply more aggressively. Others, such as Ichimada, resisted this approach. Ichimada prevailed in the early 1950s but in the process created an opening that MITI exploited to become the main driving force in Japanese development in the early 1960s.

Opposed to an aggressive expansion of bank lending to Japanese industry—what became known as overlending and led in turn to overinvestment, overcapacity, and hence the drive to export—Ichimada turned increasingly to guidelines supplied by MITI to specify the amount of loans going to various industries. These guidelines, although administrative only, took on greater significance as MITI relaxed the anitmonopoly laws of the Occupation period and Japanese city banks and enterprises began to come together again into *keiretsu*. Eventually, the close ties between the banking system, MITI, and Japanese industry provided the interventionist structure through which Japan forced postwar economic growth.

Interestingly, a trade crisis in 1953 ignited the Japanese development model. Because Japanese finance and industry had no historical inclination toward the "open market" or "free trade" (Pempel 1977, p. 742), foreign competition had to provide the spark to initiate Japanese development. Thus, without the prod of international markets, the Japanese growth experiment may have never started. And without access to such markets, of course, it could have never succeeded.

The Japanese pegged the yen at 360 to the dollar just a few weeks before the British devalued in September 1949. Not devaluing with European currencies, therefore, Japan probably began the 1950s with an overvalued exchange rate. This fact, plus the boom in imports set off by the Korean War demands, pushed the Japanese trade balance into substantial deficit by late 1953. Part of the Japanese response was traditional. The Finance Ministry and Bank of Japan cut credit and government expenditures. But MITI's reaction was anything but traditional. It slashed imports and initiated a process of targeting exports that was to characterize Japanese trade policy for the next four decades. This approach contrasted sharply with the trade liberalization approach that was under way in Western Europe and would cause long-term problems for Japan's trade relations with Europe and the United States that continue today.

The Foreign Capital Law of 1950 gave MITI control of foreign exchange. Using this authority, MITI in 1953 drastically cut imports of food, chemicals, medicine, and textiles — so drastically, as Johnson explains, that it "caused the closing of hundreds of stores in the Tokyo area dealing in consumer goods and led to the reappearance of black markets" (Johnson 1982, p. 228). Even more important, however, Okano Kiyohide, then Minister of MITI, came forward with the Okano Plan, which "reflected the view within MITI that the only way to break out of Japan's inevitable balance of payments constraints was through 'heavy and chemical industrialization'." The plan called for "building an industrial structure whose export products would have a much higher income elasticity of demand [i.e., demand for which would increase faster as foreign income rose] than Japan's traditional light industries . . . even though [such a structure] flew in the face of Japan's comparative advantage (chiefly, a large, cheap labor supply) . . . " (Johnson 1982, p. 228). The plan did not catch on immediately, but as Johnson notes, it was applied with a vengeance after 1955, especially by Ikeda, the apostle of so-called positive finance or overlending, who once again became Minister of Finance in the late 1950s. The starting point for this approach was export competitiveness. The key to exports, in turn, was the lowering of costs, and the key to that was enlarged domestic production and what Michael Borrus and John Zysman call "controlled competition" in the domestic market to achieve effective economies to scale, that is, large enough output in the domestic market to achieve lowest costs and thus export most competitively (Borrus and Zysman 1986, p. 118).

By the end of the 1950s, therefore, Japan's high growth system was in place. As different as it was from the policies pursued in Europe and the United States, it nevertheless shared two key characteristics: It was based on conservative fiscal and monetary policies that relatively encouraged savings and discouraged consumption, and it used competition in international markets to guide the reallocation of resources. The Japanese model was not built, however, on a consumer demand revolution such as that advocated by moderate as well as exclusive Keynesians in the West. (Here I differ with Johnson, who identifies a consumer revolution as a key aspect of Japan's postwar development; 1982, p. 230.) Even less, of course, was the Japanese experiment built on substantial social services, especially by contrast to Europe (see Pempel 1977, pp. 754–759.) Overlending for domestic production was a supply-side-motivated policy linked early with import protection and export promotion. Meeting competition in export markets was the overriding requirement, not consumer welfare and higher standards of living in the domestic market. Japanese development succeeded not because it ignored traditional comparative advantage, but because it benefited from a large domestic market in which it could force the pace of industrial development toward more sophisticated products and because it did not have to open this domestic market to gain access to foreign markets (see the debate on Japanese development in Borrus and Zysman 1986 and Nau 1986).

This orientation resulted partly because of unique national concerns about the vulnerability of the Japanese island economy and partly because there was no counterpressure to liberalize Japanese imports, such as the larger process of regional trade integration that was taking place in Europe. The United States opened up its domestic market to Japan unilaterally and did not require until much later in the 1960s, and then seriously probably not until the 1970s and early 1980s, reciprocal access to the Japanese market. After the war Japanese exports shifted radically from traditional markets in China and Asia to the United States (Nau 1981a), whereas Japanese imports, especially manufactured imports, remained small and static. The United States used its market to reorient Japan politically as well as economically toward the West.

In retrospect, U.S. policy toward Japan was an enormously successful undertaking. Japan today, if not thoroughly democratic, is at least a very different society from the totalitarian society of its relatively recent past. Political restructuring in Japan came, however, at considerable economic cost to the United States. It did not have to, of course. United States persistence in democratizing the Japanese economy in the early 1950s may have produced a more open and accessible Japanese market much earlier than may be taking place today, too late perhaps to save the liberal trading system (see Conclusion of this study). However, U.S. foreign policy in the 1950s and 1960s, particularly in Asia, gave priority to political and military considerations (see Chapter 5). The Japanese growth model was allowed to exploit an open international trading system, even as Japan's model of home market protection, especially if mimicked by others, posed a threat to that system. With all countries honing their competitiveness on export markets while protecting domestic markets, international markets, which are simply other countries' domestic markets, disappear. Competition and efficient economic policies are replaced by protection and political bargaining.

Italy

Italy emerged from the war with a coalition government that, as in France, initially included both communists and socialists. The policies pursued in this period led to substantial wage increases, spiraling deficit spending (which increased more than threefold from 1943 to 1947), and excess liquidity supplied by the Bank of Italy. As Bruno Foa noted, "it is not surprising that the price of [such] an economic policy . . . was paid by the nation in the shape of additional inflation" (1949, p. 30).

Already in the spring of 1947, policy began to change to rein in inflation. Alcide de Gasperi, the Christian Democrat who headed the postwar coalition in Italy, reorganized the government and removed its communist members. Professor Luigi Einaudi, Jr., was appointed Budget Minister and Deputy Prime Minister. He immediately initiated a "vigorous and ruthless dose of credit controls," establishing a system of obligatory bank reserves (Foa 1949, p.

104). In addition, he limited Treasury borrowing from the central bank. Once lectured by enthusiastic Keynesians about how to reduce unemployment, Einaudi responded that unemployment was structural, not cyclical (Kindleberger 1984, p. 11). Thus, by relying on microeconomic reforms to create new employment and credit restraint to stabilize prices, the Italian "government succeeded in stopping the inflation without resorting to currency reform." "For the next thirteen years," as Sima Lieberman points out, "Italy was to enjoy price stability" (1977, p. 103).

These developments reflected an ongoing struggle within the Italian leadership between representatives of conservative and more radical schools of economic thought. The Christian Democrats (CDU) held the balance at this point, and there is little doubt in the mind of one observer at the time that the Marshall Plan played a significant role in consolidating the CDU position and shifting Italian economic policies toward the conservative policy triad of postwar recovery. Writing in the summer of 1948, Bruno Foa concludes that

> The Harvard speech and subsequent decisions strengthened very considerably
> the hands of Premier de Gasperi, who had just reorganized his administration
> and expelled the communists from its midst. . . . it brought about the beginnings of a renewed friendship between France and Italy, and the openings of
> negotiations for a Customs Union between the two countries. Finally, it cannot be doubted that the Marshall Plan, even long before going into effect,
> clinched the outcome of the elections of April 18–19, 1948, in which the
> Communist-led Popular Front got only 31 percent of the national vote [1949,
> p. 120].

The 1948 elections gave the CDU its only absolute majority in the postwar period in both houses of the Italian parliament. From 1948 to 1953, therefore, the CDU dominated Italian economic policy and, with the help of the Marshall Plan, put Italy's postwar economic policies solidly on the path of restrained monetary and fiscal policies and freer trade. The Italians actually initiated the Customs Union discussions with France in 1947 and strongly supported the export-led strategy of European integration (Milward 1984, p. 233). The results were surprising to some analysts, who clearly favored a more exclusive Keynesian and comprehensive planning approach.

> The Italian economic boom of the period 1948 to 1955 was particularly surprising since it was characterized by rapid economic growth concurrent with
> serious unemployment without planning and without nationalization. . . .
> The first postwar decade saw in Italy, as in France, a gradual weakening of the
> enthusiasm for economic and social reform that had been developed by the
> Resistance during the years of fighting. Marshall Plan aid strengthened the
> voice of conservatism in both countries [France and Italy], and in both countries the Communist Party started losing political and economic ground after
> 1947 [Lieberman 1977, pp. 109 and 117].

Again, these macroeconomic and trade policy decisions in Italy did not rescind Italy's basic economic traditions and institutions. Italian industry was already concentrated to some extent. The industrial consortium, IRI, had been established in the 1930s to consolidate several banking operations, and it was converted progressively, beginning in the early 1950s, into a holding company to finance and direct a growing number of Italian manufacturing industries. In addition, in 1953, the CDU government created the energy consortium, ENI, to extend direct government control into the energy and chemical industries.

These operations, however, did not go as far as the *keiretsu* in Japan or the expansion of state-enterprise employment and activity that occurred later in Italy in the 1960s and 1970s. The Bank of Italy remained, to some extent, in healthy competition with the IRI banks (unlike the tight relationships in Japan that tied the Bank of Japan through the two-tiered banking structure and MITI's administrative guidance system to Japanese *keiretsu*). And the CDU government "manifested a determination to . . . run [state enterprises] according to sound business practices . . . " (Posner 1977, p. 816). Only toward the end of the 1950s, as the CDU government weakened and prepared for the "opening to the left" and coalition with social democratic parties, which came in the early 1960s, did growing divisions emerge between the conservative private business community in Italy (Fiat, Pirelli, Olivetti, etc.) and a more interventionist government. Throughout the 1950s, however, as Alan Posner concludes, the "objectives of monetary policy and . . . of economic policy in general . . . [were] designed to foster a climate favorable to business and export expansion" (Posner 1977, p. 819).

Benelux and Smaller European Countries

The Benelux and other smaller European countries in this period combined macroeconomic policies, driven more from the consumption than from the production side, with trade policies driven as much by the need to import manufactured goods as by the desire to export. Planning was popular in many parts of the political community in these countries, but the smallness of these countries and their need for trade circumscribed the ambitions of planners (Katzenstein 1985, p. 63). On the other hand, the smallness of their domestic markets meant that they could not eschew altogether a strong government role in adjustment. They, therefore, "paid more attention to industrial policy but shunned the kind of systematic, large-scale reordering of specific industrial sectors that became the hallmark of statist strategies of Japan and France" (Katzenstein 1985, p. 64). In the case of the Netherlands, for example, global macroeconomic policies, despite more highly developed mechanisms of microeconomic intervention, played the largest role in 1951 in adjustment to balance of payments deficits brought on by the Korean War boom. In general, as James Abert notes, "because of changing economic and social conditions from 1950 to 1965, the Dutch government shifted from a strong reliance on wage policy [a microeconomic instrument] to a greater use of fiscal and monetary policies" (Abert 1969, pp. 12–13).

The more interventionist structures and polices of the smaller countries do not contradict the trend in the period from 1947 to 1958 toward more market-oriented policies in all of the Western countries. Government intervention in the small countries was deliberately subjected in this period to more and more international competition through the integration of markets. As we have noted, France and, to a lesser extent, Japan also benefited from greater international competition, which counterbalanced their more interventionist domestic structures. The difference was that Japan embraced international competition on the export side but substituted a policy of "controlled domestic competition" for foreign competition on the import side. It got away with this policy, without the usual high cost of import substitution, because Japan had an enormous domestic market and a still lively system of feudalistic domestic competition. The smaller European states and even the larger ones could not have afforded such a policy and still grown as rapidly and efficiently as they did in the early postwar decades. Moreover, if all countries had imitated Japan in this earlier period, as some analysts advocate today, there would have been no growing international markets and even the Japanese experiment would have failed. It is essential not to lose sight of the overriding role that growing market competition played in the rapid growth of the 1940s and 1950s.

Developing Countries

Growth strategies in the developing world in the period after 1947 reflected a very different pattern. Instead of increasing competition, these strategies had the effect of increasing protection. Most developing countries opted in this period for policies of industrialization through import substitution. Such policies called for taking over an existing domestic market from foreign imports by protecting domestic industries that could produce the same products. While supported by infant industry economic arguments, these policies were applied in most countries in a blanket fashion across the whole spectrum of manufacturing sectors considered important for development.

Moreover, they were accompanied by an attitude of export pessimism among developing countries championed by developing country specialists such as Raul Prebisch, who headed at the time the U.N. Economic Commission for Latin American (ECLA). This pessimism counseled against relying on exports for growth because the terms of trade for developing country products — mostly raw materials and agricultural products — were said to be deteriorating as income growth in industrial countries shifted demand away from food and commodity-based products, which developing countries exported, to higher value-added manufacturing products and services, which developing countries largely imported. Instead, developing countries were called upon to rely on their own markets and seek regional cooperation with other developing countries.

There were understandable political and psychological reasons for these policies in developing countries. These countries were becoming independent for the first time and naturally placed the requirements of nation building over those of economic growth (Rothstein 1977, particularly Chapter 3). Nevertheless, the policies proved to be less than successful economically and spawned a legacy of suspicion and disinterest among developing countries toward the international economic system that hinders the adoption of more efficient economic policies today. Although a few developing countries—the so-called newly industrializing economies such as Korea, Taiwan, and Brazil—broke ranks with import substitution policies in the 1960s, those that did tended to imitate the Japanese model, turning to competition in external markets while protecting domestic markets (Nau 1985b). Without either an internal market the size of that in Japan or the degree of internal cohesion that the Japanese system has to steer industrial policy adjustment efficiently toward external market competition, these countries paid a high price for such "protected" development. Once the oil crisis hit, this price showed up in the accumulating debt burdens of the 1970s.

It is interesting though pointless to speculate about what might have happened after the war if the United States had paid more attention to developing-country policies, as it did to European policies under the Marshall Plan. Although the United States and Great Britain consulted with certain developing countries in the Bretton Woods deliberations of 1943–1945 (Blum 1967, p. 234), Richard Gardner is undoubtedly correct when he notes that Bretton Woods delegates "gave little thought to the Fund's potential impact on the less developed countries" (Gardner 1980, p. xxi). Global pressures were at work diverting U.S. attention disproportionately to Europe, and domestic pressures were at work within the developing countries themselves, leading them to ignore international opportunities. It was not until the early 1960s that the World Bank moved significantly to address third world development issues. Still, the neglect in this period of third world policies, no less than the similar neglect of Japanese trade policies, left large and unhelpful legacies that the liberal international economic order has struggled to deal with in subsequent decades.

Conclusions

From the late 1940s to the mid-1950s, the United States put in place both at home and, through the Marshall Plan, abroad the policy triad of price stability, flexible markets, and freer trade that came to characterize what we refer to in this study as the postwar Bretton Woods system (although the label is something of a misnomer, since the Bretton Woods conference in 1944 and the subsequent Bretton Woods institutions played only a small role in this period). The Bretton Woods policy triad, as suggested in Figure 1-1, coincided historically with unprecedented economic performance. Growth and productivity rose rapidly and steadily (with relatively minor recessions). Trade and exchange restrictions were dismantled. Markets flourished.

By 1960 the Bretton Woods institutions were able to assume a larger role. The IMF significantly expanded its lending, from 4 billion SDRs (Special Drawing Rights—an accounting unit issued by the IMF) from 1953 to 1963 to 18 billion SDRs from 1963 to 1973 (de Vries 1986, p. 85). In the 1960s, GATT initiated the first major global trade rounds, and the World Bank turned its attention from reconstruction to development, creating its soft-loan window, the International Development Association (IDA), to assist developing countries.

The security and economic origins of the so-called Bretton Woods system were complex, but the account in this chapter suggests that the system was stillborn until the United States and its European allies shaped a definition of shared political community that recognized the domestic dimension of international politics (namely, the difference between democratic and totalitarian definitions of private rights and public decision-making processes) and organized the diplomatic dimension to emphasize the efficient use of economic resources to defend Western freedoms. The choice of relatively conservative and efficient economic policies that followed was dictated primarily neither by social ideology and structures nor by external crisis circumstances. Within broad constraints, this choice was instead the product of voluntarist debate in each country and within the regional institutions of the Marshall Plan that eventually moved policies in all these countries incrementally toward more market-oriented standards. These policies then worked to eliminate the constraints of political division and economic restriction afflicting Europe in 1947.

The policy consensus of the late 1940s was not automatic or inevitable. It is too easy, especially in retrospect, to simplify the politics of this period. As Dean Acheson, a prominent participant, observed, the period "was one of great obscurity." "The significance of events," he noted, "was shrouded in ambiguity" (1969, p. 313).

It is also easy to sentimentalize the period as the "golden age" of American foreign economic policy. The victories in this period were won at the margins. They were fragile and always reversible. Challenges to containment and conservative Keynesian policies began to emerge as early as the mid-1950s. The competition to shape a more perfect political society continued, and that competition eventually brought about, by the end of the 1960s, significant changes in security and economic policies that ended the Bretton Woods system. The breakdown of the Bretton Woods system is the story of the next part of this study.

III

Breaking Bretton Woods

5

Blaming Bretton Woods:
Purpose and Policy Collapse

This chapter argues that there were no inherent flaws in the Bretton Woods system that brought about its collapse after the mid-1960s. Rather, deliberate policy choices by the United States and its principal allies shattered the foreign policy consensus on containment and weakened the political framework supporting the Bretton Woods economic system. Further policy choices radically altered domestic economic priorities, accelerating inflation and intervention and making it impossible to maintain the value of the dollar and the fixed exchange rate system.

From the mid-1960s on, the ideas shaping the political purpose and economic content of the postwar international economic system began to change. Under the flag of detente, France under de Gaulle and then Europe more broadly challenged the containment concept of U.S.–Soviet relations. Divisions between the United States and its allies deepened as containment not only became increasingly militarized but also extended into the swamps of Vietnam. The splintering foreign policy consensus opened early fissures in the international monetary system. De Gaulle attacked the dollar's reserve currency role. Then the United States became disenchanted with the exchange rate system and accused the allies of protecting undervalued currencies.

Simultaneously, the consensus on the content of U.S. domestic policies cracked. The moderate Keynesian postwar alternative gave way to more ambitious, exclusive, or "deep" Keynesian solutions involving a dash for full employment and substantial doses of microeconomic intervention (e.g., incomes policy, industrial regulations, etc.). Two legs of the postwar policy triad snapped—price stability and flexible markets. The third leg—freer trade—moved forward under the Dillon and Kennedy Rounds but was now threatened by increasing exchange rate crises and structural rigidities fueling protectionist strategies.

None of the changes in U.S. policies was easily explained by U.S. politics. Special interests dominated trade policy, not exchange rate or macroeconomic policy. Yet the postwar consensus persisted in trade but collapsed in money and finance. Ideas, not interests, were driving U.S. policy.

By the mid-1960s the international context of U.S. foreign economic policy was no longer permissive. Europe and Japan were growing faster than the United States, and Europe became increasingly assertive in the alliance. After unsuccessfully challenging American power in Cuba, the Soviet Union launched a major buildup to achieve strategic parity with the United States. Economic integration and interdependence had advanced far beyond initial expectations, both within the European Economic Community and, through the OECD and GATT, within the Atlantic and Western community as a whole. Vital, mobilized democratic societies had emerged in Germany and Japan and flourished in other industrialized countries, spreading diverse transnational contacts among industrial, labor, religious, student, and other societal groups. The world was no longer subject to governmental power alone, not to mention the power of one government, such as that of the United States in 1945.

Nevertheless, it would be wrong to conclude that circumstances or structural factors prompted the policy changes that occurred in the 1960s. As we observed in the previous chapter, policy choices in the late 1940s built the structures of the postwar economic system. And it would be policy debate and choices that would undermine these structures. U.S. policymakers eventually convinced themselves that the postwar system was flawed. First, it was argued, the monetary system was unsustainable because it depended upon the United States supplying liquidity through balance-of-payments deficits that over time had to erode confidence in the American dollar—the so-called Triffin Dilemma. Second, the exchange rate system was asymmetric, biased in favor of surplus countries in Europe and against the United States, whose currency was pegged in terms of gold and served as the principal international reserve currency. Third, the international trading system was unfair, reflecting residual discrimination against U.S. manufacturing products, particularly in Japan, and systematic discrimination against agriculture, most prominently in the Common Agricultural Policy (CAP) of the European Community.

Each of these arguments contained an element of truth, but as I observe in this chapter, the arguments served primarily as rationalizations for American domestic policy choices in this period that were incompatible with the existing system and in effect helped to create the alleged flaws of that system. Despite repeated condemnation of these flaws, American diplomacy in the 1960s never seriously used American power to correct the system's liquidity or exchange rate problems. Presidents Johnson and Nixon delegated foreign economic policy largely to lower-level Treasury and White House officials. The United States hesitated to act, not because it lacked power, but because it was increasingly divided internally and within the alliance on the purposes and policies to be served by American power. When it finally acted at high levels in August 1971, it did so without purpose and with highly inflationary and interventionist policies that ultimately contributed to the undermining of its power.

American international economic diplomacy in this period thus vacillated between institutions, markets, and unilateral power. The intricate balance of institutional cooperation, market-oriented policies, and U.S. aid that had characterized the Marshall Plan came unglued. From 1961 to 1968 the United States tried to manage the system through multilateral negotiations aimed at expanding liquidity, first through so-called peripheral support measures for the dollar and then through discussions in the IMF to create a new reserve currency, the SDR. But without a strong sense of shared political community (which was now weakening), the goal of a fully multilateral international economic system to include a true international monetary unit was just as utopian in the mid-1960s as it had been in the mid-1940s (when Keynes proposed the bancor), and the United States in any case never applied the power necessary to achieve such an ambitious goal.

After 1968 until early 1971, the United States retreated from multilateral initiatives and turned for the first time in the postwar period (a practice that would become quite frequent in the 1970s and 1980s) to the international marketplace to pursue U.S. objectives—the policy of benign neglect. Multilateralists, unwilling to take decisive unilateral measures to achieve their international goals, hoped that market forces would realign exchange rates—either by forcing other countries to revalue or as a last resort forcing the United States to suspend convertibility and perhaps also devalue directly. When benign neglect failed, the United States finally used its full power in 1971, but that power was completely unhinged from either international institutions or market standards. The United States acted unilaterally to cut the dollar loose from gold and to impose import surcharges as well as domestic wage and price controls.

American domestic policy choices affected the policy choices of other countries, largely through the highly interdependent markets that existed by the 1960s as U.S. officials deemphasized and ultimately, in 1971, spurned international institutions. Expansionary U.S. fiscal and monetary policies in the late 1960s created, in the fixed exchange rate system, an overvalued dollar that stimulated U.S. imports and contributed to rising prices abroad. The American trade surplus declined and protectionist pressures increased. Additionally, as tariff barriers continued to come down in the Kennedy Round (discussed later), increased domestic intervention to control prices or subsidize production raised new nontariff barriers to trade and weakened support for the free trading system, particularly as Japan, the most interventionist of the Bretton Woods governments (see Chapter 4), became a more prominent trading partner. Market pressures in both the exchange rate and trading systems began to replace to a considerable extent institutional bargaining to allocate adjustment costs.

Figure 5-1 shows the shift in domestic economic policies in this period toward more expansionary fiscal and monetary policies in the United States and toward more microeconomic intervention in Europe (see IA and IIA in Figure 5-1). Compared to the period from 1947 to 1967, average annual fiscal deficits in the United States from 1968 to 1974 quintupled and monetary growth rates (M1) in the same period more than doubled. Although U.S. macroeconomic policy had been the most disciplined among the industrial countries in the period from 1947 to 1967 (see Chapter 1), it now became the most undisciplined. Budget deficits and money growth rates in Europe, which had already been at higher levels in the 1950s and early 1960s, remained high. In addition, microeconomic intervention in Europe took off. After 1967 the

I. United States

A. Policies

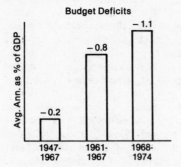

B. Performance

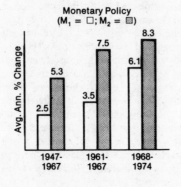

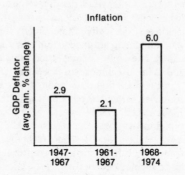

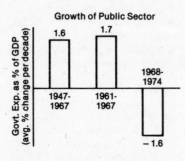

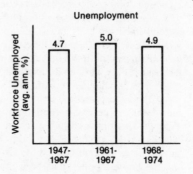

Figure 5-1 Economic policy and performance shifts after 1967. (Sources: See Appendix Tables A-1, A-3, A-4, and A-5.)

II. G-4 (France, Germany, Japan and United Kingdom)

A. Policies

B. Performance

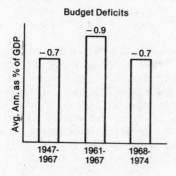

Budget Deficits

Real GDP

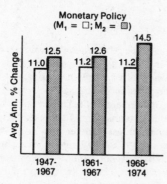

Monetary Policy
(M_1 = □; M_2 = ▨)

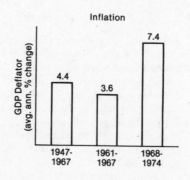

Inflation

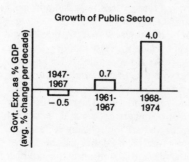

Growth of Public Sector

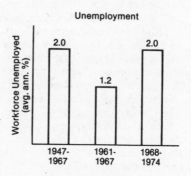

Unemployment

rate of expansion of the public sector in the G-4 countries (France, Germany, Japan, and the United Kingdom) went from a negative 0.5 percentage point per decade from 1947 to 1967 to a positive 0.7 and 4.0 percentage points per decade for the periods 1961–1967 and 1968–1974, respectively. If Japan is left out of this group, the acceleration of microeconomic intervention among the European countries alone is even more pronounced.

Economic performance after 1967 showed the consequences of these policy choices (see IB and IIB in Figure 5-1). In the United States and the other industrial countries, growth slowed and inflation doubled. Unemployment also crept up. Figure 5-1 shows clearly that these relative policy and performance shifts occurred across a wide spectrum of countries with different institutional and political features and that they occurred between the mid-1960s and the early 1970s *well before the first oil crisis*. The data at least invite the conclusion that ideas and policy, not circumstances and politics, were the main driving force behind the erosion of the postwar international economic system.

Foundations Crack: 1958–1965

As we observed in Chapters 3 and 4, the dollar emerged as the principal reserve currency of the postwar system for two reasons — containment generated a strong enough sense of shared political community to convince Congress to supply dollar aid and other countries to accept it, and conservative Keynesian domestic economic policies supported the commitment of the United States to maintain the value of the dollar at a fixed rate in terms of gold and ultimately, if gold should run out, in stably priced American assets other than gold. As Marshall Plan aid ended and trade grew in the 1950s, the United States supplied additional dollars to the system through overall balance-of-payments deficits, taking dollars earned on America's current account (sale of products and services plus unilateral transfers; see Table 1-5) and returning them overseas to finance America's growing military and foreign assistance commitments and the expansion of American multinational business.

Now two developments could undermine this dollar system. A breakdown in the shared political community or foreign policy consensus among Western countries could lead some countries to reject the American dollar for political or noneconomic reasons. That is, a country such as France could deliberately exchange American dollars for gold or refuse to accept further dollars because it disagreed with American policies around the world and sought to pressure the United States to correct its balance-of-payments deficit and reduce its controversial military or economic activities abroad. By refusing to hold the currency and thereby finance the balance-of-payments deficit of the United States, other countries could express their dissatisfaction with the political purposes being pursued by the United States and effectively limit the expansion of the international monetary system based on the dollar reserve currency.

A second development that could undermine the system was economic. Confidence in a key currency depends upon three economic aspects of the key currency country's position: its liquidity ratio, current account position (see

Table 1-5), and domestic economic policies (Bergsten 1975, Triffin 1960). The liquidity ratio refers to the net reserve position of the key currency country (i.e., the ratio of its gold reserves to short-term liabilities held by foreigners). In a key currency system, this ratio must decline as foreigners accumulate the key currency and increase their short term claims against the reserve country's gold assets[1]. As the liquidity ratio declines, however, confidence may erode as holders of the reserve currency wonder if there will be sufficient gold to cover their future claims.

Nevertheless, two other factors could continue to support confidence in the reserve currency, even in the face of a declining liquidity ratio. First, a competitive current account position of the reserve currency country could signal that country's capacity to redeem foreign liabilities not just in gold but also in price-competitive real goods and services. In that sense, the entire economy of the reserve currency country stands behind its currency. And second, the domestic economic policies of the reserve currency country could signal that country's commitment to maintain competitively priced goods and assets. If domestic policies foster growth and price stability—growth to avoid lagging productivity and eventual pressures to devalue the currency, and price stability to foster export competitiveness and preserve the key currency as a store of value—other countries can be assured of purchasing the goods and assets of the reserve currency country at competitive prices and therefore will have little reason not to hold the reserve currency. Thus, the "cardinal test of the key currency status of any national currency," as Fred Bergsten concludes, " is the degree of foreign and domestic confidence in its future price stability, relative to competing assets" (1975, pp. 172–173).

From the late 1950s to the late 1960s, consensus unraveled on both the foreign political and domestic economic policy supports of the key currency system based on the dollar. As early as the late 1950s, France had attacked the dollar system for political reasons. Foreign policy officials in the United States tried to shore up the foreign policy consensus and protect the dollar's role, but economic specialists argued that the problem was due to fundamental technical flaws in the key currency system. Robert Triffin formulated the so-called Triffin Dilemma, which focused attention on the U.S. liquidity ratio—the declining ratio of U.S. gold reserves to outstanding dollar liabilities. He saw this system leading inevitably to a crisis of confidence in the dollar. Triffin thus called for a new world currency to replace the dollar and, along with a new generation of more ambitious Keynesian economists, pressed for more stimulative domestic policies in the United States. With the dollar no longer playing the key currency role, these economists believed that U.S. domestic policies could afford to take a greater inflationary risk.

Thus, preoccupation with the Triffin Dilemma shifted attention away from the need for disciplined policies and flexible markets toward expectations of

---

[1]Otherwise, foreign claims would accumulate only as fast as the production of gold and there would be no advantage to a key currency system over that of a pure gold standard in which gold is the only widely used medium of exchange.

easier financing and liquidity. In Europe and Japan, domestic policies were already more expansionist. Higher inflation, more regulated capital and labor markets, and residual trade discrimination against U.S. exports built up powerful pressures against the timely reallocation of resources. Eventually, domestic policies in the United States also became more expansionist. From the mid-1960's on, exclusive or "deep" Keynesian solutions gained ground among U.S. policymakers (see Chapter 3 for these definitions) and eventually shattered the ability of the United States to maintain the value of the dollar or the competitiveness of its real goods and services.

According to the interpretation in this study, then, there were no technical flaws in the Bretton Woods system. Had the United States retained the confidence of its allies in both its national security and domestic economic policies, the monetary system could have been adjusted smoothly at some point from a gold exchange standard to a pure dollar standard. The requirement to convert dollars to gold could have been dropped without any basic change in the par value exchange rate system or the reserve currency role of the dollar. But this transition was possible only while a consensus on foreign policy goals and domestic price stability prevailed. By the time the United States acted in 1968 and 1971 to end convertibility, the alliance was severely weakened and domestic policy choices had already been made to abandon domestic price stability and the timely reallocation of domestic resources (i.e., flexible markets) in both Europe and the United States.

Liquidity Problems: A False Trail

In the early 1960s, however, it was not that easy to predict the durability of the dollar's role or to see clearly the underlying domestic and international political alignments that provided the crucial support for this role. The chronic balance-of-payments deficits were a new problem for the United States. It was natural for U.S. officials to assume that these deficits had to be corrected and that a new source of international liquidity besides the dollar and gold had to be created. This became all the more the case as the United States encountered stiffer trade competition in Europe and, somewhat later, in Japan. Triffin pointed to the deterioration of the U.S. current account, particularly vis-à-vis Europe, and blamed this deterioration on "creeping inflation" in the United States, a lack of "appropriate investments in research and technology," and the existence of "remaining discrimination on dollar goods and the [need for] further reduction of other obstacles to trade and payments by foreign countries, and particularly by prosperous Europe" (1960, p. 7). Europe contributed to this line of reasoning by arguing that the U.S. balance-of-payments deficits in this period fueled inflation in Europe and, because Europeans were willing to hold U.S. dollars, created excess liquidity to finance growing American direct investments in Common Market countries (Calleo and Rowland 1973; Morse 1973, pp. 216–219).

Much of this concern with excess liquidity in the late 1950s and early 1960s, however, was misplaced. Most analysts agreed then and now that U.S. balance-of-payments deficits and hence liquidity expansion in this period were not excessive (Solomon 1982, pp. 28 and 31; Bergsten 1975, p. 161). Indeed, some evidence indicated that liquidity during this period was actually inadequate and had not kept pace with the expansion of trade (Odell 1982, pp. 134–135). Triffin put the increase in total gold and dollar reserve holdings for the entire decade from 1949 to 1959 at less than $5 billion (from $36.9 billion to $41.96 billion; 1960, p. 5). Overall, from 1945 to 1965, total official reserves increased by only 50 percent, whereas trade grew almost 300 percent (Cooper 1968, pp. 51, 60). The evidence suggested, as pointed out in Chapters 3 and 4, that the postwar Bretton Woods system was clearly tilted toward relatively rapid adjustment of domestic policies and not toward generous financing or liquidity that encouraged inflationary or interventionist policies.

Similarly, European arguments about the export of U.S. inflation and direct investment in the early 1960s puzzled knowledgeable experts at the time (Cooper 1968, p. ix). United States inflation was practically nonexistent in this period; European inflation, by contrast, was considerably higher. From 1961 to 1967, for example, inflation as measured by the CPI averaged 3.8 percent in the G-4 countries (France, Germany, Japan, and the United Kingdom) compared to only 1.7 percent in the United States. Moreover, U.S. capital was being attracted to Europe largely by tighter monetary policies and higher interest rates in Europe, which were the consequence of accumulating surplus reserves in European countries. By refusing to expand money supply in tandem with reserve accumulation, European monetary authorities were in effect lending short to the United States; the United States, in turn, was lending long to Europe through the expansion of American direct investments. Given restrictions on European capital markets and higher savings preferences in Europe, the New York financial markets were performing a vital intermediation role for Europe, taking short-term Eurodollar deposits and converting them into long-term industrial investments in Europe (Cooper 1968, pp. 133–134). Restrictions in Europe on imports from the United States, in the form of both restrictive demand policies and prospective trade diversion from regional integration within the Common Market, compounded these financial flows (e.g., by encouraging U.S. direct investment to substitute for U.S. exports).

The real economic problems in this period, therefore, were reserve accumulations, tighter macroeconomic policies, and relatively restricted trade and capital markets in Europe (and Japan), not inflation in the United States or abuse of the U.S. reserve currency position through direct investments in Europe. The fact that Europe was experiencing higher inflation than the United States, despite relatively tighter monetary policies, suggested that microeconomic or structural rigidities, such as labor and capital market regulations and interventionist policies in agriculture, transportation, and other sectors, may have been an important cause of inflation in Europe. At this point, Europe, not the United States, needed to adjust, and the adjustment had to involve not only macroeconomic stimulus but also, if inflation was not to go even higher, structural liberalization. Unfortunately, by allowing the problem in this period to be defined as a liquidity problem focusing on U.S. balance-of-payments deficits rather than an adjustment problem focusing on European domestic economic and exchange rate policies, the United States played into Europe's hands.[2]

By defining the problem inappropriately, U.S. responses were inappropriate, at least from the point of view of fostering the further development of an open, efficient international economic system. Instead of encouraging Europe to liberalize financial markets, for example, the United States progressively restricted its own financial markets. Operation Twist in the early 1960s sought to drive up U.S. short-term interest rates and to stem the outflow of short-term capital to Europe while lowering U.S. long-term rates and encouraging more capital investment in the United States. The Interest Equalization Tax in 1963 simultaneously discouraged long-term foreign borrowing in U.S equity markets. In the middle to late 1960s, the United States applied quantitative controls, initially voluntary and then mandatory, to staunch the outflow of U.S. capital. Had these efforts succeeded, the results ironically would have been to reduce liquidity available for international finance just as the problem was being defined as the need for more liquidity. In another sense, of course, the reduction of U.S. dollar outflows would have accelerated pressures to create new liquidity in the form of SDRs.

As it turned out, capital controls did not work. United States banking simply moved offshore, and the Eurodollar markets burgeoned throughout the 1960s. Liquidity increased sufficiently, but adjustment pressures built up, especially on the exchange rate system. What was lost was the opportunity to address early the continuing domestic policy adjustment requirements of the postwar system, particularly structural adjustments, such as the liberalization of labor and capital markets in Europe, or, in another context, the liberalization of import and financial markets in Japan. These structural rigidities in labor and capital markets had been the primary targets of Marshall Plan programs in the 1950s. But in the 1960s and 1970s they slipped from the agenda of international economic priorities. When they were rediscovered in the 1980s, the expansion of the government's role in microeconomic areas that had occurred in the meantime made them much more difficult to overcome (see Chapter 6).

Foreign Policy Conflict Weakens Monetary Consensus

Contrary to the arguments posed by the Triffin Dilemma, then, the attack on the dollar in the late 1950s and early 1960s did not have its roots in economic factors. United States net reserves were still more than adequate (U.S. gold reserves roughly equaled U.S. dollar liabilities in 1959 — approximately $20 billion each); America's deteriorating current account position was arguably more the consequence of residual import barriers in Japan and inflexible microeconomic policies in Europe than relative price inflation in the United

---

[2]As Richard Cooper notes, the conflicting arguments set out here are "not easily distinguished empirically," but indirect evidence seems to refute the European argument (1968, pp. 134–136).

States; and U.S. domestic economic policies, particularly the much greater flexibility of U.S. labor and capital markets, were exemplary, especially compared to labor markets in Europe and financial markets in Japan. The attack on the dollar derived instead from a breakdown of the implicit international political consensus underwriting the broader purposes of the dollar system. Europe and particularly France started to dump American dollars for noneconomic reasons.

As soon as he came to power in 1958, Charles de Gaulle acted to redress the privileged position of the dollar. He devalued the franc and began to blame the inflow of dollars, not French domestic economic policies, for the rise in French inflation. Underlying his actions was a long-simmering animus toward Anglo-American dominance of the postwar economic system. His economic advisers, principally Jacques Rueff, had convinced him that the "United States . . . was benefiting from the dollar exchange standard by being exempted from the rules of international exchange" (Morse 1973, p. 217). Unlike other countries that had to pay for their balance-of-payments deficits in foreign currencies or gold, the United States could simply print its own currency to cover external debts. This privilege gave it the capacity to finance any foreign policy activities it might choose, overwhelming other countries' political objections in the process and exporting a surfeit of unwanted dollars and hence inflation to those countries as well. "By returning to the gold standard," de Gaulle believed, "the United States would be prevented from enjoying this exceptional position, and France and the other European countries would gain if they could maintain surpluses in their balance of payments" (Morse 1973, p. 217).

Here, then, was the political justification for the surplus or mercantilist economic policies that France and other European countries began to pursue in the late 1950s (and of course Japan pursued for other reasons). Europe accumulated surpluses of U.S. dollars and instead of investing these reserves in domestic restructuring and expansion, began to redeem them for U.S. gold. Form 1949 to 1957, U.S. gold holdings had declined by only $1.7 billion, or 5 percent of the total. But from 1958 to 1960 they fell $5.1 billion, $2.3 billion in 1958 alone. The economic reasons for this drain were not self-evident. As Robert Solomon points out, there may have been a downward secular adjustment of postwar European demand for U.S. goods, weakening the U.S. current account. But as he goes on to conclude,

> there must have been a political element in the changed attitude toward the dollar. With strength and self-respect restored, it was only natural that continental Europeans would react to their earlier position of weakness and subservience to an all-dominant America by finding fault with America's currency. Renewed strength brought with it a desire for greater symmetry in international monetary arrangements [Solomon 1982, pp. 28–29].

These political divisions among the allies widened throughout the first half of the 1960s, first with the American–French dispute over British entry into the Common Market and then with de Gaulle's challenge to the whole structure of Cold War relations and the Western alliance. They frustrated effective U.S. actions to achieve an orderly expansion of international reserves.

In the early 1960s American officials, including President Kennedy, were still firmly committed to the dollar role. These officials—led by C. Douglas Dillon, Secretary of the Treasury; Robert Roosa, Under Secretary of the Treasury; and George Ball, Under Secretary of State—sought to preserve the role of the dollar by gaining allied consensus to expand short-term lending to the United States through gold pools, swaps, dollar guarantees, and other mechanisms that Robert Roosa called "peripheral support" for the dollar (Odell 1982, p. 102). The allies pooled gold reserves to combat speculative increases in gold prices, which momentarily in October 1960 broke through the $35 range and reached as high as $40 per ounce before retreating again; they also intervened for the first time since the war to influence foreign exchange markets, negotiated a network of bilateral swap agreements in which central banks created reciprocal short-term credits to support exchange market interventions, refinanced these short-term credits when necessary through medium-term Treasury bonds denominated in foreign currencies (so-called Roosa bonds), expanded IMF lending through the General Arrangements to Borrow (GAB), and sought to issue future dollars against a basket or "bouquet" of foreign currencies held by the United States. Over time, U.S. officials expected to share some of the responsibilities for managing the world reserve system with other leading currency countries (Odell 1982, p. 104).

France and Europe, however, wanted more immediate and direct control. They insisted on a veto over loans under the GAB. In 1963 the French government proposed a new international reserve currency to replace the dollar, the Composite Reserve Unit (CRU). It would be distributed on the basis of existing gold holdings and thus end the privileged position of the United States in creating new liquidity.

By 1965 Europe moved from a simple critique of American purposes and policies underwriting the monetary system to the articulation of an alternative set of goals that conflicted with U.S. objectives. This conflict had three parts. First, it concerned the formula for shared political community in central Europe. De Gaulle and Europeans more generally had never fully shared the containment approach to East–West relations that separated economic and political systems in central Europe. Closer to Eastern Europe and the Soviet Union historically, they placed less emphasis on the domestic dimension of East–West relations—the role of fundamental values in interstate conflict—and more on the historical and diplomatic dimension (see Table 1-3).

Detente as it emerged in Europe put the emphasis back on the way relationships in international politics are structured, rather than on differences in basic values among individual nations. In its more extreme versions, detente represented a return to the commercial–legal view of Cordell Hull and Henry Morgenthau. If relationships could be structured around economic and political activities, peace would follow. In its more practical version, detente accorded a much greater role to diplomacy and active negotiation to link economic and political developments with the gradual resolution of basic conflict. Peace did not follow automatically from the encouragement of economic and political cooperation; instead diplomacy engineered the construction of peace. As it was later adopted by Henry Kissinger, detente did not end geopolitical conflict between the United States and the Soviet Union, but it downplayed the domestic moral dimension and focused diplomacy on managing U.S.–Soviet differences by wrapping them in a "web of interdependence" and systemic constraints (Kissinger 1979, Chapter V, and 1982, Chapter VII).

A second aspect of alliance conflict over East–West relations concerned the extent to which Western security interests were at stake in the third world. Was peace indivisible? Did Soviet-abetted aggression in developing countries, especially through guerrilla-led national wars of liberation, threaten the central balance in Europe? In the 1950s, the attack on South Korea had been interpreted by the alliance as a potential threat to Western Europe. But in the 1960s, the war in Vietnam would prove to be a different case. Europe would come to feel that containment outside Europe ultimately weakened the basis for containment in Europe. It would see, therefore, much less of a role for conventional or even guerrilla military power in the third world and much more of a role for economic assistance and basic human needs (food, health, etc.). If economic problems could be solved in poor countries, peace and political development would follow. Once de Gaulle extracted France from Algeria in 1962, the differences with the United States over military and economic aspects of third world policies became increasingly apparent.

Finally, and related to the second aspect, the alliance divided on the question of whether the West should support any type of government in the third world, if foreign policy interests required it, or only certain ones. Europeans, with stronger socialist and in some countries communist parties involved in their own domestic affairs, did not share America's aversion to left wing governments in the third world, nor did they share America's affinity at times for right wing governments. Inevitably, European governments saw wars of liberation more as struggles to overcome poverty and oppression in authoritarian societies, whereas the United States saw these conflicts as communist inspired revolutions and totalitarian threats to the values of free societies in the West. These differences toward third world issues mirrored differences toward U.S.–Soviet relations. Europe saw the principal sources of conflict in unresolved economic and social problems and diplomatic relations; the United States traced these conflicts to more fundamental domestic values. Eventually, diplomacy in Europe obscured continuing differences in domestic values between East and West, while U.S. policy in Vietnam lowered the appeal of Western values by excessively militarizing the conflict and by backing one totalitarian society, South Vietnam, to defend against another, North Vietnam.

Thus, the foreign policy conflict that tore the alliance apart in the middle and late 1960s began in the early 1960s. This conflict by itself may have eventually led other countries to reject the dollar reserve currency role. But de Gaulle's power peaked in 1968, if not 1 or 2 years before, and thereafter economic pressures on the dollar became more important than political ones. Inflation in the United States rose dramatically after 1965 and ultimately eroded foreign confidence not only in the gold backing of the dollar but also in the backing provided by price-competitive American goods, services, and capital assets. These economic pressures had their roots in the erosion of the domestic economic policy consensus in the United States to preserve price stability. That consensus too began to fragment as early as 1960.

Ambition Undermines Price Stability

As we saw in Chapters 3 and 4, the consensus that eventually dominated U.S. and Western economic policy in the 1940s and 1950s set moderate goals for employment and fiscal policy. It called for maximum, not full employment, proposed a budget surplus at high employment to retire national debt rather than running any deficit necessary to achieve full employment, and looked to monetary policy to maintain nominal demand consistent with price stability. It also viewed skeptically further extensions of governmental intervention in microeconomic areas.

In the late 1950s, however, economic policy thinking became more ambitious. Although U.S. growth was faster in the 1950s than it had been in the past and the three recessions in this decade were shorter and milder than previous ones, Europe and Japan were growing faster. The United States was seen as paying a domestic price in the form of slower growth in order to play the role of international central banker. Maintaining confidence in the dollar, it was thought, "caused, or at least reinforced, policies of tight money and fiscal restraint . . . costing perhaps tens of billions in lost growth" (Gardner 1980, p. xxxvii; see also Van der Wee 1986, pp. 67–68). The new ideas that emerged in the early 1960s were much less willing to accept these perceived domestic costs to preserve the role of the dollar.

The new economic thinking originated in the university and policy elite layers of the nongovernmental arena that I have called the cocoon of consensus building. Herbert Stein describes the roots of this new thinking:

> There were two elements in the intellectual underpinning of the new liberalism. One was basically Keynesian and reflected what by then was the mainstream of academic economics. By 1960 the young economists who had been hypnotized by Keynes when they were graduate students in the late 1930s were mature enough to be advisers to Presidents and pundits to the nation. This included such people as Paul Samuelson, Walter Heller and James Tobin. . . . They regarded the postwar consensus . . . as progress but still far from the true gospel. They had much more ambitious goals for unemployment . . . , they were much more confident of their ability to forecast and manage the economy, and they were ready and eager to put into practice the stereotype of Keynesian macroeconomic policy that . . . Walter Heller later invented a name for . . . "fine-tuning."

> The second strand . . . was what might be called Galbraithianism. . . . The Galbraithian view . . . had little regard for "consumers' sovereignty"—a staple of free market economics . . . [and] scoffed at the idea that the American economy was governed by competition which yielded a high degree of efficiency. . . . So [Galbraithians] were prepared for, and eager for, a large degree of government regulation of the economy and government redistribution of income [Stein 1984, pp. 95–97].

The new ideas represented a shift from conservative to exclusive Keynesianism and from a skeptical attitude toward microeconomic intervention to a planner's penchant for it (see definitions in Chapter 3). The combination became what I call in this study "deep" Keynesianism.

Kennedy's advisers saw the conflict in classic New Deal dimensions. A task force organized during the transition and chaired by George Ball, who subsequently became Under Secretary of State in the Kennedy administration, concluded that conservative European bankers were using gold purchases from the United States to thwart the prospect of more expansionary domestic policy in Europe aimed at reducing unemployment and getting the economy moving again (Odell 1982, p. 106). These were the same central bankers that Keynes feared would throttle postwar recovery and expansion. The "new" New Deal enthusiasts sought maximum latitude to eradicate domestic unemployment and poverty. They did not consider inflation an evil; indeed, their analysis based on the Phillips curve, which predicted a certain trade-off between inflation and unemployment, led them to expect more inflation as unemployment went down. One could choose the balance desired, and the Kennedy economists left no doubt that they gave priority to reducing unemployment. In the end, if it was required, inflation could be managed with incomes policy, more direct and targeted government intervention to control prices and wages.

The desire to pursue more ambitious domestic goals gave rise to misgivings about the dollar's reserve currency role. Robert Triffin's concern that confidence in the dollar would inevitably decline stemmed in part from his interest in more vigorous domestic expansion. In testimony before Congress, he worried about the British experience of the 1920s in which domestic expansion had been held hostage to the British pound. "If and when we feel reassured about our internal price and cost trends," he stated, "we may wish to ease credit and lower interest rates in order to spur our laggard rate of economic growth in comparison not only with Russia, but with Europe as well." But when we do that, he fretted, "we may then be caught . . . exactly as the British were in the 1920s between these legitimate and essential policy objectives and the need to retain short term funds in order to avoid excessive gold losses" and maintain the value of the dollar (1960, p. 10).

As we noted earlier in this chapter, maintaining the value of the dollar required above all price stability, yet by the early 1960s sentiment was clearly shifting in favor of more growth and employment, even if that meant some anticipated inflation based on the Phillip's curve. Richard Cooper, who served on the Council of Economic Advisers in the early 1960s, subsequently offered the view that balance-of-payments requirements in the 1950s and early 1960s "inhibited the use of expansionary monetary and fiscal policies which, on domestic grounds alone, would have been desirable before 1965" (Cooper 1968, p. 42). "Domestic grounds" were now gaining on international obligations as the standard for U.S. domestic economic policies. As this became evident, although there was little actual inflation in the early 1960s, financial markets began to anticipate a weakening dollar. These were some of the psychological currents that spooked the markets in the 1960 election (Odell 1982, p. 87) and dogged American efforts in the early 1960s to shore up "peripheral support" for the dollar. Confidence in the American commitment to price stability was the first of the economic factors supporting the role of the dollar that was cast in doubt.

Initially, Kennedy remained ambivalent toward the new ideas of his advisers. He tended to agree with members of the postwar foreign policy establishment, such as Dillon and Roosa in Treasury, who defended the dollar's role (Collins 1981, p. 178). He once lumped dollar devaluation together with nuclear war as the two events that frightened him most (Odell 1982, p. 108). Yet he also represented the "New Frontier," and his economic advisers were eager to cross a new frontier in domestic economic policy. They advocated a major tax cut program to spur the economy, the kind of supply-side fiscal reform that had taken place in Germany in the early 1950s but had been generally spurned by other industrial economies. Walter Heller, in particular, undertook a massive educational effort to convert Kennedy to a deeper version of Keynesianism. In 34 months Heller's Council of Economic Advisers produced more than 300 memoranda to convince Kennedy that a tax cut was needed and would not harm the balance of payments (Kettl 1986, p. 97).

Kennedy yielded in January 1963, though he did not live to see his tax program enacted. Johnson secured passage of the tax bill in early 1964 and went on to add substantial new social programs to attack poverty, illiteracy, disease, and old age needs—the Great Society programs. The initial results seemed to prove the "new" New Dealers right. Nineteen sixty-four and 1965 were banner years for the economy. A more ambitious Keynesianism, in fact, together with a dollop of microeconomic intervention to eradicate poverty, "reached its finest hour—in terms of self-confidence and popular acceptance, as well as achievement—in 1965" (Stein 1984, p. 27). Unemployment dipped to 4 percent in December 1965, the U.S. trade surplus was the largest since 1947, and inflation continued to run below 1 percent per year.

In 1965, however, the more ambitious domestic economic program merged with the U.S. decision simultaneously to escalate the war in Vietnam. This decision added enormous new defense expenditures to an already overloaded budget and sharpened foreign policy conflicts with the allies, weakening the sense of shared political community supporting the dollar's role. When Johnson decided not to finance the war and Great Society programs through new tax increases for fear that Congress would modify his Vietnam and domestic policies, the stage was set for an eventual clash between domestic programs and the dollar's fixed value in terms of gold.

Ideas Not Politics Drive Change

What happened after 1965 can be viewed, in retrospect and somewhat irreverently, as a massive attempt at self-delusion and academic rationalization to disguise the fundamental incompatibility between the budget policies pursued by the United States and the dollar's role in the Bretton Woods international economic system. The dilemmas the United States confronted in the years after 1965 were not consequences of a faulty Bretton Woods system, whether in the areas of liquidity, exchange rate adjustment, or trade (and theories were spawned to identify technical defects of the system in all these areas — see further discussion). They were the creation of U.S. domestic economic and foreign security policies, which weakened the U.S. balance of payments and diminished the will among the allies to cooperate on shared political objectives.

The force behind change in the United States in the early 1960s was not pressing circumstances or pluralistic politics; it was ideas. New ideas emerged that challenged the existing purposes and policies of the postwar system and eventually forged new political alignments to support different policies. (Odell's analysis also strongly supports this emphasis on the role of ideas; 1982, pp. 124–130.) Charles de Gaulle did not come to power in 1958 on a groundswell of anti-American sentiment in Europe; he did much to create it. Kennedy had not been elected on the basis of the more ambitious ideas of his economic advisers. As Stein points out, there was no massive economic problem in 1960 compared to that of 1932. Kennedy and Johnson had "to identify their problems and persuade the American people of the seriousness of the problem" (Stein 1984, p. 89).

The environment was relatively permissive. International monetary and domestic macroeconomic policy were not areas of visible or divisive partisan debate in the United States, as domestic macroeconomic policy would become in the late 1970s or international monetary issues became in the 1980s. The impulse for change came from a relatively small number of foreign policy and economic experts and reflected ideas generated largely in the quasi-governmental and nongovernmental circles of American universities and think tanks. In short, the foundations of the international monetary system cracked because of new ideas spawned by de Gaulle and his academic advisers in Europe and by Kennedy and his academic advisers in the United States. It is powerful testimony to the critical role of policy ideas, intellectual debate, and leadership in shaping foreign economic policy outcomes, as opposed simply to the influence of structural or political forces.

Trade policy was one arena in which special interest politics did begin to play a larger role in the early 1960s. Ironically, however, policy in this arena did not change significantly, largely because of the continuing force of basic policy ideas associated with comparative advantage and free trade. As imports went up, beginning with Japanese cotton textiles in the late 1950s, U.S. congressmen began to demonstrate greater sensitivity toward affected domestic interest groups. When the Kennedy administration sought legislation in 1962 to launch the Kennedy Round (the Trade Expansion Act), Congress mandated a new trade policy organization in the Executive Office of the President, the Office of Special Trade Representative. Congress sought to link trade policy more closely to domestic political interests while not separating it entirely from continuing free trade and foreign policy concerns. The White House, it believed, could represent domestic interest groups better than the State Department and still aggressively pursue free trade objectives. On this basis, Congress approved U.S. goals and participation in the Kennedy Round, the most ambitious effort to date to advance multilateral freer trade. Thus, commitment to comparative advantage and the nondiscriminatory trading system received new impetus, despite the growth of interest group pressures in this area. Meanwhile the commitment to the international monetary regime weakened even though there were comparatively few interest group pressures in this area.

Superstructure Sways: 1965–1971

From 1965 to 1971 America's international economic diplomacy tried to hold on to and even add to the superstructure of the Bretton Woods monetary system while the foundations of that system progressively disintegrated. From 1965 to 1968 American officials negotiated to put in place the vaulted ceiling of a true international banking system complete with a nonnational monetary reserve unit, the SDR. The crumbling foundation, however, supported only the thinnest layer of new SDR issues. Thereafter, U.S. officials hoped that increasing structural pressures would bring about a spontaneous realignment of monetary and exchange rate relationships. What diplomatic architecture could not achieve by design, "benign neglect" would accomplish by default.

But policy choices in this period were never stable enough to define a predictable set of exchange rate relationships. Macroeconomic policy in the United States was now on a roller coaster, going from stimulus in 1965–1967 to restraint in 1968–1969 to stimulus once again in 1970–1971. Inflation and unemployment ratcheted up simultaneously, and microeconomic intervention—wage and price controls—became the preferred way to deal with the inflation predicted by the Phillip's curve.

New circumstances were said to justify new policies. Unwilling to recognize the roots of instability in domestic policies, the United States blamed the international system. Serious flaws were found in the exchange rate and trading systems, in addition to those in the monetary system (i.e., Triffin's Dilemma). Yet, surprisingly, the United States made few serious attempts to correct these flaws. It seemed to recognize subconsciously that to address them would require new discipline on U.S. domestic policies.

The SDR: Paper-Thin Diplomacy

By 1965 the defenders of the dollar's role in the Bretton Woods system were exhausted. The assault of the Triffinites, who called for a new world currency to replace the dollar, had finally penetrated the Treasury Department. (The following discussion relies heavily on Odell 1982, Chapter 3.) Secretary Dillon and Under Secretary Roosa had left the Department, frustrated by growing European resistance to "peripheral support" measures for the dollar. The Europeans, especially de Gaulle, continued to insist on U.S. demand adjustment, urging cutbacks in U.S. domestic and foreign expenditures. Unwilling to do that, the United States had two choices: suspend convertibility and/or devalue the dollar, or create a new reserve unit to replace the dollar.

Support for the multilateral system, if not for the specific role of the dollar, was still strong. Floating exchange rates were equated with isolationism, and academic acceptance of floating rates was still limited. Hence, moves to suspend convertibility or devalue the dollar at this point were not live options, especially because both suspension and devaluation would require the cooperation of the Europeans to succeed. Suspension without devaluation would imply going to a pure dollar standard in which the United States would be able to run balance-of-payments deficits and issue dollars free of any constraints in terms of gold or any other standard. Unless Europeans agreed to this step, they could have simply refused to accept additional dollars, forcing the United States to curb overseas expenditures or devalue. Devaluation, on the other hand, with or without suspension, might have been offset by competitive devaluations by European countries. So fundamental changes in the exchange rate system seemed at this point too complicated. Given the desire to continue to work through the system—that is, multilaterally—it was easier to create a new reserve unit than to adjust exchange rates.

This choice was helped by the fact that Henry Fowler, the new Treasury Secretary, was a Triffinite, as was Francis Bator, President Johnson's new economic adviser in the White House. In the spring of 1965 they briefed President Johnson on the dollar. Subsequently, in June, Johnson sent Fowler a secret memo ordering him to find a new way to create liquidity. One wonders whether the decision to escalate the war in Vietnam played a subconscious role in these deliberations. McGeorge Bundy, Johnson's National Security Adviser, had commented in late 1964 that if the war required it, devaluation of the dollar might have to be considered (Odell 1982, p. 141). Devaluation, however, as discussed earlier, was too complicated. Maybe the same running room could be obtained for the U.S. balance of payments by creating a new international reserve currency.

In the end, it is doubtful that such reasoning played a significant role. If it had, President Johnson would have shown a greater interest in the subject and ensured that adequate influence was brought to bear to persuade the Europeans to accept the generous creation of new liquidity. As it turned out, the talks to create the SDR dragged on desultorily for 2.5 years. They were conducted largely by lower-level officials under the general direction of the Secretary of Treasury. There is no evidence that President Johnson took an interest in these deliberations until the sterling devaluation of November 1967 precipitated a rush of gold sales and eventually led in March 1968 to the closing of the gold market for private sales. Thereafter, with government no longer buying or selling gold in the private market, official gold reserves were essentially frozen. No new nondollar liquidity could enter the system except through the creation of a new reserve currency.

Simultaneously, therefore, the major industrial countries, working through the Group of 10 minus France,[3] agreed in March 1968 to create a new unit

---

[3]The Group of 10 included Belgium, Canada, Germany, France, Italy, Japan, the Netherlands, Sweden, the United Kingdom, and the United States. Subsequently, Switzerland associated itself with the Group, making 11 members, but it continued to be called the Group of 10.

called the SDR. The decision, though acclaimed by President Johnson as the "greatest forward step in world financial cooperation in the 20 years since the creation of the IMF," proved to be a very small step at best (Solomon 1982, p. 143). The Europeans sought to restrain the creation of new liquidity and thus keep the United States' feet to the fire to reduce overseas expenditures. They insisted on a collective veto for the European countries over the issue of SDRs and a requirement that borrowers pay back 30 percent of any drawings (the United States having opposed any obligation to repay). France did not attend the meetings that closed the gold window because it refused to participate in decisions that weakened the role of gold, which France saw as the ultimate sanction on U.S. domestic and foreign policy profligacy. The first issue of SDRs agreed to in 1969 amounted to a mere $9.5 billion over the 3-year period from 1970 to 1972. The world hardly noticed this addition to what became massive new liquidity in the 1970s created by inflation and the oil crisis.

While the United States was adding this modest piece of superstructure to the Bretton Woods system, domestic policies deteriorated rapidly. The tax reduction in 1964 and large new social and military programs beginning in 1965 imparted a powerful fiscal jolt to the U.S. economy. The pressure on monetary policy was severe. William McChesney Martin, Chairman of the Federal Reserve Board, was the last holdout of the old school resisting Triffinism. He worried about inflation and the fate of the dollar. In December 1965 he confronted Johnson with a hike in the discount rate. As Donald Kettl notes, however,

> Johnson was in no mood to slow the economy. He had been won over on the
> benefits of stimulative federal spending, and even with the Vietnam war build
> ing he did not want to put his Great Society programs on the line. They were
> his "children," his "babies" as he used to describe them [Kettl 1986, p. 106].

The dispute was patched up, but the Fed continued to "lean against the wind" through the spring and summer of 1966. Finally, in late 1966, Martin agreed to ease monetary policy when Johnson announced his intentions to propose an income tax surcharge. The latter did not go into effect until 1968, however, by which time the economy was substantially overheated. In that environment the surcharge proved to be an insufficient brake. The economy headed into uncharted waters, where monetary and fiscal restraint was never sufficient or sustained enough to arrest price increases. Eventually the next Fed chairman, Arthur Burns, would become convinced that new microeconomic medicine — wage and price controls — was needed to cure the resulting inflation.

Benign Neglect: Market Tremors

The macroeconomic policies put in place from 1965 to 1968 predictably weakened the American balance of payments. The merchandise trade surplus, which had leveled off at around $5 billion from 1960 to 1965, dropped steadily after 1965 to lows of $600 million in both 1968 and 1969. Meanwhile net capital outflows doubled from $5.7 billion in 1965 to $11 billion in 1968 and $11.6 billion in 1969 (*ERP* 1976, pp. 366–367). Had the United States controlled inflation in this period, it is possible that the balance-of-payments problems might have corrected themselves. A Brookings Institution study in 1965 projected that Europe would inflate faster than the United States and that under these circumstances the basic U.S. balance-of-payments deficit would disappear, at then prevailing exchange rates, by 1968 (Odell 1982, p. 146). The crucial requirement was and always had been to maintain price stability in the U.S. economy.

With the commitment to domestic price stability increasingly in doubt, however, efforts to cope with the balance of payments shifted progressively from the futile capital controls of the early and middle 1960s (see earlier discussion) to exchange rate adjustments and possible trade restrictions (these four measures being the only immediate government policy options available; see Table 1-4). Advocates of trade restrictions were growing, but economists recommended this choice only with great reluctance (see, e.g., Cooper 1968, p. 273). The theory of comparative advantage still held sway, and economists would hold off until the 1980s before they proposed trade theories that justified more restrictive trade policies (Helpman and Krugman 1985).

The fixed exchange rate system, however, was less sacrosanct. The argument now materialized that the adjustment process, as well as the liquidity arrangements, of the Bretton Woods system was faulty. The adjustment process was asymmetrical. It put great pressure on deficit countries to adjust, because their capacity to borrow was in the end limited (even the capacity of the United States to borrow, if you accepted the Triffin argument about an inevitable loss of confidence in the dollar). But it put little or no pressure on surplus countries to adjust. They could maintain existing exchange rates, which became increasingly undervalued, and continue to accumulate reserves. If these reserves were sterilized (i.e., countries offset the sale of their currencies by other monetary operations to absorb excess liquidity), surplus countries could avoid expansion of the domestic economy, while holding on to export markets, and thus sustain surpluses indefinitely. In this way, it was argued, Bretton Woods encouraged mercantilist policies. The old scarce currency clause of Bretton Woods did not help much in this case (see Chapter 3). It allowed for discrimination against the scarce currency, but this now implied U.S. trade restrictions against Europe and Japan. Applying this provision would have wrecked the free trade system, and no one was ready for that just yet.

The issue was not only how to get surplus countries to inflate or revalue their currencies. Principal deficit countries were also reluctant to change their exchange rates—witness the defiant efforts of the British Labour Party to support the pound from 1964 to 1967, which the United States encouraged because of the feared consequences of a depreciation of the pound on the dollar. Moreover, when deficit countries finally did adjust, it was generally believed that they overadjusted (Solomon 1982, Bergsten 1975). Thus, over time deficit countries acquired undervalued currencies just as surplus countries did.

Now, the argument continued, the United States was the one country in the system that could not change its exchange rate. As other countries acquired undervalued exchange rates, the U.S. currency became progressively overvalued. The price of foreign products undercut U.S. prices in international markets, and the U.S. balance of payments and particularly its trade surplus continued to deteriorate. Thus, it was argued, through no fault of its own but because of the multiplying flaws of the Bretton Woods system, the United States was unable to correct its balance-of-payments accounts.

What identifies this argumentation as rationalization is the complete lack of conviction with which it was applied. From 1968 right up to the dramatic unilateral actions of the United States in 1971, the Nixon administration made no substantial diplomatic attempt to negotiate a realignment of exchange rates or to offer new rules for more balanced adjustment procedures between creditor and debtor countries, such as those it included in its plans for international monetary reform in September 1972 (see the following discussion). Why, if the system was so faulty, did it not merit at least one, even half-hearted, attempt to change it? Nothing was tried, not even an effort comparable to the meager one to create the new reserve currency unit known as the SDR.

Instead, the United States pursued a concerted policy in this period of pulling back from multilateral diplomatic initiatives. It reacted defensively to a series of exchange rate adjustments first in Great Britain in November 1967, then in France and Germany in August and October of 1969, respectively, and finally in Canada in May 1970. (Canada actually went to a floating rate.) The United States pursued a deliberate policy of what became known as benign neglect, a policy to let international market forces work to pressure surplus countries to revalue. The principal market forces at work, of course, were inflationary U.S. domestic policies. Under the fixed-rate system, accelerating U.S. inflation in the period after 1965 put pressure on surplus countries to inflate or to revalue. If they wished to avoid revaluing, they had to sell their currency for dollars, increasing the domestic money supply and risking potential inflation. The United States hoped they would choose to revalue, but except for a modest revaluation in Germany in October 1969 (4.3 percent in traded goods terms; Solomon 1982, p. 164), other countries refused to revalue. Instead, the principal industrial countries in this period chose to risk inflation.

Not only did the Nixon administration show little diplomatic enthusiasm for realigning exchange rates, but it is not at all clear that actual market developments in the period from 1947–1967 required it. Just as in the case of the Triffin Dilemma, the exchange rate system may not have been technically flawed.

Exchange rate alignments have to do, among other things (e.g., supply conditions), with differential tolerances for unemployment and inflation among various countries. What seems to have happened in the mid-1960s is that U.S. preferences shifted much more rapidly from emphasizing the control of inflation to eliminating unemployment than similar preferences in other countries, even in the United Kingdom, where, as we noted earlier, a Labour government maintained relatively severe domestic measures from 1964 to 1967 to support the pound. This relative shift would have, under a more flexible exchange rate system, led to a lower value of the dollar compared to the currencies of other countries. In other words, the dollar, by remaining fixed, became overvalued. But this is not the same as arguing that the dollar was already overvalued before U.S. inflation took off. That depends on what was happening in the United States and other countries before 1968.

The IMF data show that from 1948 to 1967, prices in the United States increased by a factor of 1.4. In some 18 other industrial countries, prices increased within the same period by a factor of 1.4–2.0, slightly higher than that in the United States. Now, using a strictly purchasing power parity calculation of exchange rates, some real devaluation of the currencies of these 18 countries against the dollar was justified because they had experienced higher inflation. The question is how much devaluation actually occurred. Because nominal devaluations in other industrial counties during this period were on average less than the internal price inflation in these countries (cumulative devaluation of approximately 40 percent compared to price inflation of 100–200 percent), it is reasonable to conclude that the real devaluation of other currencies, vis-à-vis the dollar, was not excessive (for data in this section, see de Vries 1986, pp. 59, 62–65).

On this evidence the dollar cannot be said to have been seriously overvalued, as some economists claimed already in the early 1960s (see Odell 1982, p. 91). Because the preceding calculations average across a large number of countries, they do not preclude serious overvaluation of the dollar with respect to individual currencies, such as the German mark or the yen. And they do not account for productivity and other changes in the real economic situation among these countries[4]. Nevertheless, the argument in the late 1960s was about a general bias in the exchange rate system overall. The argument pointed to excessive, across-the-board devaluations in 1949 and then a process of adjustment, as we have reviewed, in which both surplus and deficit countries systematically undervalued their currencies vis-à-vis the dollar. The data do not seem to support this argument.

The U.S. dollar, therefore, may not have been in a disadvantageous position until U.S. inflationary policies of the late 1960s put it there. In short, the dilemmas of the dollar were the doings of U.S. domestic policy, not the fault of a biased, fixed exchange rate system (Bergsten 1975). Exchange rates changed very little in the period from 1948 to 1967 because inflation rates among the major industrial trading countries were relatively uniform and low throughout this period (see Figure 5-1). United States inflation performance, in particular, was exemplary and undoubtedly underpinned relatively low inflation throughout the system (except, of course, in some developing countries, where inflation was excessive and nominal devaluations were often greater than internal inflation rates; see de Vries 1986, pp. 63–64). It is true that in this period

---

[4]On the basis of productivity increases, which were generally higher in Europe than in the United States during this period, European currencies should have been revalued. That they were not revalued adds to the case that they were undervalued vis-à-vis the dollar.

exchange rate changes were not all that easy to make, the European Monetary System (EMS) did not exist to facilitate realignments between the French franc and the German mark, and the IMF had inadequate machinery for exchange rate surveillance. But the exchange rate system was not so faulty or out of whack by 1967 that relatively minor adjustments might not have sufficed to improve it substantially. Such domestic and exchange rate policy adjustments in the early 1960s would have been particularly timely. But this was precisely the moment when adjustment dropped from the U.S. international economic policy agenda, not only in exchange rate policy but, even more important, in domestic macro- and microeconomic policy.

Creating the Dollar Dilemma: Domestic Policy Choices, 1969–1971

Putting these arguments about systemic constraints in perspective helps to focus attention on the critical domestic policy choices the United States made in the period from 1969 to 1971. In 1969 the Nixon administration initiated classic adjustment policies to rein in the overheated domestic economy. It tightened macroeconomic policies, both monetary and fiscal (maintaining the tax surcharge passed by the Johnson administration in the summer of 1968), and dismantled capital controls that President Johnson had imposed to stem the dollar outflow. In short, the Nixon team took the policy steps called for by the earlier postwar domestic economic consensus under Truman and Eisenhower, but it appeared to do so more by instinct than by conviction and without a clear road map of where it wanted to go internationally. Indeed, it was retreating internationally to reliance on market forces alone, including more and more talk about flexible exchange rates. Once markets imposed political costs, the Nixon team would back off from tighter macroeconomic policies and undertake some of the most interventionist domestic and international policies of the postwar period, including trade restrictions and wage and price controls as part of the August 1971 decisions. In some sense, as David Calleo argues, "Nixon's first two and a half years in office witnessed, in effect, the last act of the Kennedy–Johnson economic policy" (1982, p. 5).

Unemployment reached a new low of 3.3 percent in 1969, but the restrictive macroeconomic policies adopted in 1969 eventually pushed the country into a mild recession in 1970, the first such setback since 1959. Unemployment rose to 5 percent with 5 percent inflation. The public reacted very unfavorably. Grown accustomed during the 1960s to the longest expansion since World War II, the voters were unforgiving at the polls. The Republicans lost ground in the 1970 elections, and Nixon resolved that the experience of 1960, when Kennedy had defeated him in part on the promise of a more rapidly expanding domestic economy, would not be repeated again in 1972. As Stein notes, Nixon was "in tune with the conventional wisdom that the country valued continuous high employment above price stability" (1984, p. 135). In December 1970 he appointed John Connally as Secretary of the Treasury, and the outlines of a new approach emphasizing intervention in both the domestic and international economy began to take shape.

Arthur Burns had taken over at the Fed in January 1970. John Ehrlichman's notes from this period reflect clearly the sentiments, if not instructions, with which Richard Nixon dispatched Burns to the Fed. At a meeting in the Oval Office soon after the appointment, Nixon told Burns, "You see to it: no recession." Later Nixon reiterated to Ehrlichman and Bob Haldeman, "The Fed *must* loosen — it must risk inflation" (Kettl 1986, pp. 120–121; emphasis in original notes). The close ties between Burns and the Nixon White House subsequently spawned a host of stories about the rigging of the money supply. The stories are probably too facile to be true and are not corroborated by convincing evidence (Kettl 1986, p. 114; Woolley 1984, p. 155; see also Hibbs 1987a, 1987b). Nevertheless, the story of monetary policy in the period from 1970 on is one of reluctant but steady submission to an uncompromising and unconstrained fiscal policy (a story that may be repeated in the early 1990s; see Chapter 9). After gaining and sharing a coequal role with fiscal policy in the 1950s and 1960s, monetary policy clearly became the junior and accommodating partner in the 1970s. Exclusive Keynesianism, as we noted in Chapter 3, played down the role of monetary policy. It would take a sharp shift to the right, to monetarism, in the late 1970s to retrieve a role for monetary policy. And in the early 1980s, monetary policy would go on to dominate domestic policy, driving interest rates to unprecedented heights as fiscal policy careened even further out of control.

In his first meeting of the Federal Open Market Committee in early 1970, Burns opted for a looser policy (Kettl 1986, p. 120). Like most economists of the period, he had expected that 12 months of monetary restraint would squelch inflation. When it did not, he concluded that circumstances had changed and that monetary policy was no longer the appropriate instrument to rein in inflation. Already in May of 1970 he began to urge a new policy of wage and price controls. Burns believed that excessive Keynesian stimulus beginning in 1964–1965 had embedded inflation so deeply in the economy that normal Keynesian remedies no longer worked (Burns 1978, p. 55). Inflation was due to cost–push pressures, not monetary phenomena (Burns 1978, p. 144). Wages accelerated, even as unemployment rose. What was needed was an incomes policy that "could change the psychological climate, help to rein in the wage–price spiral, squeeze some of the inflation premium out of interest rates, and improve the state of confidence sufficiently to lead consumers and business firms to spend more freely out of the income, savings, and credit available to them" (Burns 1978, p. 131).

Nixon did not care for Burns' income policy or for what he viewed as continued Fed restraint on the money supply. The two sparred throughout 1970. Unemployment rose. Public dissatisfaction grew, with only 19 percent of the people expecting a prosperous year (Odell 1982, p. 126). Congress dared the President to be more aggressive, having passed the Economic Stabilization Act in August 1970, which gave the President authority to impose wage and price controls. Business increasingly supported such controls. Eventually, as Donald Kettl explains, the "high unemployment, pressure from the White House and Congress, complaints from outsiders like [Walter] Heller, and serious talk of credit controls convinced Burns and his colleagues to push their fears about worsening inflation to the background and to ease the money supply" (1986, p. 126). In early 1971, therefore, even though unemployment had leveled off and inflation was coming down, the Fed opened the money spigot once again. From January to August of 1971, M1, the monetary aggregate that included currency plus demand deposits, grew at an annual rate of 10.8 percent, compared to 5.2 percent for the entire preceding year.

This decision proved to be fateful for the economy and the dollar. Although it is difficult to identify the point of no return in the American breakout of Bretton Woods, early 1971 seems as good a point as any. As Herbert Stein points out, the "recession in 1970 was not very deep." By the end of the year, "there were already signs of recovery" (1984, p. 157). Unemployment was flattening out, and inflation, although it fluctuated, was down from 6 percent to 5.5 percent in 1970 and was to go down to 3.7 percent in the first 8 months of 1971 (Shultz and Dam 1977, p. 68). The evidence of recovery was ambiguous, as always. But a "reasonable judgment of this policy," Stein concludes, is "that [the policy] was working in the sense that it was getting the inflation down at the price of a recession which, at least by later standards, seems moderate" (1984, p. 158).

Alas, it was not enough to meet the rising expectations of the public and Congress. Moreover, political time was running out. With elections approaching and without deep convictions, such as those that held Ronald Reagan on course in 1982 despite similar congressional election pressures (see Chapters 7 and 8), the Nixon administration threw caution to the winds and inflated the sails of both monetary and fiscal policy.

Having started with a reliance on markets, the Nixon team ended up vitiating markets in a way that even Johnson's economists never contemplated. The costs of domestic adjustment were simply perceived as being too high. A good bit of economic analysis supported this view. C. Fred Bergsten, who served in the early Nixon administration as a staff member of the National Security Council, was preparing at the time a detailed analysis demonstrating that for a country like the United States, with a relatively small international sector yet large absolute role in world markets, domestic adjustment was by far the most costly route to restored balance-of-payments equilibrium (Bergsten 1975). This analysis confirmed what had been assumed by a growing number of economists and policymakers ever since the recommendations of the early Kennedy advisers, namely, that the United States should not accept international constraints on a domestic policy to achieve full employment. That this view was held so widely and so strongly in 1971, 10 years after the longest peacetime expansion in postwar history had reduced unemployment only from 5.4 percent in 1960 to 3.3 percent in 1969 (and back up to 4.8 percent in 1970) attests to the almost ideological fervor with which the proponents of exclusive Keynesianism clung to their economic outlook. It would take another decade to rid this school of thought of the assumption that inflation was the acceptable consequence, rather than the enemy, of higher employment. And it is not surprising that in 1981 some of these same individuals would be the most outspoken critics of the Reagan administration when it accepted the costs of domestic adjustment to reduce inflation and temporarily at least (because it failed to complete the course; see Chapters 8 and 9) restore the value of the dollar (Bergsten 1981). These critics would make the argument that "all post-war American administrations" had endorsed and pursued the Bretton Woods policy goals of low inflation, market incentives, and freer trade (Bergsten 1985, p. 140). But the different content of the macro- and microeconomic policies pursued in the late 1960s and 1970s compared to those pursued in the 1940s and 1950s makes this argument difficult to accept.

Conclusions

The real tragedy of the period from the mid-1960s to 1971 is that, although the United States retained the international instincts of the Bretton Woods system and still had the power to implement these instincts, its purposes and policies shifted and it failed to act. This failure, according to John Odell, "proved to be one of the more critical miscalculations of the decade" (Odell 1982, p. 241). It happened because the United States lost its sense of national purpose in foreign policy and exacerbated conflicts with the allies in Vietnam that weakened the consensus underlying the dollar exchange system. The obsession with unemployment at home went on to sever its domestic economic moorings as well.

By the time the United States did act in the summer of 1971, it acted without a purpose or policy, except to seek national autonomy free of international commitments (Gowa 1983). As Robert Solomon put it, Secretary of Treasury John Connally, who engineered the decision in August of 1971 to end the dollar's convertibility to gold and who symbolized the new U.S. attitude toward the international economic system, had "no vision of how to improve the economic welfare of his own country or the world" (Solomon 1982, p. 191). The day after the announcements of the new measures on August 15, he asked his advisers, "What do we do next?" (Odell 1982, p. 262). The lack of vision and content in American actions in 1971 had a lot to do with the subsequent erosion of American power. As Robert Keohane has noted, the "collapse of international stabilizing arrangements . . . were not results of a *natural* decline of American power so much as consequences of American policies that rapidly and unnecessarily reduced American power" (Keohane 1982, p. 11). United States policy had come full circle. From comprehensive plans for world economic order in 1945 with no commitment of power to apply them, the United States committed its power in 1971 with no plans for where it wanted to go. The uncoupling of American power from American goals would continue to characterize U.S. policy throughout the 1970s.

6

Ending Bretton Woods:
Cooperation Without Content

The decade of the 1970s witnessed a veritable explosion of international economic interdependence and institutional cooperation. The oil crises ignited a massive mobilization and integration of financial markets, as well as an endless round of summits and U.N. conferences. Yet this chapter demonstrates that such international activity was increasingly unstable and hollow. Expanding to include North and South as well as East and West, the international community lost a sense of shared political purpose and abandoned the minimum domestic economic policy consensus necessary to sustain an open, stable, and efficient world economy. Without agreed direction or content, the world economy drifted, economic growth and trade slowed, and international conferences and institutions became increasingly irrelevant to the real economic forces of inflation and microeconomic inflexibility crippling world markets.

In the 1970s the United States pursued domestic economic policies that consistently weakened its power and foreign security policies that accommodated the resulting diffusion of power. A world of many powers emerged, not only the five-power world with which the decade began—the United States, the Soviet Union, China, Japan, and the European Community—but also in the course of the decade the Organization of Petroleum Exporting Countries (OPEC) and the Islamic fundamentalists in Iran and the Middle East. For the first time perhaps in history, the world political community became truly global. With the oil crises the world economy also expanded. The surplus revenue OPEC states became important financiers of world trade and investment and recycled petrodollars that fueled a growing economic interdependence between North and South, as more advanced developing countries—known as the newly industrializing countries (NICs)—increasingly borrowed capital and expanded manufactured exports in northern markets.

Politically, the world looked more like the diverse universalist community envisioned in 1945; economically, it had become far more complex, with highly integrated trade and financial markets. In North–South relations, economic interdependence seemed to expand faster than the sense of shared political community, leading to sharp clashes over the need for a new international economic order (NIEO); a political process of detente seemed to outpace the viable economic opportunities for trade between East and West. More than ever, the world community needed firm guidelines that could maintain a proportion between political aspirations and economic realities.

In this environment, no longer as permissive as that in 1945 yet equally obscure, American policy lost its moorings. In the early 1970s foreign policies of detente and retrenchment consciously stripped away the moral or ideological (domestic political) dimension of international politics in favor of a purely geopolitical view of great power balances, yet these same policies simultaneously diminished U.S. military power so essential for the traditional exercise of geopolitical balance. Without an assertive moral or military dimension to U.S. foreign policy, economic and arms control diplomacy filled the vacuum. By mid-decade, U.S. political and economic diplomacy was more active than it had been since the Marshall Plan days. But the enthusiasm for international cooperation, in contrast to the Marshall Plan days, was now unmatched by an equally clear and inspiring definition of America's national purpose and was increasingly inconsistent with America's domestic economic policies. America's national purpose defined a world of such complex diplomacy and interdependence that the general public became increasingly disenchanted, and U.S. domestic economic policies disrupted complex, interdependent world markets in ways that made substantive international cooperation increasingly difficult.

As a result, international cooperation did little more in this period than accommodate growing political and economic diversity. The international monetary system was manipulated to absorb the consequences of a major shift in domestic economic policies, not only in the United States but also, in part through the U.S. example, in other industrial countries. Practically all countries shifted toward greater macroeconomic stimulus and microeconomic intervention to pursue the twin objectives of full employment and a higher quality of life. International rules constraining the use of exchange rate changes and trade restrictions had to be relaxed, and international liquidity exploded, particularly after the oil crisis of late 1973.

In contrast to the Bretton Woods standards, the emphasis shifted decisively from adjustment to finance. Adjustment problems accumulated. By the end of the decade, the result was less employment, rampant inflation, and an interventionist, no-growth philosophy that put the open international economic system itself at risk. Interventionist policies in domestic economies sparked a new debate over trade policy focusing on export subsidies and nontariff barriers. This debate elevated a new model of trade policy to center stage—the Japanese model—in which home markets are protected while export markets are exploited. This model generalized, of course, meant an end to the open multilateral trading system.

International institutions could not cope with the stresses. Despite a decade of intense international diplomacy—U.N. special conferences, industrial country summits, and North–South conclaves—progress was marginal. Cooperation lacked content. Indeed, international cooperation was frequently designed precisely to accommodate the absence of common standards. The floating exchange rate system was praised because it relaxed international constraints and allowed more room for diverging national economic preferences and performance with respect to inflation, government intervention, and import substitution. There was less need, it was argued, for common standards to guide the mutual adjustment of domestic policies. The IMF accords at Jamaica in 1976 did little to provide for serious domestic policy surveillance in industrial countries, and the only subject off limits in the North–South dialogue of the 1970s was the domestic policies of the developing countries. Thus, the nations of the world wrestled to establish common external arrangements even as they acknowledged that they shared fewer and fewer common standards for the management of their policies at home (Calleo 1982).

Bretton Woods Destructs: 1971–1973

There is no doubt that the principal as well as proximate cause of the decisions that the United States took in August 1971 to suspend the dollar's convertibility into gold was the incompatibility between the new deep Keynesian consensus on domestic economic policy and the dollar's role as the linchpin of the fixed exchange rate system. The fixed-rate system could not accommodate a policy of endless debasement of the dollar's value. Confidence would clearly decline at some point, whether liquidity was perceived to be excessive in terms of gold or not. Indeed, by the late 1960s, liquidity was not the problem. Dollar holdings abroad in 1969 were only $300 million higher than 5 years earlier. "Despite domestic inflation and the erosion of its trade surplus," Robert Solomon concludes, "the United States had stopped adding to world reserves through its balance of payments" (1982, p. 106). Confidence in the dollar now depended on U.S. domestic economic policies, and for good reason, this confidence steadily weakened.

The economic policy consensus in the United States ruled out domestic adjustments to preserve the value of the dollar but did not determine whether the United States would resort to suspension and devaluation of the dollar, a rash of domestic and trade policy controls, or both sets of measures. The United States could have simply floated the dollar. That was in fact the final outcome of its policies from 1971 to 1973. But it was as if there had to be a demonstration of domestic policy discipline at the very moment that that discipline was being irrevocably abandoned. "Ending gold convertibility would go down better," Herbert Stein noted, "if it was packaged as part of an independent, positive American economic policy . . . that looked strongly anti-inflationary" (1984, p. 166). Extensive wage and price controls were the result. In the end the United States also imposed trade controls. Striking at unfair trade practices abroad was a convenient way of demonstrating that the fault lay in the international trading system and with other countries' policies, not with U.S. domestic policy.

Ironically, the United States ended the Bretton Woods era by standing it on its head. Domestic and trade policies were made less free, whereas exchange rates and ultimately capital flows were made more free. Intervention invaded the real economy, and free markets took over the financial economy.

Unilateralists Take Control

John Connally, Secretary of the Treasury, and Peter Peterson, the President's chief international economic adviser in the White House, were both appointed in early 1971. Both were strong advocates of restrictive actions on trade. Peterson headed the newly established Council on International Economic Policy. In one of the first acts of this new council, he issued a comprehensive report that emphasized the decline of American economic power and catalogued the grievances of the United States against its various trading partners (Peterson 1971). Following the end of the Kennedy Round in 1967, trading issues had dropped off the U.S. agenda, as U.S. policy wrestled with liquidity and exchange rate issues (see Chapter 5). Meanwhile, U.S. domestic policies had shifted greater adjustment burdens to the trading sector (as they would do again in the early 1980s, see Chapter 7). United States inflation made it more attractive to import and less easy to export. The U.S. trade surplus disappeared in 1971 for the first time in this century. In that year the protectionist Burke–Hartke bill was introduced in Congress; supported by labor, it called for ceilings on nearly all imports based on their average level for the previous 4 years (Destler 1986, p. 114).

U.S. officials were in no mood to acknowledge that growing protectionism had its roots in U.S. domestic policy. Instead, Japan became a symbol of all that was wrong with the trading system (as it would again in the 1980s). Following up campaign promises, Nixon moved against Japan on textiles. The protracted and bitter negotiations ended with a voluntary restraint agreement and a broadening of the multilateral marketing arrangement dealing with textiles (Destler et al. 1979, Aggarwal 1985). These negotiations offered an early showing of the ugly face of bilateral trade disputes and the use of so-called gray-areas measures outside GATT to impose discriminatory trade relief (circumventing Article XIX of GATT, which required nondiscriminatory multilateral application of relief measures). They also spotlighted the growing awareness and concern with nontariff barriers. Japan was considered to be a country rife with all sorts of subtle and hidden trade restrictions and subsidies. These restrictions, it was thought, made a mockery of trade liberalization through tariff reductions. The United States, with adversarial arm's length relationships between government and business, had few defenses against market penetration once tariffs were lowered. By contrast, it was argued, Japan simply retreated to a multilayered defense of nontariff barriers while it subsidized its exports even more vigorously to capture market share in the United States.

There was much truth to these arguments, but Japan might have made the same case against the trade-distorting effects of the interventionist and inflationary policies then dominating in the United States. The decline of tariffs and the shift of the domestic economic policy consensus toward excessive macroeconomic stimulus and microeconomic intervention were beginning to reveal the incompatibility between liberal international markets and highly interventionist domestic economic policies. The initial postwar consensus had imposed clear limits on domestic economic policies so as to foster exchange rate stability and the liberalization of trade. Now the shift in this consensus threatened both the stability and openness of international markets.

The steady drumbeat of complaints in the United States about foreign countries consummated the intellectual rejection of Bretton Woods. As John Odell argues, the most decisive factor explaining U.S. policy in 1971 was the "disappearance of old attitudes toward the Bretton Woods system" (Odell 1982, p. 240). The feelings of cooperation that prodded the United States to stay with the multilateral approach to liquidity problems in 1965–1968 and to rely on market forces rather than unilateral actions to revalue currencies in 1968–1971 were now being replaced by a sense that the United States had been betrayed by ungrateful allies who exploited a system of trade and monetary relations badly skewed against American interests. Under the assertive style of John Connally, a head of steam was building up to strike back at the allies. "Foreigners are out to screw us," said Connally; "Our job is to screw them first" (Odell 1982, p. 263).

There was still reluctance to suspend the convertibility of or devalue the dollar. Arthur Burns, as the good central banker, argued until the bitter end to do neither. He sought aggressive actions domestically, including wage and price controls, but no action internationally except to suspend convertibility as a last resort if the domestic package failed to reverse market expectations (Odell 1982, p. 259). Paul Volcker, then Under Secretary of Treasury, reluctantly accepted nonconvertibility but expected a devaluation of the dollar, which he hoped to negotiate immediately as part of a general realignment of exchange rates. Ironically and perhaps indicative of the loss of long-term direction in this period, no one anticipated that exchange rates would end up floating. The idea of floating rates, although increasingly accepted in academic circles, where the ideas of Milton Friedman began to replace those of Robert Triffin, was still considered too risky. But Connally, Peterson, and initially Nixon were in no hurry to realign rates. They wanted the allies to understand that a new era had begun in which the allies would have to bear a greater share of the liquidity, adjustment, and trade burdens of a revamped international economic system. If floating helped to make that point, U.S. leaders were prepared to float, at least temporarily.

Suspending the convertibility of the dollar without devaluing was still a possibility. Despite the volatile financial situation from January through mid-August 1971, U.S. gold outflows totaled only $845 million. Other countries in the end "did not try to push the United States off gold" (Odell 1982, p. 216). Declining confidence in the dollar was much more a function of the persistent rise in prices for U.S. goods and assets than it was of vanishing U.S. gold reserves. But U.S. policy would now leave no doubt that domestic prices would rise still further.

The Camp David Package: A Time Bomb

The domestic economic package put together at Camp David in August of 1971 contained a time bomb. Responding to the widespread priority to reduce unemployment, the package called for rapid increases in government expenditures in 1972, with restraint to follow in 1973. The rationale was pure Keynesian. As Herbert Stein explains:

> The deficit for fiscal 1972 was going to be large anyway, and there would be no complaints about making it larger, as long as the budget for next year, fiscal 1973, would be in balance, at least on a high-employment basis. This was an old FDR trick, to combine expansion with the appearance of fiscal prudence by making this year's deficit so large that future deficits would look moderate by contrast [1984, p. 183].

Conveniently, of course, 1972 was an election year; therefore the joy of pump priming could be enjoyed during the election campaign while the pain of adjustment would only come afterward. How, though, was one to avoid an embarrassing debasement of the dollar? A domestic stimulus package would encourage imports over exports and lead to further pressure on the dollar. If convertibility were suspended, the dollar would probably decline. A decline of the dollar in turn would exacerbate inflation, already spurred by the budget stimulus. In addition, contemplated trade restrictions would increase prices. Alas, the way to contain these price pressures was through wage and price controls. The solution to excessive macroeconomic stimulus was microeconomic intervention. An "incomes policy" would demonstrate the necessary commitment to domestic discipline and counter speculation against the dollar. Ironically, such a policy was perceived not only to counter the effects of stimulus but also to permit it. As Stein notes, "now that they were about to impose controls they thought they had more room for more stimulus" (1984, p. 178).

As it turned out, the budget deficit for fiscal year 1971, which had ended June 1971, jumped to $23 billion from $2.8 billion in 1970. The deficit for fiscal year 1972 was an equal $23.4 billion (*ERP* 1976, p. 339). These two budgets recorded the largest deficits since 1945, except for the budget-busting year of 1968 ($25.2 billion). Meanwhile, the monetary supply grew more rapidly from 1970 to 1973 than for any 3-year period since the end of World War II (Kettl 1986, p. 114). The combination of unprecedented domestic stimulus and comprehensive wage and price controls set the fuse for the time bomb and an eventual explosion of prices and collapse of the dollar once the controls were removed. The United States had fallen for the classic economic elixir for which it earlier had criticized other countries and later would criticize many more (e.g., Brazil in 1986). It was now only a question of time as to when the bomb would go off.

While controls held, the economy looked great. Nineteen seventy-two was a banner year. The GNP grew at an annual rate of 5.7 percent and inflation at only 3.3 percent. Nixon's reelection was assured. Flush with victory Nixon then lifted controls in January 1973. The bomb went off in March. Inflation soared to 8.8 percent in 1973 and 12.2 percent in 1974, while unemployment remained stuck around 5 percent. "By the middle of 1973 at the latest and possibly earlier," as Herbert Stein notes, "the United States was in the grip of a classical demand pull inflation against which the controls [which Nixon sought to reinstate in June 1973 and elaborated after the oil crisis in October 1973] were powerless" (1984, p. 185). The dollar floated, as did other major currencies, and the world entered a new era of highly unstable and increasingly protected international markets. All of this happened, it should be remembered, before the oil crisis of October 1973.

Fluttering U.S. Monetary Diplomacy

The deteriorating domestic economic situation puts U.S. diplomacy in this period in proper perspective. That diplomacy became more active after August 1971, ending the policy of benign neglect from 1968 to 1971. But now, much like a newly decapitated chicken, U.S. diplomacy fluttered without its head, having cut any links between international initiatives and the necessary domestic policy reforms in the United States. Without such reforms, U.S. diplomacy stood no chance of rescuing the fixed exchange rate system, let alone of negotiating a new multilateral agreement providing for greater symmetry of adjustment responsibility between surplus and deficit nations. The United States tried both of these initiatives, no doubt quite sincerely and earnestly. But the initiatives were hollow. The epicenter of the earthquake that shook the world economic system after 1971 was in the United States, not abroad. Rationalizations, of course, continued to disguise this fact. In 1972–1973 the Friedmanite doctrine of floating rates achieved the same exalted status as the Triffinite doctrine of eventual dollar inconvertibility. Floating rates, it was thought, would salvage national autonomy where a new reserve currency (i.e., the SDR) had not. Nations would be able to pursue whatever domestic policies they chose and exchange rates would float to balance external accounts. Analysts predicted the end of the dollar's role and a system "much more akin to a multiple reserve currency system" (Bergsten 1975, p. 401).

In addition, an old nemesis from the postwar period returned to fuel justification for floating rates—the speculators and money managers. Offshore banking had grown substantially in the late 1960s and early 1970s, and in 1969 the Nixon administration discontinued capital controls. However, with national capital markets in Europe and Japan still relatively restricted and with government intervention in the real economy increasing throughout the industrial world, prospects for long-term investment declined and the movement of short-term money became increasingly important and disruptive. Eventually, policymakers convinced themselves, as many remain convinced today, that "hot money" precludes the maintenance of stable, relatively fixed exchange rates. Fixed rates, it is true, can not be maintained without an international framework of rules. But, as John Odell argues, "if such a framework had been in place and government had carried out its rules, the massive currency movements of early 1973 would have been less likely" (1982, p. 302).

International cooperation in this period therefore was a sideshow—but not if you read the newspapers or listened to American officials. In the venerable, vaulted halls of the Smithsonian Institution building, President Nixon proclaimed the new agreement reached in December 1971, which realigned and once again fixed exchange rates, the "most significant monetary agreement in the history of the world" (*New York Times*, December 19, 1971). In reality, the agreement proved to be inconsequential, like the SDR agreement in 1968. The Smithsonian realignment lasted less than 15 months and represented little more than a momentary intrusion of great power politics into the tedious world of international money. Henry Kissinger, who was preoccupied with global strategic and political objectives, became convinced in late 1971 that Connally's approach was threatening the alliance. Kissinger prevailed upon Nixon, who met with French President Georges Pompidou in mid-December and patched up the exchange rate dispute.

The Smithsonian agreement devalued the dollar by about 10 percent against the major industrial countries (Solomon 1982, p. 209). Although the United States had sought a 12–15 percent devaluation and would later claim that a second devaluation was justified to achieve this target, the numbers were not critical. The United States was not willing or able to support the new rates. At the Smithsonian it decided not to restore the convertibility of the dollar (putting the world in effect on a pure dollar standard) and its inflationary domestic policies ensured that anything less than a continuously depreciating dollar would be insufficient to staunch the U.S. balance-of-payments hemorrhage, particularly after wage and price controls were ended.

In 1972 Nixon and Kissinger went back to great power diplomacy, and Connally went back to private life. George P. Shultz replaced Connally and quickly became the economic czar in the Nixon administration, being appointed an assistant to the President in December 1972 along with his cabinet post—the equivalent role in domestic and international economic policy that Kissinger enjoyed from 1973 to 1975 in foreign policy. Shultz was more of a multilateralist than Connally and more interested in the content of domestic policy. He disliked wage and price controls and generally leaned toward market solutions for both domestic and exchange rate problems (Shultz and Dam 1977). Shultz, therefore, was torn in different directions by his predilections. He wanted to maintain cooperation with the allies but did not share their insistence, particularly French insistence, on maintaining a system of fixed rates. Shultz in effect wanted cooperation not to cooperate.

The next flutter of U.S. diplomacy revealed the contradiction of such a formula. In September of 1972, the United States tabled an elaborate proposal to reform the international monetary system and launched a 2-year process of diplomatic negotiations within the IMF to secure agreement on such reform. The U.S. proposal called for the greater use of SDRs as international reserve assets and for upper and lower limits on total foreign exchange reserves as indicators of the need for both surplus and deficit countries to adjust (Solomon 1982, p. 243). The desire for SDRs to play a greater role was consistent with the desire to reduce the dollar role, but the entire focus on reserves to indicate adjustment requirements (as opposed to domestic policy variables, which emerged later, tentatively, in the 1976 IMF Jamaica Accords and then more forcefully in the mid-1980s; see Chapters 7–9) seemed inconsistent with the preference of at least some key policymakers, such as Shultz, to move toward more frequent and ultimately flexible exchange rate changes. Reserves were less important in a system of floating rates because they exerted little discipline on surplus or deficit countries as long as these countries were willing to accept, respectively, appreciating or depreciating currencies. The Europeans probably sensed that the reserves issue was a diversion. They continued to insist that the United States adjust through domestic policy measures. They preferred maintaining fixed rates and were even moving themselves toward a zone of fixed rates within the European Community, which initiated in March of 1972 an arrangement known as the "snake in the tunnel," whereby European currencies were maintained within a closer range with one another than with the U.S. dollar.

The IMF reform talks dragged on from 1972 to 1974 with no substantive outcome whatsoever. The only outcome was to perpetuate the reform process itself through the establishment of the Interim Committee of the IMF. Meanwhile, money growth rates expanded more rapidly in this period than either before or after (Keohane 1982, p. 13). The process of international monetary cooperation was being robbed progressively of any meaningful outcomes by the steady deterioration of domestic economic policies.

The end came with a rush. Once U.S. wage and price controls were lifted in early 1973, speculation broke loose as markets anticipated higher inflation in the United States. In preparing for a second devaluation, the United States undertook another flurry of diplomacy, this time with multilateralists like Shultz and Volcker determined not to repeat the nonconsultative, high-handed style of the Connally devaluation. Volcker traveled to all the allies in February and negotiated new exchange rate alignments. The dollar was devalued by another 8 percent. The exercise, however, did not matter. The new rates held less than 3 weeks. Meeting in Paris on March 16, the major industrial countries finally declared an end to what they rationalized as the "crisis-prone" system of fixed exchange rates.

Economic conditions were hardly favorable for a new initiative in trade. Floating exchange rates and domestic inflation both discouraged trade. Nevertheless, security factors apparently pushed trade legislation to the top of the second-term Nixon agenda and helped launch a new round of multilateral trade negotiations known as the Tokyo Round. According to George Shultz, Nixon was influenced in February of 1973 by advice from Edward Heath, then prime minister of the Conservative government in Great Britain, who argued for a new trade initiative to salvage alliance cooperation (Shultz and Dam 1977, p. 138). Having neglected the allies in his first term in favor of triangular diplomacy with the Soviet Union and China, Nixon now launched his "Year of Europe" exercise. Trade legislation became top priority on the White House legislative calendar; with considerable expenditure of waning political capital by a Watergate-debilitated administration, a new trade bill was passed in 1974

(Pastor 1980). The Tokyo Round was under way, but volatile macroeconomic policies and new microeconomic issues (i.e., nontariff barriers) offered poor prospects for speedy or substantive outcomes.

Domestic Purpose and Policy Shift: Pluralism and an Expanding Public Sector

United States power had declined by the early 1970s. The interpretation in this study does not deny that. The world was more competitive and complex. As early as the 1960s, Europe and Japan had challenged aspects of America's purposes and policies for the world economy. But in the 1970s American power was still enormous, particularly among the industrial countries (47.6 percent of OECD GNP; see Table 2-3); and what power it had lost to the world as a whole (from 39.3 percent of world GNP in 1950 to 30.2 percent in 1970; see Peterson 1971, vol. II, Chart 1) it had regained in part through the politically more homogeneous (i.e., strengthening of democracies in Germany and Japan) and economically more interdependent industrial world that existed by 1970. The decline of American power is not a sufficient explanation for the events of the 1970s. More important appears to be the loss of legitimacy and effectiveness of American power. And that loss had to do with the crisis in America's national purpose following Vietnam and the persisting choice of inflationary and inter- ventionist domestic economic policies incompatible with the openness and stability of international markets.

Pluralism Substitutes for Purpose

In 1973 the United States extricated itself from Vietnam but began the long, dark descent into the Watergate miasma that would hobble U.S. leadership for the rest of the decade. America contracted an identity crisis. United States security policies in Vietnam had ultimately negated the domestic ideals of political freedom and economic competition that America sought to secure. The domestic dimension of containment, it was argued, had led America astray, and America would now have to accept the moral ambiguities of tradi- tional great power diplomacy. Nixon and Kissinger were avid practitioners of great power diplomacy and sought to base its use on the material requirements of peace, rather than on the moral requirements of freedom. This was all the more necessary, they believed, in a world of growing nuclear parity between the United States and the Soviet Union in which public opinion was becoming increasingly aware of the intolerable premises of the balance of nuclear terror.

Counseling America to become an ordinary power had its costs. Kissinger himself recognized and was sympathetic to the view that America could not sustain a purposeful presence in the world just to maintain the balance of power, any more than it could sustain such a presence in a moral crusade for freedom (Kissinger 1982, p. 242). Moreover, the world was now deeply divided between fundamentally different political or moral cultures (Judeo-Christian,

Islamic, Confucian) in a way that was not true of the traditional balance of power in Europe. In a sense the world needed a peace that permitted a debate among diverse national purposes, not a peace that assumed these differing purposes did not matter. In any event, while Nixon and Kissinger tried to rescue America from one moralism, that of militarized containment, America succumbed to another, that of demilitarized detente.

Detente focused new attention on restructuring diplomatic relationships to overcome or at least make less dangerous continuing domestic political differences. It gave priority to arms control and economic interdependence. In Congress it quickly led to a reduction of U.S. defense expenditures. From 1967 to 1974, total U.S. government expenditures remained nearly constant at 20.5–20.9 percent of GNP, while defense expenditures dropped from 9.0 to 5.4 percent of GNP and nondefense spending rose from 11.5 percent to 15.5 percent of GNP (*ERP* 1987, p. 337). The peace dividend in effect was used to finance the Great Society programs. It would be debated for the rest of the decade whether this reallocation of resources was a reasonable response to the end of the war in Vietnam and the apparent attainment of a more stable nuclear balance through the limitation of strategic arms. What is important for our story is that the reduction of defense expenditures coincided by the middle of the decade with a world view that tended to downplay the role of military power and to celebrate growing political and economic diversity. This view emphasized the fragmentation and separateness of interests and issues that made it increasingly difficult for national governments to influence policy events or to maintain the confidence and loyalty of their citizens.

The new definition of America's national purpose that arose in the early years of the Carter administration emphasized six "apparent" trends that together described a world physically unified but politically pluralistic:

1. An emerging global economy whose management depended on strengthening institutional cooperation, recognizing the blurring of traditional distinctions between foreign and domestic interests, and treating issues in a fragmented and disjointed manner.
2. Growing physical interdependence reflecting the closed system of world resources, population, geography (oceans, atmosphere, outer space), and society (terrorism, nuclear proliferation, etc.).
3. Declining American power.
4. The multiplication of important actors to include transnational and transgovernmental elites as well as traditional state leaders.
5. The diminishing utility of military force.
6. The loss of public consensus and confidence in governments.

This view was described at the time as "quite conventional" and "not in serious dispute" (see Allison and Szanton 1976, p. 45). Its thrust was to treat problems separately and cooperatively, much as the commercial–legal perspective had sought to treat problems in 1944–1945 between the Soviet Union and the United States. In addition, however, the new view encouraged maximum participation of political interest groups and countries, as well as technical experts. Although it saw national governments as increasingly incapable of reconciling diverse interests, it expected somewhat naively that international institutions and expertise would facilitate trade-offs and compromise (Allison and Szanton 1976, pp. 47–48, 52–54).

Under the influence of this world view, institutions proliferated in the 1970s at both the national and international level. In the United States, foreign economic policy acquired its own President-chaired Council on International Economic Policy (CIEP), putting foreign economic issues on the same level in the White House as national security issues which were handled by the President-chaired National Security Council (NSC). Economic summits emerged among the heads of state and government of the major industrial countries to parallel NATO security meetings. And numerous U.N. special sessions and conferences addressed such global issues as food, water, desertification, science and technology issues, and the problems of the "new international economic order" (NIEO).

Although all of this activity was said to reflect the new importance of global and economic issues, it also reflected the loss of importance and priority of these issues. If foreign economic policy had been critical for the United States in 1971, it would have remained linked with security and foreign policy concerns, as it was in the late 1940s, or it would have been subordinated to domestic economic policy, as it became in the early 1980s. Instead by being separate from both foreign policy and domestic economic concerns, foreign economic policy lost out consistently in the trade-offs between security and domestic economic policy. It neither benefited from the motivation of shared political community that security interests might have supplied nor coupled initiatives for international economic policy cooperation with the necessary domestic economic policy reforms without which these initiatives could not succeed.

Economic Policy Vacillates

Keynesianism expended itself in the course of the 1970s, but nothing replaced it. There was a sporadic tendency throughout the decade to move back toward the conservative end of the economic policy spectrum but no consensus on what the new mix of more efficient policies might be. The Ford administration moved in 1974 to reverse the highest inflation rate in the United States since 1919 and invoked once again the old-time religion of "tight money, fiscal restraint, and narrowly targeted relief for those most suffering from the recession" (Kettl 1986, p. 135). The apparent hope was to reestablish the premise of fiscal and monetary policy restraint while accepting a larger government role for cushioning the effects of recession on the poor. But the Ford administration "made little effort to fix upon the national mind any general principles of policy that would prevent the return of inflation" (Stein 1984, p. 215). Congress and the public regained their appetite for further expansion and government intervention as soon as inflation subsided — from 12.2 percent in 1974 to 7.0 percent in 1975 to 4.8 percent in 1976. Thus, whether Ford had been reelected or not, the "basis for the persistence of such a [noninflationary] course had not been laid" (Stein 1984, p. 215).

President Carter made unemployment once again the centerpiece of U.S. domestic economic policy. Fiscal policy became highly stimulative in 1977, and monetary policy followed closely behind, playing its now subordinate role of accommodating domestic expansion at the expense of internal and external price stability. For the 3-year period 1977–1979, the most commonly watched monetary aggregate, M_1, grew faster than in any previous 3-year period of the postwar era, topping the expansion from 1970 to 1973 (Stein 1984, p. 218). The dollar fell rapidly in the course of 1978, precipitating a series of dollar rescue operations. Initially, Carter also carried forward the government intervention- ist theme. Microeconomic intervention reached new heights (or depths) in Carter's National Energy Plan, through which, it was said by later critics, U.S. policymakers sought "to regulate every BTU that flowed through the U.S. economy" (Stockman 1986, p. 38). The growing tendency to regulate expressed the neo-Malthusian "limits to growth" theology: What could no longer grow had to be reallocated.

In retrospect, as Herbert Stein concludes, U.S. policy from 1974 to 1979 "relapsed into the form of conventional, *ad hoc*, pragmatic fine tuning" (1984, p. 210). It no longer had the ideological zeal of the deep Keynesianism of the early 1970s, but it retained the general direction. Economic policy reflected the growing uncertainty in economic theory that in the late 1970s became "envel- oped in a thicker fog of intellectual confusion than earlier" (Odell 1982, p. 328).

Throughout the 1970s, therefore, the U.S. economy got the worst of both worlds of countercyclical demand management and microeconomic interven- tion. Economic policy pursued the prizes of low unemployment and low prices (especially energy prices), even as demographic trends swelled the work force with new entrants (with women and minorities entering the U.S. work force in unpredicted numbers) and the postponement of domestic policy adjustments made the economy less able to absorb new employment without higher and higher inflation.

Metastasis: Deep Keynesianism Abroad

Its easy both to underestimate and to overestimate the impact of U.S. domestic policy in this period on the rest of the world. On the one hand, the United States remained the largest market and the most observed (i.e., respected and criticized) society in the Western world. It was still the pacesetter for economic policy ideas as well as practical economic developments in the world communi- ty. On the other hand, it did not determine the course of events in other countries in the late 1960s and early 1970s, any more than it did in the late 1940s. And its influence in the later period was much more indirect, largely through markets and forces of interdependence rather than through interna- tional institutions or direct negotiations and foreign aid. Yet the United States still played a critical role in legitimizing policy trends throughout the world.

Europe and Japan, as we have noted in Chapter 4, started the postwar period with stronger policies and institutions for both macro- and microeconomic intervention in the domestic economy. They would have persisted in the use of these unique instruments regardless of U.S. influence. But, at the margins, U.S. policy under the Marshall Plan clearly leaned against these tendencies. Had U.S. policy persisted in this direction in the 1960s, further structural change would have undoubtedly been achieved in Europe and would have begun at least a decade earlier in Japan. But in the late 1960s, U.S. policy turned in exactly the opposite direction. It reinforced policy thinking that emphasized fine tuning of macroeconomic policy and sectoral programs for industrial development and regulation. In so doing, it led Europe, Japan, and much of the developing world into a decade of non-market-oriented policy changes and frequently postponed structural adjustment.

To attribute the developments after 1973 primarily to systemic complexities and oil shocks is implausible. This reasoning reflects the attempt to excuse domestic policy for the shaping of international events. As a major OECD study in 1977 concluded, domestic policy errors contributed significantly to international outcomes in this period. "The major error of policy," according to this study, "was the overly expansionary policy at the beginning of the 1970s, especially monetary policy." The drafters of the study, chaired by Paul McCracken, former chairman of the Council of Economic Advisers under President Nixon, were so persuaded by the evidence that they added: "We do not think it debatable that the massive increase in monetary aggregates beginning in 1970 contributed significantly to the excessive speed of the 1972–1973 boom, the build up of inflationary pressures, and stubborn inflationary expectations" (OECD 1977, p. 102).

The same study also went on to note the marked rise in government microeconomic intervention prior to the oil crisis, with public expenditures rising at a rate of four full percentage points in the OECD countries as a whole over the decade prior to 1974 (OECD 1977, p. 210). Although this expansion of the public sector enjoyed political support, it significantly raised economic costs. "Efficient resource allocation in the public sector," the study pointed out, "always involves special problems because of the absence of market discipline and competition"; and "beyond some threshold, an over-rapid increase in public expenditure can so preempt resources from other uses, including investment, that it will have adverse effects on economic growth and, concomitantly, this competition for resources can be a source of inflationary pressure" (OECD 1977, p. 168).

As this book argues, therefore, policies have something to do with creating circumstances, and macro- and microeconomic policies in the OECD prior to the oil crisis certainly helped to create the supply and demand conditions in the early 1970s that OPEC exploited to quadruple oil prices.

The pervasive shifts that occurred in domestic policies throughout the OECD have already been recorded in Figures 1-1 and 5-1 of this study. The shifts had different origins in different countries. As early as 1961 Britain experimented with the National Economic Development Council (NEDC, or

Neddy), taking its "first halting step toward the interventionist policies that would draw the government into detailed intervention with the affairs of specific sectors and individual companies in the 1960s and 1970s" (Zysman 1983, p. 179). Neddy called for voluntary tripartite (labor, government, and business) planning to spur economic growth but until 1967 took a back seat to Labour government efforts to defend the pound. Then in 1967 the Labour government overturned traditional reliance on global macroeconomic policies and government "neutrality" and sought "to institutionalize a price and incomes policy as an integral element of its economic strategy" (Blank 1977, p. 707). Briefly, in 1970, a new Conservative government tried to reverse course—opposing incomes policy, tripartite planning, and the interventionist policies of the new Ministry of Technology created in 1968. But as in the United States, the Conservative government's commitment to the moderate Keynesian policies of the 1950s was not very deep. By 1972 the Conservative government and then in 1974 the new Labour government stepped once again on the twin accelerators of stimulus and intervention. As Douglas Ashford reports, Britain joined the "big spenders" among the welfare states in the 1970s (1981, p. 97).

Planning was an important element of French economic policy in the late 1940s and 1950s. But Monnet modernizers had used the planning process to move French industry into the bracing economic waters of international competition, particularly through the European integration movement. De Gaulle, however, feared that French industry would not be up to such competition. From 1958 to 1968, therefore, he "relied more heavily on public sector controls and instruments" to consolidate French industry and develop "national champions" in key sectors such as electronics and computers (Ashford 1982, p. 153). French targeting strategies proved much less successful than those of Japan. "The bureaucrats," according to John Zysman, "came to have less confidence in their prescience and began to realize that their intervention often produced new troubles without resolving the old ones" (1983, p. 149). After 1968 and particularly after 1974 France moved cautiously toward less *dirigiste* policies. France was now somewhat out of sync with the rest of the OECD countries, as it would be again in the early 1980s, when it lurched back toward interventionism. Nevertheless, French policies in the 1970s "did not disengage the state from specific interventions but only shifted the priorities and techniques of state policy" (Zysman 1983, p. 151).

If French policies after the mid-1950s moved sooner toward interventionist practices than other industrial countries (with the exception perhaps of Japan), German policies moved later. Even in the 1970s, Germany resisted extensive subsidies to industry (Zysman 1983, p. 257). Yet relaxation of anticartel legislation as early as 1957 fostered a corporatist industrial structure in Germany that privately allocated credit to German industries. By 1966, when the Social Democratic Party (SPD) entered the government, the German social market economy had also overcome its reluctance to employ traditional Keynesian demand management policies. Moreover, German social programs were already the most extensive among the industrial countries. Thus, German policy paradoxically pursued relatively moderate micro- and macroeconomic policies on the surface, while it preserved and extended communitarian structures within the underlying society. After the first oil crisis, German economic and industrial policies became more aggressive, but the state's role, by comparison to other industrial countries, remained modest (Katzenstein 1987). Germany's greater success in managing growth and inflation, compared, for example, to Gaullist France, may have been a function of both the continued pervasiveness of private ownership in industry and finance in Germany (creating more immediate costs for business failures and hence earlier adjustment) and the more consistent respect for market standards in German government policy interventions.

Japan was always the most interventionist of the industrialized nations but moved progressively after 1960 to implement explicit strategies of industrial targeting, protected domestic competition, and aggressive export. Italy began the slide toward greater intervention in the early 1960s. The "opening to the left" in 1962 made the coalition government more sympathetic to labor demands, the expansion of social services, and state intervention in industry (Posner 1977, p. 817). The small states in Europe were also more interventionist to begin with and "granted [in the 1960s and 1970s], in relative terms, about twice as much public subsidies to enterprises as did the large industrial countries" (Katzenstein 1985, p. 65).

There were common forces at work in each of the industrialized countries after the mid-1960s, and it is tempting to conclude that either structural circumstances—the disappearance of favorable postwar conditions for growth—or broad historical trends toward more socialist ideologies were driving the shift of government policies toward greater intervention. But this conclusion conceals ideological preferences. It appeals to structure or history to resolve disagreements that are in fact a matter of continuous contest among different schools of thought that seek to move policy ideas in one direction or the other. The shifts toward more interventionist policies in the late 1960s were tentative, especially in Britain and Germany. Had the United States counteracted these shifts by maintaining reasonable macroeconomic policies and avoiding wage and price controls, it is conceivable that the OECD countries would not have slid as far as they did in the 1970s toward inflation and intervention. Indeed, by assertive action to undo the excesses of inflation and intervention after 1979, Britain and the United States were still able in the 1980s to lead the OECD countries back toward general disinflation and deregulation. Thus, it is unreasonable to argue that U.S. policy choices in the late 1960s, and for that matter similar policy choices in other industrial countries, had nothing to do with outcomes in the 1970s.

What happened after the mid-1960s, according to the interpretation of this study, is that new ideas came along in the United States and other industrial countries about how to achieve a higher quality of life through more aggressive government intervention and more comprehensive planning. Structural factors may have made these new ideas possible, but they did not make them necessary. As Herman Van der Wee explains,

> During the 1960s the solid confidence in the operational growth of the mixed economy led most politicians and economists to judge that the problem of

economic crises had been largely solved and that attention should be turned to
the problem of social and economic welfare. The application of Keynesian
macro-economic principles . . . planning and associated policy instruments
seemed to guarantee a climate of long-term economic growth. So the consolidation of the modern welfare state became the central plank of government
policy.

This shift in emphasis had the effect of switching creative attention away
from the concept of technological efficiency and towards that of social efficiency [1986, p. 79].

The shift from economic to social efficiency was encouraged by the decline
of larger purposes among the Western countries justifying an emphasis on
growth. The need to use scarce resources efficiently to rearm as well as rebuild
the West, which had spurred growth policies and trade liberalization in the
early 1950s, was now less urgent. Detente and the collapse of America's leadership in Vietnam brought new participants into the shared political community
underlying the world economy at the same time that these developments
robbed this community of its common values and hence its political intensity.

The first oil shock reinforced but did not initiate these developments. If
expansive monetary policies preceded and significantly contributed to the oil
shock (see earlier discussion), expansive fiscal policies and microeconomic
intervention followed it, especially in Europe (U.S. fiscal policy having expanded dramatically well before the first oil shock; see Figure 5-1). Between 1973
and 1975, the government's share of total expenditures in OECD countries rose
on average by six percentage points. "Not matched by a similar expansion of
revenue," as an OECD study observes, "it was largely in this period immediately after the first oil shock that the general trend towards budget deficits . . .
began" (OECD 1985b, p. 31). This trend was not cyclical, a consequence of the
1974–1975 recession, but secular, amounting to a pincer movement of government policy that increasingly stimulated demand at the macroeconomic level
while intervening in the economy and making it less flexible at the microeconomic level. Macroeconomic stimulus ran up against less flexible labor and
capital markets and pushed up prices. Even when macroeconomic restraint was
exercised, microeconomic factors prevented prices from declining. The ratchet-
ing up of inflation convinced policymakers, as it had Arthur Burns in the
United States (see Chapter 5), that circumstances had changed, that modern
economies no longer responded to market signals. Policymakers stepped in to
intervene still further, with energy and more general price and wage controls; in
the process they only compounded market rigidities and inflation.

Public Invasion of Domestic Markets

It is worth taking a moment to investigate further the growing government role
in the 1970s, particularly in microeconomic intervention. This study argues
that this role slowed growth, not necessarily because governments were doing
the intervening but because they often gave priority to noneconomic goals and
thus made labor and capital markets less responsive to market signals. The

OECD has now compiled convincing evidence that government behavior in this period departed from the economic premises that fostered incentives and flexibility in labor and capital markets in the 1950s and 1960s. In the 1950s and 1960s, as the OECD notes, "investment, output and employment responded so strongly to the signals coming from domestic and international demand . . . because the capabilities . . . could be mobilized through factor [labor and capital] markets that were markedly more efficient than in earlier years." In the 1970s, however, in factor as well as product markets, "a broad range of microeconomic policies acted . . . to reduce the efficiency of individual markets — and slow the recovery from internal and external shocks" (OECD 1988, pp. 18–33; quotes from pp. 20 and 25, respectively).

Table 6-1 shows the expansion of the public sector during this period. From 1968 to 1980 the public sector grew by 16 percentage points in Belgium, 8 in Canada and France, 21 in Denmark, 15 in Germany, 13 in Italy and Japan, 18 in the Netherlands, 12 in Norway, 22 in Sweden, 7 in the United Kingdom, and 4 in the United States. In every case except the United States this growth was faster — in most cases substantially faster — than growth during the preceding 12 years from 1956 to 1968 (see first and second columns of Table 6-1). Germany's public sector, for example, grew five times as fast in the later period; Italy's grew twice as fast. Other countries had rates of growth ranging from 1.1 to 1.6 times as fast. Only Canada's public sector grew at roughly the same pace as that in the earlier period. Except for the United States, therefore, the data clearly show that the late 1960s represented a watershed period in the growth of

Table 6-1 Government Expenditures as Percent of GDP, 1955–1982

	1955–1957*	1967–1969*	1974–1976*	1978–1982†
Belgium		35.6	43.0	52.0
Canada	25.1	33.0	39.4	41.3
Denmark	25.5	35.5	46.4	56.3
France	33.5	39.4	41.6	47.8
Germany	30.2	33.1	44.0	48.8
Italy	28.1	35.5	43.1	48.6
Japan		19.2	25.1	32.8
Netherlands	31.1	42.6	53.9	60.3
Norway	27.0	37.9	46.6	50.3
Sweden		41.3	51.7	63.8
Switzerland		25.0	33.5	
United Kingdom	32.3	38.5	44.5	45.8
United States	25.9	31.7	35.1	35.3
OECD average‡	28.5	35.5	41.1	45.3

Sources: OECD (1978), pp. 14–15, for 1955–1957, 1967–1969, and 1974–1976.

OECD (1985b), pp. 41–42, for 1978–1982.

*Three-year averages.

†Five-year average.

‡Unweighted and for 1955–1976, excludes Greece, Ireland, New Zealand, Spain, and Switzerland.

government intervention. Moreover, even Japan joined the trend. Its public sector expanded the most proportionately, increasing by two thirds over its 1968 level. By the end of the decade Japan and the United States had public sectors comprising about a third of GDP, whereas Europe had public sectors ranging from 45 to 60 percent of GDP.

Not only the size but the composition of public expenditures shifted after 1968. Table 6-2 shows the growth of public sector expenditures by category for the period from 1955 to 1982. After 1968 the largest increases in public expenditures by far came in the category of transfers and subsidies (T & S in Table 6-2), principally payments to households for income maintenance, public health, and education expenses. From 1968 to 1980 (data for 1967–1969 and 1978–1982 in Table 6-2), this share increased by at least seven percentage points in every country (including the United States) and by as much as 18 percentage points in some countries (Netherlands and Sweden). Calculated on a percentage growth basis, these increases represented jumps of more than 50 percent for every country and as much as 100–200 percent for some countries (Canada, Denmark, Japan, Netherlands, Sweden, and the United States). By comparison, in the 12 years preceding 1968 (data for 1955–1957 and 1967–1969 in Table 6-2), shares going into transfers and subsidies increased in the countries listed from a low of 0.7 percentage points in the case of Germany to a high of 8.8 points in the case of the Netherlands. Even though this expansion occurred on a smaller initial base, the growth of shares in this period on a percentage basis in no case exceeded 100 percent and in most cases was less than 50 percent.

The shares of government expenditures going into consumption after 1968 (C in Table 6-2) increased much less than shares going into transfers and subsidies. Nevertheless, except for the United States and Norway, this growth was substantially more than that during the preceding 12-year period (for Denmark, for example, 10 percentage points after 1968 compared to 4.6 percentage points before 1968). Meanwhile shares going into investment after 1968 (I in Table 6-2) generally declined—for the OECD countries as a whole from 4.7 percent of government expenditures to 4.5 percent after rising from 4.0 to 4.7 percent in the years before 1968. The only exceptions were Germany, Italy, and Japan.

Thus, after 1968, government spending not only increased rapidly in comparison to the period before 1968, but also shifted decisively from investment to transfers and subsidies and, somewhat less so, to consumption. In real terms, the mean of the government's share of gross fixed investment in the OECD economies dropped from 16.0 percent in 1960 to 15.8 percent in 1970 to 13.9 percent in 1982, declining steadily after 1965 (OECD 1985b, p. 62). Government was spending less of a growing public share of the economy on investment, which might be expected to boost production, and more on transfers and subsidies, which might be expected to boost consumption. Even then, transfers and subsidies contributed less to consumption than other government purchases. As George Shultz and Kenneth Dam note, "although estimates differ as to the degree of stimulus of the two kinds of expenditures, purchases are considerably more stimulating than are transfer payments" (1977, p. 34). Thus,

Table 6-2 Government Expenditures by Category as Percent of GDP, 1955-1982

	1955-1957*			1967-1969*			1974-1976*			1978-1982†		
	C	T&S	I	C	T&S	I	C	T&S	I	C	T&S	I
Belgium	11.5	10.5	—	13.8	15.1	3.3	16.2	19.3	3.3	18.5	25.3	2.5
Canada	13.2	6.2	3.3	17.2	8.2	4.1	19.7	11.8	3.6	20.2	18.5	2.9
Denmark	12.6	7.4	2.9	17.2	11.8	4.9	24.0	15.8	4.4	27.2	25.6	4.2
France	14.1	15.0	2.3	13.7	19.2	3.9	14.4	21.9	3.7	15.5	28.8	3.5
Germany	12.5	12.5	2.7	14.4	13.2	3.5	20.3	16.9	3.9	19.9	23.3	5.1
Italy	11.9	10.9	3.2	13.5	17.1	2.8	13.7	21.5	3.6	17.4	26.8	4.6
Japan	9.7	4.0	—	8.4	5.2	5.0	10.7	8.4	6.0	10.0	15.3	7.5
Netherlands	15.1	9.3	3.5	16.0	18.1	5.0	18.0	27.3	4.0	18.0	37.0	5.0
Norway	11.3	11.1	3.1	15.4	16.0	4.5	16.7	22.3	4.7	19.4	26.6	4.2
Sweden	15.6	8.2	—	20.2	12.3	6.1	24.8	19.3	4.4	29.2	29.8	5.3
Switzerland	9.4	6.0	—	10.3	8.8	4.5	12.1	13.6	5.0	12.2	16.3	—
United Kingdom	16.6	7.9	3.5	17.7	11.3	6.1	21.5	14.7	4.8	21.2	20.5	3.5
United States	16.7	4.5	2.1	19.2	7.1	2.8	18.8	11.2	2.1	18.4	14.8	1.3
OECD average‡	13.0	8.8	4.0	15.3	12.2	4.7	18.0	16.1	4.5	17.7	22.3	4.5

Sources: OECD (1978), pp. 14-15, for 1955-1957, 1967-1969, and 1974-1976.
OECD (1985b), pp. 41-42, for 1978-1982.

*Three-year averages.

†Five-year average.

‡Unweighted and for 1955-1976, excludes Greece, Ireland, New Zealand, Spain, and Switzerland.

Note: C=consumption; T & S=transfers and subsidies; I=investment.

177

government expenditures after 1968 not only were buying less investment, but may have been buying less consumption. Although it is difficult to know precisely the consequences of these shifts for economic output, it is unreasonable to argue that there were no such consequences.

Expanding government spending, of course, had to be paid for out of rising taxes. Although revenues did not generally keep up with expenditures after 1968, leading to growing budget deficits, total taxes as a share of GDP increased sharply. From 1967–1969 to 1978–1982, taxes as a share of GDP within the OECD as a whole increased from 30.1 percent to 40.6 percent. This increase was about 100 percent more than that for the preceding 12-year period from 1955–1957 to 1967–1969 (OECD 1978, p. 42; OECD 1985b, p. 44). In Europe alone tax receipts went up from roughly one third of GDP in the early 1960s to 38 percent in the early 1970s to 45 percent in the early 1980s (Drouin et al., 1985, chapter 6, p. 9). Indirect taxes in the OECD countries increased only modestly, but direct and social security taxes grew rapidly. Direct taxes increased from 8.9 percent of GDP in 1955–1957 to 10.4 percent in 1967–1969 and then to 13.8 percent in the period after 1968, an increase over the entire 24-year period of nearly 200 percent (OECD 1978, p. 42). The larger tax take came primarily from private households (income and social security taxes), although an OECD report in 1978 noted that "corporate taxation as a proportion of corporate profits has been increased in most countries" (OECD 1978, p. 47).

The evidence compiled by the OECD does not indict the growth of the public sector per se. Nor does the argument in this study do so. But the OECD evidence and this study do indict the degree and type of public sector intervention that characterized the 1970s. If anything, this evidence understates the degree of intervention that occurred because it excludes costs that were forced indirectly on consumers and industries through price supports and regulations rather than increases in government expenditures and taxes. The most egregious example of such indirect costs was in agriculture. Price support programs in Europe and Japan added substantial costs to the price of food. The OECD estimates that in the period 1979–1981 consumers paid $35 billion of a total of $56.9 billion in agricultural costs in Europe and $16.7 billion of a total of $26.9 billion in Japan (whereas consumers in the United States, which relies less on pervasive price support programs, paid $7 billion of a total of $26.4 billion in agricultural costs; OECD 1988, p. 189). Indirect costs due to government regulation also accelerated in the industrial sector. The OECD lists three developments in particular that distorted labor markets: (1) "a fairly widespread compression of occupational and industrial wage differentials [that] occurred as a result both of collective bargaining and of government policy" and that reduced incentives to shift labor resources from declining to advancing industrial sectors (compression of wage differentials occurring much less in the United States than in Europe and probably accounting for the much better U.S. job performance in the 1970s and 1980s); (2) "legislative and regulatory restrictions" that actually prohibited layoffs of workers, notably in larger firms; and (3) "restrictive work practices" that made it more difficult to redeploy labor even internally within industries (OECD 1988, pp. 23–24).

By any reasonable standards, the evidence suggests that the expansion of the public sector in the 1970s went too far. At the time, there was much reluctance to acknowledge this fact. Evidence and data in OECD studies, such as the McCracken Report in 1977 (see earlier discussion in this chapter), were dismissed as conservative ideology and poor policy science (Keohane 1978). It would take the experience and perspective of the 1980s to suggest that economic outcomes are not just a consequence of circumstances outside the control of policy (the oil shocks of the 1970s, for example) or partisan, political decision making that serves narrow ideological or social interests. As this study argues, outcomes are also a consequence of economic policy choices, some of which are more efficient than others. The economic policy choices of the 1970s that discounted inflation and encouraged government intervention to regulate labor, product, and other economic markets proved to be very costly in terms of economic efficiency, particularly at the margins. As an OECD study in 1988 pointed out, "the costs of increasing the size of the public sector, as well as the gains from reducing it, are significantly larger than the average economic cost of public sector spending" (OECD 1988, p. 364). Thus, the marginal surge in government spending in the OECD countries after 1968, recorded in Tables 6-1 and 6-2, must have been particularly costly in terms of employment and economic output, especially in Europe. Even studies that pay less attention to this comparative data on public spending before and after 1968 and place more emphasis on the shocks of the 1970s as the cause of employment and output losses cannot avoid the conclusion that "increasing structural rigidities in European labor markets . . . " were a significant factor in lowering economic performance in the 1970s (Lawrence and Schultze 1987, p. 42).

International Process Compensates:
Markets and Institutions at Cross-Purposes

The conventional argument is that circumstances changed in the 1970s, markets became less efficient, and governments intervened both domestically and internationally to correct market imperfections. The argument here is that government purposes and policies changed, destabilizing international markets and making it increasingly difficult for national or international institutions to influence economic events. New purposes emphasizing pluralism and participation encouraged a more activist international economic diplomacy than ever before (sometimes as a substitute for traditional military diplomacy), whereas new policies emphasizing the government's role in domestic economies made that diplomacy less and less effective in restoring noninflationary growth. This argument is not against international institutions or international cooperation per se, any more than it is against government intervention per se. It is simply that international cooperation cannot succeed in nurturing reasonably stable and expanding world markets if domestic purposes and policies increasingly accommodate conflicting concepts of shared political community (e.g., as the Bretton Woods agreements in 1944–1945 tried to do between the United States and the Soviet Union) and any degree or type of government intervention (e.g., as the Bretton Woods agreements tried to do between the United States and Great Britain). The pluralist appeal for participation (e.g., East–West, North–South) is not enough, and the persistent preference for public control of the domestic economy is too much to maintain open, efficient international markets. Under these policy choices, international markets will either consistently escape effective international influence or close down as national institutions reassert control. The 1970s witnessed both—the apparent failure or at best marginal influence of one international conference after the other and the emergence of creeping protectionism.

Trade and Finance in the 1970s

International markets continued to grow in the 1970s but on the basis of increasingly weak domestic policy foundations. Trade volume grew at a rate only one quarter that of the preceding decade (see sixth column of Table 1-1), despite an enormous growth of international finance. Trade among the OECD countries was increasingly squeezed between the volatility of domestic and international (i.e., exchange rates) prices driven by macroeconomic stimulus and the rigidity of labor and capital markets exacerbated by spreading microeconomic intervention. Japan and a few of the developing countries, the NICs, seized on the deterioration of competitive conditions in Europe and North America to expand exports to these markets, particularly in labor-intensive and then intermediate products—textiles, footwear, consumer electronics, steel, automobiles, and so on. The United States and Europe responded with increasing import restrictions, particularly in the form of voluntary export restraints and orderly marketing agreements. By one estimate, the OECD countries increased some form of nonmarket control over trade from 36.3 percent of all traded goods in 1974 to 44.3 percent in 1980. The figure in manufactured goods alone quadrupled from 4.0 to 17.4 percent (Anjaria et al. 1985, pp. 22, 102; see also trade policy indicators in Table A-2). Quantitative restrictions reemerged in international markets after having been virtually eliminated in Europe and North America in the late 1950s.

Meanwhile, Japan and the NICs did much less to liberalize imports and either contribute to the growth of international markets in the case of Japan or lower import costs and enhance the efficiency of exports in the case of the NICs. While some NICs liberalized imports from the mid-1960s until the first oil shock, liberalization either leveled off (e.g., Korea) or declined after the early 1970s (e.g., Brazil, Singapore, the Philippines, Argentina, and Indonesia; see composite liberalization index in World Bank 1987b, pp. 96–97). Developing countries, more broadly, continued and even accelerated traditional policies of import substitution, price controls, and extensive public enterprises. By 1983, tariffs in developing countries were on average four times higher than tariffs in industrial countries, and nontariff barriers (NTB) were almost three times higher. (The NTB measures are, of course, only approximate; see Laird and Finger 1986.)

Public sectors in developing countries also expanded rapidly. Central government expenditures alone (excluding local government expenditures) rose in some 60 middle-income developing countries from 18 percent of GDP in 1970 to 26 percent in 1980, a growth of 44 percent compared to a 29 percent increase in similar expenditure ratios in industrial countries (World Bank 1983b, p. 47). State-owned enterprises (SOEs) mushroomed. Over the decade, the contributions of SOEs to GDP in developing countries grew from 7 to 10 percent, whereas this figure in industrial markets rose only from 9 to 10 percent. The growth in numbers was even more revealing. From 1960 to 1980, SOEs grew in Brazil from a few more than 100 enterprises to almost 500, in Mexico from under 200 to over 500, in Tanzania from less than 100 (1967) to 400, in India from around 40 to 180, in Pakistan from 50 to 175, and so on (World Bank 1983b, p. 76). These institutions dominated budget and credit requirements in many countries. Although in theory state-owned enterprises may have operated efficiently, in practice this was not generally the case. A World Bank study noted the conflict between political and economic imperatives:

> As a commercial entity, an SOE must sell in the marketplace. As a public organization, it is given other objectives and is exposed to pressure from politically powerful sectoral interests. SOEs are often operated as public bureaucracies, with more attention to procedures than to results; and ready access to subsidies can erode the incentive of managers to minimize costs [World Bank 1983b, p. 50].

Macroeonomic policies in the developing countries in the 1970s compounded the inefficiencies. Planning institutions established after the war "generally failed to fulfill the high hopes placed in them in the 1950s and 1960s and [were] often limited to assembling public investment programs with only weak links to budgets and policymaking" (World Bank 1983b, p. 64). Central budget institutions either did not exist or were too weak to control expenditures. "Examples abound," as the World Bank reported, "of public expenditures being out of control because the central authorities [were] not aware of the spending programs of different public agencies" (World Bank 1983b, p. 71). Finance ministers generally controlled central banks, and monetary policy was dominated by short-term questions of financial liquidity, rather than by long-term development objectives.

With foreign markets growing more slowly and becoming more competitive and domestic markets becoming increasingly rigid and inefficient, a new trade policy model emerged in the 1970s symbolized by Japan and the newly industrializing countries. It was based not on reciprocally negotiated freer trade but on unilaterally initiated targeted trade. It called for protecting home markets through tariff and especially nontariff barriers to achieve the production experience and economies of scale to export. Then it called for systematic and heavily subsidized targeting of export markets. The model assumed that competitiveness was no longer a function of relatively fixed, traditional factors of comparative advantage, such as land, labor, and capital, but increasingly the consequence of highly mobile technology. Whereas governments could do little, except over a longer period of time, to influence traditional factors of comparative advantage, they could promote the rapid advance of technology within relatively short periods of time and thus "create" comparative advantage. Trade policy was no longer a matter of reducing government intervention and allowing competitive market forces to work. It was now a matter of increasing government intervention to force competitive markets. The new trade policy model fit quite well with the interventionist ideology of the 1970s. (For a discussion and critique of this model, see Rushing and Brown 1986, and especially my article in that volume, Nau 1986).

The new trade policy model was seen, and still is seen by many today (see Conclusion of this study), as a successful response to changed circumstances, especially intensified export competition and the exhaustion of the postwar backlog of technological innovations. The argument here, however, is that these changed circumstances were themselves created by government policies and that targeted trade policies were among the factors in the 1970s that weakened international markets and slowed technological innovation. Domestic policy shifts, we have observed, created more volatile macroeconomic conditions and more pervasive microeconomic intervention and protectionism in world markets. Growth and trade slowed, reducing market incentives for technological innovation. Targeted trade policy approaches compounded these trends by intensifying competition for export markets while contributing little to the expansion of import markets. These approaches worked only as long as *some* foreign markets remained open (i.e., not every country pursued the new model), and domestic markets in the country pursuing targeted trade remained reasonably efficient despite protectionism (the case, for example, with the large, competitive domestic market in Japan). During the course of the decade, both foreign and domestic markets weakened. Accumulating debt and ultimately the debt crisis were in considerable part a reflection, rather than a cause, of the declining opportunities for exports and the escalating costs of import protection and domestic policy intervention.

Thus, the explosion of financial markets in the 1970s and the inevitable debt crisis of the 1980s were, according to the argument in this study, in considerable part a consequence of the implosion of real markets, that is, the deterioration of macroeconomic, microeconomic, and trade policies in both industrial and developing countries that in effect rescinded the postwar policy triad of price stability, flexible markets and freer trade. The oil crisis of 1973, as we have already observed, would not have occurred with the suddenness and sharpness that it did if it had not been for the prior deterioration of price stability in the OECD countries. The multifold increase in the price of oil, in turn, generated the enormous petrodollar revenues that subsequently fueled the expansion of international bank lending. Not accidentally, this bank lending circumvented the conditionality of IMF adjustment programs. In the 1970s no one wanted to focus on domestic policy adjustment (see next section). Domestic conditions, consequently, adjusted very slowly, if at all, to the effects of the oil price increases on the real economy. The absence of adjustment characterized both borrowers and lenders. In the United States, for example, price controls encouraged oil consumption and discouraged domestic oil production, leading to steadily rising imports from 1974 to 1978 and setting up world markets for the second oil price shock in 1979 (Nau 1981b).

On the one hand, therefore, the oil shock increased pressures for adjustment in domestic economies that were already suffering from deteriorating policy conditions. On the other hand, it generated the international liquidity that allowed such adjustment to be postponed. This is not to say that some adjustment and growth did not occur in the 1970s, particularly in Japan and the NICs. The strong export-led growth of the NICs was particularly gratifying. Much of the success of the postwar system in redistributing world income (see discussion in Chapter 1) was due to the acceleration of exports and growth in these countries as well as a group of second-tier, newly exporting countries (NECs), including Chile, Colombia, Peru, Thailand, Malaysia, the Philippines, and Pakistan (Havrylyshyn and Alikani 1982). But this growth was not sustainable, given real economic conditions in the industrial world and in the middle-income developing countries themselves. In some sense, growth and trade went as far as they did in the 1970s only because the enormous expansion of liquidity drove them.

Neither the OPEC nor the borrowing oil-importing developing countries can be blamed for the market trends of the 1970s. Primary responsibility rests with the industrial countries, especially the United States. Industrial countries also set the patterns with respect to international cooperation in this period, but here the developing countries played a more aggressive role, not always to their own advantage.

International Institutions:
Summitry and North–South Dialogue

The decade of the 1970s was better off for the international cooperation that occurred than it would have been without it. For a while, at least, international dialogue can hold a community together politically even as that community declines economically. But in many ways, international cooperation in this period was unhelpful. It deliberately ignored the central causes of economic instability in world markets, namely, less efficient domestic economic and trade policies, and it promoted international solutions that sought to regulate markets on the assumption that markets were now too complex to be efficient. International activities, in short, served too often as an excuse for international intervention. Unlike the Marshall Plan institutions that worked to encourage market linkages, international institutions in the 1970s worked at cross-purposes with markets. There was a willingness to recognize that market interdependence required coordination of domestic policies but a refusal to recognize that existing domestic policy tendencies of stimulus and intervention were inconsistent with growing market interdependence.

The inconsistency between domestic intervention and market interdependence became most evident in the trade negotiations of the 1970s. The Tokyo Round, the seventh postwar round of multilateral trade negotiations, succeeded in further reducing tariff levels among industrial countries, but it was much less successful in dealing with nontariff barriers (domestic subsidies, procurement policies, health and safety standards, etc.), which were rapidly becoming the major source of trade conflicts in world markets (Winham 1986). Nontariff barriers became a problem in part because of the success of postwar negotiations in lowering tariff barriers. Mac Destler explains:

> As long as trade policy involved tariffs—a distinct, separable instrument—nations could reconcile barrier reductions with activist policies at home. They could be "liberal" on cross-border transactions and interventionist within the home market. But their "domestic" economic actions had considerable impact on trade, and the lowering of tariffs made this impact more visible. Inevitably, American producers began to focus less on tariffs and more on other nations' domestic steps: the subsidies benefiting Europe's state-owned steel companies, or the buy-Japanese policies of the government telecommunications agency in Tokyo [1986, p. 34].

Thus, the increasing importance of domestic policy actions in world trade competition would have become a problem at some point in any case, requiring some agreement on the limits of domestic intervention if "free and fair" trade rules were to be preserved. Nontariff barriers became a more immediate and serious trade problem, however, because the consensus in the 1970s sanctioned interventionist policies, unlike the consensus in the 1950s and 1960s, which restrained such policies, at least at the margins. As government intervention became increasingly critical for more aspects of trade competition, it seemed natural to regulate trade through more direct government bargaining. By the end of the 1970s, despite the partial progress of the Tokyo Round in negotiating a series of codes for nontariff barriers—affecting subsidies and countervailing duties, procurement policies, customs valuation, licensing, and technical standards—the participating countries seemed to be marching off in the direction of greater government management of international trade.

The developing countries, which stood to gain most from open and expanding world markets, cooperated with and even welcomed this trend toward greater intervention and politically managed trade. In the North–South dialogue spawned by the oil crisis, they appealed for a new international economic order (NIEO), a set of integrated institutions to manage internationally, on a one-nation, one-vote basis, both trade and financial markets. Their priorities called for cartelizing commodity markets (e.g., UNCTAD's integrated commodity program), securing special and differential treatment for developing countries' manufacturing exports (e.g., the Generalized System of Preferences [GSP]), declaring moratoria on debt obligations, and organizing major international transfers of technology and other economic resources.

Their preference for redistribution, rather than growth, of resources and for control of, rather than participation in, open markets was understandable (Krasner 1985). But it was also unlikely to succeed or to benefit the rest of the world, and in the meantime, the NICs and other developing countries with increasing influence in world markets forfeited their leverage to keep industrial markets open by bargaining for reciprocal access in manufactured products or by pressing for liberalization in new sectors such as agriculture, where food-exporting developing countries had much to gain from altering the industrial country practice of dumping protectionist-generated food surpluses on world export markets.

North–South discussions in the 1970s were largely irrelevant, at least in an economic sense. The areas of greatest dynamism in this period—manufactured trade and commercial bank lending—remained outside the influence of the North–South dialogue. And the domestic policies of both developing and industrial countries were explicitly off limits. Thus, although the rhetoric of North–South discussions emphasized comprehensive, integrated approaches to development, culminating in an appeal at the end of the decade for Global Negotiations—the ultimate conference to integrate all subjects (see Chapter 7)—the most important economic policies and developments in the real economy were largely ignored or dealt with only piecemeal.

World Bank programs in this period illustrated the piecemeal approach. Singling out poverty, especially rural poverty, the Bank expanded its aid dramatically (elevenfold) in the course of the decade to finance thousands of individual projects to improve basic human needs in the poorest countries—health, literacy, and child nutrition programs (Ayres 1983). Although these efforts were hardly irrelevant and contributed to the strong improvements in basic human needs in low-income countries in the postwar period (see Chapter 1), they were also isolated from general macroeconomic and trade policy conditions of the period in both developing and industrial countries. By the end of the decade, despite its good work (its interventionist approach being justified to combat deep poverty but nevertheless contributing to non-market trends in the 1970s), the World Bank was largely irrelevant to the looming debt crisis and spent the first half of the 1980s languishing in the shadows of its sister institution, the IMF (see Chapter 9).

Industrial countries also intensified the process of economic cooperation in the 1970s but again without significant consequences for real economic developments. The first economic summit among the five major industrial countries at Rambouillet, France, in November 1975 confirmed the new system of flexible exchange rates but did not agree on any meaningful rules to govern the relationship among domestic economic policies. The agreement reached 2 months later at the IMF meeting in Jamaica ended for all practical purposes any commitments to defend exchange rates, except under extreme circumstances of "disorderly market conditions." Because exchange rates, which are international prices, can be defended on a sustained basis only by defending domestic price stability, particularly in the key currency country (i.e., the United States), the Jamaica Accord in effect threw away any rule governing domestic prices—the first leg of the postwar policy triad. Flexible exchange rates, it was thought, would compensate for diverging and generally rising prices. The steady depreciation of the dollar throughout the 1970s was a logical consequence of this abandonment of domestic price stability.

The economic summits of the 1970s brought an orientation to the world economy that gave priority to process and participation rather than to a con-

sensus on the substantive policy requirements of open, stable, and efficient international markets (see de Menil and Solomon 1983, Henning 1985,
Putnam and Bayne 1984). International coordination was necessary not to
achieve a consensus on domestic policy but to cope with three procedural
changes that had occurred in the international system. These changes
included:

1. The increasing entanglement of foreign and domestic politics that flowed
   from economic interdependence.
2. The waning of U.S. hegemony that had undergirded the long period of
   postwar prosperity.
3. The bureaucratization of international relations, a trend that some political leaders found particularly frustrating.

Summits offered a means to deal with these procedural changes — "to reconcile
international economics and domestic politics, to supplement and perhaps
supplant hegemonic stability with collective management, and to restore political authority over bureaucratic fragmentation and irresponsibility" (Putnam
and Bayne 1984, p. 8). According to this interpretation, summits were an
institutional or regime response to changed structural factors, in particular the
decline of American power and the increasing pluralization of both domestic
and international politics.

The counterinterpretation of summits offered in this study does not deny
the need to deal with procedural complexities of international coordination
once market openness and interdependence have become extensive and well
established. But it does not view these procedural realities as being independent
of substantive domestic and trade policy choices. The structures of market
interdependence in the 1970s were, as shown in Chapter 4, created in good part
by domestic policy choices in the 1940s and 1950s that emphasized specific
substantive criteria to moderate macroeconomic policies, restrain microeconomic policies, and accept freer trade. Instabilities in these structures in the
1970s were in turn a consequence of significantly different substantive policy
choices made after the mid-1960s. Thus, the substantive content of domestic
policies is all-important for the creation and maintenance of open, stable, and
growing international markets.

The process or procedural perspective on summits contends that "policy
coordination does not necessarily mean the adoption of identical or even similar policies, but rather the adoption of mutually reinforcing policies or at least
the avoidance of mutually inconsistent policies" (Putnam and Bayne 1984, p.
2). Yet it might be asked, Mutually reinforcing or at least not mutually inconsistent with respect to what substantive goals? Stability cannot be expected if
nations pursue significantly diverging unemployment and inflation goals (e.g.,
France accepting a higher level of inflation than Germany). And fairness cannot be expected if governments are permitted to intervene in domestic affairs
without some common limits, as when Japan pursues targeted rather than freer
trade. And without stability and fairness, skepticism will grow that markets
could or should be kept open.

Thus, some commonality of domestic policy priorities is absolutely essential to maintain open and efficient markets. Theoretically, these priorities could converge around high rates of inflation, more inflexible labor and capital markets, and targeted trade policy approaches. But if this were the case, the system, though open and stable, would be much less efficient. And if the system is less efficient, one may well ask why keep it open? There may be political reasons to do so, but the principal economic justification for open markets, namely, greater competition and efficiency, would no longer apply.

The history of summits in the 1970s ultimately revealed the contradiction between open, stable, and efficient international markets, on the one hand, and inflationary and interventionist domestic policies, on the other. The Bonn Summit of 1978, which, it is argued, "represented a rare, perhaps even unique, case of comprehensive policy coordination among the seven major economies of the world," demonstrated the risks of trying to convince every country to accept higher levels of inflation (Putnam and Henning 1989). The Germans agreed at this summit to accept a greater dose of fiscal stimulus, but subsequently regretted this decision when the second oil crisis reinforced inflationary pressures throughout the West. To argue that the German decision was a good one except for the occurrence of the second oil crisis is a bit like arguing that U.S. decisions in 1971 were good ones except for the occurrence of the first oil crisis. The fact is that the Nixon administration in 1972–1973 and the Carter administration in 1977–1978 were determined to stimulate domestic and international demand regardless of the short-term consequences for prices. The German decisions in 1978 thus considerably weakened the world economy's defenses against the inflationary impact of the second oil crisis, just as U.S. decisions had done in 1972 before the first oil shock.

The particular orientation of summits in the 1970s may in fact contain an inherent bias toward inflation and intervention. It is hard to conceive of these meetings taking tough collective decisions to restrain inflation or to rein in government intervention. Active summitry tends to focus on monetary and exchange rate policies and generally pushes these policies toward more accommodating stances and hence lower interest rates and potentially higher inflation (the case in 1977–1978 and again from 1985 to 1987). Indeed, coordinated exchange market intervention may be particularly useful to pressure central banks, which are relatively independent in the United States and Germany, to adopt somewhat looser monetary policies than they might otherwise, or at least to refrain from tightening money policy when they might wish to do so (Funabashi 1988, and see also Chapter 9 in this study).

Meanwhile, fiscal and structural policy issues play a somewhat lesser role in modern summitry and, when they do, the thrust is once again generally in a stimulative or interventionist direction. At Bonn in 1978, the Germans agreed to stimulate their economy in return for a U.S. commitment to decontrol energy prices (Putnam and Henning 1989). The latter was a rare case of summits reducing interventionist policies, although subsequent Reagan administration actions to accelerate the decontrol of energy prices demonstrated that international agreements may be unnecessary to undertake microeconomic reforms. More commonly, measures to restrain fiscal stimulus or promote structural change have been frozen out of the summit process because highly mobilized and divided interest groups at home generally oppose bargaining away the benefits of domestic spending and intervention (Aho and Levinson 1988, pp. 57–58). Why advocates of summitry, as it was practiced in the 1970s and again in the mid-1980s, often assume that international meetings and institutions will help to break these domestic deadlocks when stronger domestic institutions have been unable to do so at home is never clear. More frequently, international meetings merely add international differences to domestic ones and become highly visible media events for grandstanding to domestic publics.

More effective than this type of international summitry might be decisive political actions first to resolve economic deadlocks and disagreements at home and then to influence policies and institutions abroad through competitive performance outcomes in world markets (see Chapter 2). This was the type of international cooperation that the Reagan administration and some other summit countries (e.g., Great Britain) turned to in the early 1980s.

Conclusion

New policy ideas consolidated in the 1970s around pluralist political and deep Keynesian economic perspectives. These ideas encouraged the process of international cooperation, even as they introduced policy content that made international markets more unstable and international cooperation less effective. Growth and international trade slowed down, whereas international finance and debt expanded. Vast amounts of money moving within a real-world economy that was increasingly less open and more rigid gave rise to more unstable exchange rates and more short-term-oriented capital markets. International institutions, ignoring the roots of instability in the real domestic economy, tried valiantly to preserve open and stable markets increasingly through short-term exchange rate and monetary policy coordination. The world community had an economic tiger by the tail, heading toward the inevitable day of reckoning when international liquidity would level off and domestic adjustment would have to begin.

IV

Restoring Bretton Woods?

7

Recession 1979–1982:
Domestic Policy Adjustment
and International Conflict

This chapter traces the early efforts of the Reagan administration to redirect international economic diplomacy away from the process and institution-oriented approach of the 1970s to a more domestic policy and choice-oriented approach that sought to recapture the policy triad of the earlier Bretton Woods system — price stability, flexible markets, and freer trade. In the late 1970s the United States began efforts to redefine American purpose and policy and to close the gap between American diplomacy and power. The Soviet invasion of Afghanistan in December 1979 and the appointment of Paul Volcker as chairman of the Federal Reserve Board in October 1979 marked turning points in both American security and economic policies. Thereafter, in 1981, the Reagan administration propelled the nation in new directions, revitalizing a sense of pride and purpose in American democracy and setting forth a sweeping program for economic reform.

In theoretical outline, Reagan initiatives charted a path toward enhanced U.S. and Western security and more efficient economic policies to support security — that is, the Bretton Woods policy triad. In this sense they offered the prospect of restoring the essential elements of postwar peace and prosperity and adapting these elements to a more pluralistic and interdependent world community. After a blustering foreign policy debut and punishing recession, U.S. policies actually moved circumstances in this direction. American power surged. Defenses bristled, inflation dropped, investment accelerated, and manufacturing productivity rebounded sharply. American diplomacy became more self-confident. The Western alliance deployed new nuclear weapons, and the United States and Soviet Union embarked on unprecedented arms control negotiations to reduce large numbers of nuclear weapons.

In practice, however, America's domestic economic and foreign security policies remained both incomplete and unintegrated. Although the gap between America's purpose and power narrowed, coherent policies to bridge this gap and to rally a new political consensus around common security and economic goals never materialized. Instincts rather than doctrine guided both U.S. relations with the Soviet Union and efforts to restore an efficient U.S. and world economic order. In too many instances, ideology–anti-Communism and free markets — substituted for intellectual leadership. America was moving but it was not clear where. Improvements in performance lacked both context and explanation. They neither inspired new followers nor integrated support among existing ones. Progress was attributed to "the luck of the president" or his personal popularity, not his policies (Mandelbaum 1985).

The economic revolution became stuck. Massive tax cuts could not be matched by proportionate spending cuts. Fiscal policy went into gridlock, and unprecedented levels of federal debt accumulated. Monetary policy had to compensate, tightening sharply in the early 1980s to wring out inflation, then expanding after 1985 to bring the dollar down, tightening again in 1987 to stabilize the dollar, and fluctuating thereafter to avoid both recession and inflation. Fiscal imbalances shifted enormous pressures to the trading and financial sectors, fueling protectionism and compounding the international debt crisis. America itself became the world's largest debtor country, increasingly dependent on foreign capital flows. World financial markets grew shaky, and in October of 1987 stock markets crashed around the world. The world economy seemed perched on a knife's edge, teetering unsteadily from expectations of recession in late 1987 to revived growth and fears of inflation in 1988 and back to concerns about slower growth in late-1989. Ironically, after a free market Reagan administration, the United States faced developments that portended less free trade, more regulation of markets, potentially higher prices, and less growth. The instinctive quest to restore the Bretton Woods policy triad appeared to have come full circle.

American economic diplomacy reflected the lurching fortunes of domestic policy. In the late 1970s and early 1980s, diplomacy mirrored the deep divisions in domestic policy priorities among countries, with some countries giving priority to bringing down unemployment (e.g., France) and others giving priority to lowering inflation (e.g., the United Kingdom and the United States). After 1983 domestic policy priorities among the major industrial countries converged, but American policymakers never launched a vigorous and consistent international economic diplomacy. American diplomacy remained weak at a time when U.S. domestic policy reforms projected powerful and, on balance, helpful influences through the international marketplace to disinflate the world economy and to revive market incentives and growth (Marris 1985, p. 74). Although U.S. diplomacy paid more attention to domestic microeconomic and trade policy reforms, downplaying the heavy emphasis of policy in the 1970s on exchange rate and monetary policy coordination, it failed to generate a sense of community in the world economy in a period of difficult economic transition and to translate its domestic ambitions patiently into a broadly accepted design for international economic reform. In the void, U.S. policy came across as unilateral and ideological.

In 1985 U.S. diplomacy became more active, but it began to lose touch with its domestic roots. United States diplomatic initiatives to lower the dollar, manage the debt crisis, and stimulate broader international growth substituted for, rather than complemented, efforts to redress domestic imbalances. The administration largely abandoned domestic efforts to reduce the budget deficit and take the lead in drafting new trade legislation, steps that might have alleviated growing pressures on monetary policy and protectionist sentiment in Congress. Instead, it placed all its bets on exchange rate coordination to lower the dollar and on revenue-neutral tax reform to sustain the recovery.

American diplomacy resting on massive fiscal stimulus and precarious monetary policy began to look more and more like the failed diplomacy of the 1970s. It had more initial credibility and less immediate adverse impact because it began in a worldwide situation of low inflation and high unemployment, unlike the synchronous boom of the early 1970s. But its directions were unmistakably similar, and the world in 1988–1989 seemed increasingly nervous as the U.S. and some other economies approached levels of full employment and confronted either slower growth or potentially higher prices or conceivably both.

Purpose and Policy Turns in the United States

By 1979 the interlude with political pluralism (detente) and deep Keynesian economic policies had peaked. The United States and its allies began to regroup to face new external Soviet challenges and to attack the domestic roots of stagflation. The NATO two-track decision in 1979, the failure to ratify SALT II after the Soviet invasion of Afghanistan in December 1979, and even the ill-fated Iranian hostage rescue attempt in April 1980 reflected a reawakening of U.S. policy to the realities of military power and competition in a world of fundamental political differences. Similarly, the Carter administration's retreat from the stop-and-go economic policies of 1977–1978 and its belated concern for price stability and the value of the dollar marked a renewed awareness of the link between internal discipline and external stability.

Initial steps to reverse security and economic policies were tentative, to be sure. Security policy was a subject of intense and sometimes bitter debate in the election campaign of 1980. Inflation continued to climb from 11.3 percent in 1979 to 13.5 percent in 1980, and unemployment in 1980 still stood at 7 percent, higher than in all but 2 years (1975–1976) since 1947. But public sentiment was clearly shifting. As William Schneider reported, "in the early 1980s, the preponderance of public sentiment toward the Soviet Union was once again negative, although not quite as negative as it was in the 1950s" (1987, p. 50). Moreover, public concern about domestic economic problems, which had dominated all the elections after 1972, escalated to a postwar peak in 1979–1980 (Hibbs 1987a, p. 128).

The Reagan administration had a unique opportunity for leadership. It offered a more realistic view of the cooperation that could be achieved between totalitarian and democratic societies. And it recalled the virtues of a more cautious, longer-term approach to economic prosperity based on price stability and market-oriented competition rather than on the more radical pump-priming and interventionist policies of the 1960s and 1970s.

A New Realism

The prevailing definition of America's national purpose shifted after 1979 in two ways. Realist voices in the Carter administration grew stronger and injected both a sharper human rights (domestic) dimension and a stronger military or geopolitical dimension in U.S. policy toward the Soviet Union (Brzezinski 1983, especially Chapter 9). In summer 1978 the United States protested the treatment of human rights activists in the Soviet Union and eventually won the release of several prominent dissidents. This demarche coincided with a modest tightening of the administration's economic policies toward the Soviet Union, placing some oil and gas equipment to the Soviet Union under license (whereas earlier the administration advocated energy cooperation with the Soviet Union; see Chapter 10).

More important, the United States moved after the Soviet invasion of Afghanistan in December 1979 to strengthen the military dimension of U.S. diplomacy. Although real U.S. defense expenditures began to increase again in fiscal year 1976, the Carter administration had made a number of highly visible decisions to cancel or delay major weapons programs (e.g., the B-1 bomber, MX missile, neutron bomb), and it was torn internally between the commercial–legal approach to foreign policy reminiscent of Wilsonianism and the more traditional geomilitary approach. The Afghanistan invasion finally resolved the issue in favor of a more assertive military posture, as reflected in the failed Iranian rescue mission and the resignation of Secretary of State Cyrus Vance, the leading advocate of the commerical–legal view (Vance 1983). If containment had gone too far in militarizing U.S. security policy, it was now recognized that detente had gone too far in demilitarizing it.

The Reagan administration reinforced tendencies to strengthen both moral and military dimensions of U.S. foreign policy. National purpose now included a new sense of pride in American and Western freedom and a greater will to use military force on behalf of asserting and defending that freedom. The Reagan administration subordinated U.S.–Soviet relations to defense and alliance initiatives (e.g., military build-up and the NATO two-track decision) and launched more assertive policies to promote democracy in the third world, including the use of U.S. and proxy military forces (the Reagan Doctrine).

As the decade progressed, Reagan policies produced initial dividends in unprecedented arms reduction agreements with the Soviet Union and more even-handed attitudes toward totalitarian governments, on both the right (Philippines, Panama) and the left (Nicaragua, Cuba). At the same time, the United States avoided to some extent both the moral and military excesses of Vietnam—that is, supporting one totalitarian regime to defeat another (though many Americans questioned the democratic bona fides of, for example, the Nicaraguan contras) and using military force so as to become entangled in protracted and unwinnable military conflicts. The greater moral substance and militant style of American policy defined a shared political community with like-minded countries, including now a growing number of democratically oriented developing countries, that potentially offered sharpened motivations for efficient international economic policies—both to rally prospects for freedom in the third world (as containment rallied prospects for freedom in postwar Western Europe) and to use resources more efficiently for defense (as the Korean War stimulated liberal trade policies in the 1950s).

Nevertheless, the Reagan administration's definition of national purpose never achieved a coherence or context that could inspire a broadened and sustained consensus within the United States or the Western and third world. In the early 1980s, administration efforts frequently vacillated between ideological belligerence and military, including nuclear, posturing, on the one hand, and marked indifference to foreign policy when it seemed to clash with domestic priorities, on the other. By the mid-1980s the defense build-up was still uncoupled from a widely recognized or accepted strategic rationale, and in 1985 real defense expenditures began to decline once again. Although there was some greater bipartisan support for the idea of negotiating arms control agreements from a position of strength and leverage, defense support weakened further in 1988 as a major defense procurement scandal unfolded. The Iran–Contra affair also discredited much of the administration's leadership against terrorism and totalitarianism in the third world.

Despite its many failings, however, Reagan's concept of national purpose provided new direction. It neither shattered the sense of shared political community in the West, as militarized containment in Vietnam had done, nor dissipated it too broadly across nations with fundamentally different domestic systems, as some versions of detente and pluralism had done. There was a common basis for revitalized community in the West, as strong U.S. security relations with France suggested. Indeed, European and third world friends of the United States complained in the 1980s less about America's sense of national purpose than about the economic policies the administration chose to implement this purpose. These economic policies had more to do with both enhancing and undermining the prospects for restored prosperity and stability in the world economy than American security or military policies.

Carter's Tentative Economic Turn

In August 1979, Carter appointed Paul Volcker as the new Chairman of the Federal Reserve Board. He did so reluctantly, offering the job first to Tom Clausen, then President of the Bank of America, who was seen as more of a team player (Greider 1987, pp. 22, 45–46). Nevertheless, Carter acquiesced in October 1979, when Volcker engineered the technical changes that permitted interest rates to rise and eventually rein in double-digit inflation (Greider 1987, p. 121). Monetary policy broke free once again, as it had in the early 1950s, from the clutches of fiscal policy and short-term interest rate management. The results were predictable. The federal funds rate, the overnight interest rate for interbank borrowing, soared from 12 percent in October 1979 to 18 percent in

March 1980. The prime rate, which banks charge their best customers for short-term funds, peaked at 20 percent.

Carter's 1980 budget, however, remained stimulative, and credit availability and inflation continued to rise. Facing reelection, Carter invoked the little known Credit Control Act of 1969 to restrain credit. The economy slowed precipitously, with GNP plunging 9.6 percent in the second quarter of 1980. By June the credit controls had been removed, and the Federal Reserve accelerated the growth of bank reserves. "By the fall," as David Calleo notes, "the rapidly growing money supply was pushing the Fed back to a tight monetary policy, despite the election, high unemployment and depressed business profits" (1982, p. 149). When the Fed did tighten in October, Carter, on the campaign trail, openly criticized it (Kettl 1986, p. 179).

The economic turnaround in 1979–1980 was highly tentative. Although the Carter administration gets credit for appointing Volcker, its commitment to disinflation was suspect (McKinnon 1984, p. 48; Melton 1985, pp. 61–62). As Herbert Stein concludes, "The Carter–Volcker policy of 1979–1980 was an attempt to feel a way through the difficulties of the transition to a less inflationary world by extremely cautious, tentative and reversible steps. . . . But the policy was not acceptable or credible, especially to people who considered themselves conservatives" (1984, pp. 229–232). The Reagan administration would supply both the conservative political coalition and the policy persistence to make the Volcker program work. Disinflation would become one of its most notable achievements, less by virtue of integrated and consistently implemented policies than by way of clear and unchanging priorities.

Reaganomics: Dissensus in the Cocoon

While inflationary and interventionist policies continued under the Carter administration, ideas in the nongovernmental arena (the cocoon of consensus building; see Chapter 1) were shifting in new directions. The new ideas stressed the old policy triad of price stability, flexible markets, and freer trade, but they did so in a piecemeal and unintegrated fashion. Monetarists saw price stability as the single most critical key to prosperity, supply-siders emphasized tax cuts, budget balancers sought to reduce the welfare state, and laissez-faire free market types championed freer trade. The pieces were never integrated, as they had been in 1946–1947, and Reagan policies reflected warring intellectual camps within the administration as well as outside in the larger arena of nongovernmental relations.

By the end of the 1970s, deep Keynesian economic policy in the United States was in full retreat. Mainstream economists had jettisoned the conviction that employment was more important than price stability and that more jobs could be created with higher and higher inflation. The country seemed headed "back to the future," to recapture some of the basic economic themes that had characterized the late 1940s and 1950s. As Herbert Stein explains:

> By 1979, as the country wound up for a presidential election, the stage was set
> for a change in economic policy, and the direction of the change was clear.

There would be a shift of national priorities—toward greater price stability, faster growth and greater freedom for individuals in the use of their own money and management of their own affairs, and away from higher employment, the redistribution of income and the promotion of particular industries and uses of the national income as the primary objectives [1984, pp. 226-27].

The monetarist school of economic policy centered around the writings of Milton Friedman and other economists working from base camps at the Universities of Chicago, Rochester, California at Los Angeles, and Miami and at Texas A and M and Virginia Polytechnic Institute. The monetarists had convinced increasing numbers of congressmen that the demand for money over time was relatively stable and that control of monetary aggregates was a better way to influence aggregate demand and employment than Keynesian policies of fiscal stimulus and monetary policy accommodation. In 1975 Congress passed a resolution calling for more attention to monetary aggregates, and in 1978 it passed the Humphrey-Hawkins Act (Full Employment and Balanced Growth Act) mandating annual targets for money and credit, as well as semiannual reports by the Fed to Congress on progress in meeting these targets. The monetarists gained another apparent victory in the change of Fed policy in October 1979. This change targeted one element of banking reserves known as nonborrowed reserves as a means to influence borrowed reserves and hence short-term interest rates. Monetarists, to be sure, advocated targeting total reserves and rejected any role for interest rates (Melton 1985, pp. 47-57). Nevertheless, the change in Fed practice did represent a partial, albeit temporary, move toward monetarism.

Advocates of supply-side tax cuts operated from different bases in the nongovernmental arena. They enjoyed only selective support in the academic community (e.g., Robert A. Mundell at Columbia University and Arthur B. Laffer at the University of Southern California) but drew more support after the mid-1970s from Wall Street and a small but growing band of members in the U.S. Congress (e.g., Jack Kemp, William Roth, and Phil Gramm) (Wanniski 1978, Roberts 1984). Their main concern was with reducing tax rates to improve incentives to work, save, and invest, and they crafted the Kemp-Roth tax proposals, which were initially introduced in Congress in 1977. The Kemp-Roth bill eventually became the centerpiece of the Reagan economic program.

Other perspectives figured in the growing economic policy debate of the late 1970s. The traditional conservative Keynesian economists, such as Herbert Stein and Alan Greenspan, were still around, but they were too compromised by the Republican fling with deep Keynesianism in the early 1970s to exert a significant influence. A new generation of fiscal conservatives shared the traditionalists' concern with budget control but believed that this control must be achieved at much lower levels of government involvement, that is, lower levels of both taxes and expenditures. This generation, unlike the earlier one, was unwilling to live with the interventionist social programs of the 1960s and 1970s and, as David Stockman put it, to "become the tax collector for the welfare state" (1986, p. 53). Their commitment to expenditure reduction, however, along with the monetarists' commitment to tight money, made the Reagan program look too much like the traditional "root canal" economics of conservative Republican administrations. Thus, the new generation of fiscal conservatives gladly embraced the supply-siders' logic that simultaneous tax cuts would spur investment, growth, and enhanced revenues (Anderson 1988, p. 116). Whether they fully accepted the extreme claims of supply-siders that revenue gains from higher growth would more than offset revenue losses from lower taxes is beside the point. (They all vehemently deny accepting these claims, except for Don Regan 1988, p. 174; Anderson 1988, pp. 140–163; Stockman 1986, pp. 10–11; see also Roberts 1984, p. 40.) Their commitment to both tax and expenditure cuts made them dependent on an extremely difficult political balancing act, if expenditures and taxes were to be reduced simultaneously, or susceptible to more extreme economic and political arguments, if expenditures could not be reduced in line with taxes.

Stockman's positions reflected the dilemma. Initially he was identified with the supply-siders and their extreme expectations regarding revenue growth. Once the budget slipped into deep deficit, however, he articulated the view, propagated already in 1983 by his earlier mentor, Senator Daniel Moynihan, that the deficits were political, deliberately designed to maximize congressional pressures to trim the welfare state (Greider 1981; Moynihan 1983, 1988a, and 1988b). Either way, Stockman and the new fiscal conservatives came off as extremists. Their only escape would have been to make some part of the tax reductions contingent on expenditure cuts (Niskanen 1988, p. 27). But then they might have wound up with policies that looked just like the traditionalists' fiscal policies (i.e., budget control but at higher levels of social spending).

Taken together, the new policy ideas clearly represented a struggle back toward the more conservative end of the economic policy spectrum. The desire for a less cyclical approach to fiscal policy, a stronger role for monetary policy, and a need to restrain if not reduce levels of government intervention and social spending were all elements of moderate Keynesianism in the late 1940s and 1950s (see Chapters 3 and 4). The problem in 1980 was that the public sector was now much larger than it was in the late 1940s. Unlike the moderate Keynesians of the late 1940s and 1950s, the new conservatives did not want to settle for merely leveling off the expansion of microeconomic intervention; they sought to roll back the interventionist legacy of the 1960s and 1970s. Hence, they were more willing to take chances, even at the risk of fiscal imbalance. They experimented with a still more conservative alternative, such as that which had inspired Ludwig Erhard's policies in the late 1940s — sharply lower taxes, radical deregulation of commerce, and draconian control of the money supply (Stockman 1986, p. 39).

In theory, monetarism and supply-side economics remained complementary (Niskanen 1988, p. 19), but in practice, each school was prepared, if necessary, to do without the other. Supply-siders talked about monetary policy restraint, and some, such as Congressman Jack Kemp, even espoused a return to gold or a fixed-quantity standard of money. But in contrast to monetarists, supply-siders seldom called for a contraction of the money supply. In 1981, for example, supply-siders "did not want any tight money prior to the tax reductions" because they felt that anticipation of the tax cuts and hence deferral of income to the future when taxes would be lower would already slow the economy temporarily (Roberts 1984, p. 118). And "after the tax reductions had improved the profitability of work and investment," supply-siders believed that "the economy would demand faster money creation because of the faster real growth of goods and services" (Roberts 1984, p. 142). Somewhere in between the supply-siders might have accepted monetary restraint. (Roberts favored tighter money in April 1981 when volatile money policy was perceived to be driving up interest rates; 1984, pp. 140–141.) But because they were the most optimistic that disinflation could be brought off without any increase in unemployment, they clearly rejected monetary restraint if it was associated at any point with an increase in unemployment. In the end, the supply-siders on the right were not unlike the exclusive Keynesians on the left. For both, monetary policy had to accommodate expansion, even though each advocated significantly different fiscal measures to achieve expansion.

The monetarists, by contrast, were more concerned with inflation than with growth. They were less optimistic that disinflation could take place without higher unemployment. But to shed their bankers' image, they were willing to play along with supply-side logic. They advocated a gradual reduction of the money supply—50 percent over 6 years—and subsequently criticized the Fed for providing 75 percent of that reduction in the first year alone. Nevertheless, when they had to choose between higher money growth and lower GNP, they chose the latter. In June 1981 the Reagan administration realized that its nominal growth forecasts were too high given projected reductions in the money supply and historical data on the velocity of or demand for money. The monetarists favored lowering the growth forecasts; the supply-siders opted for higher money growth. This conflict, even before the recession began, suggested the tension between the two schools that would doom any prospect of reconstructing within the Reagan administration the moderate Keynesian consensus of the late 1940s.

The differences among these various economic perspectives within the administration and outside the official policymaking process in what we have called the cocoon of consensus building made it certain that, even if the Reagan program succeeded, there would be no coherent explanation for it. Reaganomics was a modern mosaic that could mean anything to individual people in particular and nothing to everyone in general. A more sensible integration of the major elements of the Reagan program was possible. It would have called for a return to the stable money policies of the 1950s and early 1960s by whatever technical means the Fed might have found appropriate. The Fed used no rigid money rule in the earlier period ("leaning against the wind," it was called) and did quite well maintaining price stability. The new integration would have also advocated tax reduction and reform to revitalize longer-term capital formation in American business and promote savings among American consumers. It would have pressed deregulation to make labor and capital markets more flexible after the expansion of public services and regulations in the 1970s (see Chapter 6). And it would have asserted a rule for fiscal policy somewhere between chronic and large deficits (which ultimately undermine long-run capital formation by consistently diverting a greater share of total investment from the private sector) and a constitutionally mandated balanced budget. In short, it was not inevitable in 1981, any more than it is in the 1990s, that no consensus could be reached on domestic economic policy. In the short run, as the experience of 1981 suggests, such a consensus may not have been necessary to implement important and helpful steps to improve the economy. But in the long run this consensus was essential to complete, consolidate, and sustain these improvements. The Reagan administration was not able to construct such a consensus in 1981 or afterward.

As a result, the administration was particularly vulnerable to the buffeting of political forces in 1981. These forces, both inside the administration and outside, eventually transformed the Reagan program from one advocating new limits on domestic economic policy into what Herbert Stein subsequently called the "economics of joy" (1984, Chapter 7). The tax reduction program became a free lunch for industry and others, the new monetary policy was applied with such alacrity that it facilitated the worst recession of the postwar period, and the expenditure reduction program quickly disintegrated amidst the competing claims of defense, welfare, and entitlement programs. But it would be a mistake to see these outcomes as the consequence of special interest or bureaucratic politics alone or primarily. The Reagan program, despite its lack of intellectual integration, contained all the elements of a coherent and comprehensive attack on the major economic conditions that retarded growth in the early 1980s. Special interests pulled the program out of line here and there (the excesses of the 1981 tax bill, which were retrieved in large part in 1982 and 1984 tax revisions) and ultimately contributed to gridlock in the budget area. But the President and the administration, as this chapter and the next make clear, had a chance to pull the program off. They failed not because of politics but because they did not address the need to construct a coherent and consistent rationale for their program that might have integrated the individual elements and shaped a larger political coalition to support it.

The President's Choices

The initial Reagan program put forward four basic elements or pillars of domestic economic reform:

(1) A budget reform plan to cut the rate of growth in federal spending.

(2) A series of proposals to reduce personal income tax rates by 10 percent a year over three years and to create jobs by accelerating depreciation for business investment in plant and equipment.

(3) A far-reaching program of regulatory relief.

(4) And, in cooperation with the Federal Reserve Board, a new commitment to a monetary policy that will restore a stable currency and healthy financial markets [Executive Office of the President 1981].

The program did not mention trade policy and was silent on other aspects of international economic policy. At this point, supply-siders, monetarists, and the new fiscal conservatives all agreed that domestic reform took priority. Everyone expected that what was good for America would be good for the rest of the world, and in retrospect, much of the domestic policy reform undertaken in 1981–1982, unlike the domestic policy steps of the early 1970s, were beneficial for the rest of the world after 1982 (see discussion in Chapter 8). But the international purposes and consequences of the domestic reforms were only weakly envisioned and explained in the first year of the new administration.

The tax and expenditure bills submitted to implement the administration's program passed the Congress in the summer of 1981. They left a huge gap between revenues and expenditures. The budget went radically out of control and remained so for the rest of the decade. Revenues leveled off at around 20 percent of GDP while expenditures soared as high as 25.8 percent of GDP (IMFc 1988). Chronic budget deficits raised the gross federal debt from 33.5 percent of GNP in fiscal year 1980 to 53.5 percent in calendar year 1988 (*ERP* 1988, pp. 338–339, 248; OMB 1989).

Meanwhile, money policy turned sharply downward. Although larger monetary aggregates, such as M_2 and M_3, grew above their target ranges in 1981, raising some questions about the overall tightness of money (Melton 1985, p. 62), narrow monetary aggregates, such as M_1, decelerated dramatically (to zero in October 1981; *ERP* 1982, p. 303). Moreover, nominal GNP growth suggested tightening, decelerating from 20.5 percent in the first quarter of 1981 to −1.4 percent in the first quarter of 1982 (*ERP* 1984, p. 221). For 1982 as a whole real GNP dipped 2.5 percent. Inflation dropped from an annual rate of 13.5 percent in 1980 to 10.4 percent in 1981 and 6.1 percent in 1982 (*ERP* 1988, p. 316).

What happened? The advocates of each element of the Reagan program had their favorite scapegoat. Monetarists blamed excessively tight and volatile money; supply-siders pointed to the reduction and delay of the tax cuts. (The 30-percent, 3-year reduction effective January 1, 1981, had been reduced to 25 percent effective October 1, 1981.) The fiscal conservatives blamed politics. As David Stockman explained, "soup kitchen" politics on Capitol Hill involving raw, parochial horse trading "caused the budget reduction package to shrink and the tax cut package to expand," leaving the yawning gap between revenues and expenditures (1986, p. 251). In the end, the new fiscal conservatives, such as Stockman, came to agree with the traditional fiscal conservatives, such as Stein. The government's involvement cannot be reduced. The welfare state exists because democracy wills it. All that is left to do is to make democracy pay for it (Stockman 1986, pp. 391–392).

The intramural fault finding among the various economic theologians within the administration was amusing but not enlightening. Reaganomics neither failed totally as a package of unintegrated ideologies as some critics alleged (e.g., Stockman 1986) nor would have been more successful if only one of the ideologies had dominated. Tax cuts without tight money would have probably ignited hyperinflation; tight money without the tax cuts would have precipitated a still deeper recession; and a balanced budget without tax reform or monetary discipline would have perpetuated stagflation. The separate parts of the program needed each other. But because they had not been integrated intellectually or politically, the President ultimately had to assign priorities to them. With nothing to guide him between economic ideologies, on the one hand, and political pragmatists, on the other, he shaped the program with his own instincts. In this sense, the program was well named. Reaganomics was very much the President's personal program. The view that the President was out of touch with the facts (Stockman 1986, p. 375) or that the "President himself had very little to do with the invention and the implementation of the policies and mechanisms that encouraged [the] remarkable increase in the nation's wealth and general well-being" (Regan 1988, p. 143) is simply not persuasive. In the end Ronald Reagan got exactly what he wanted from his economic program.

And what Ronald Reagan wanted most of all was the tax cut. Even Stockman acknowledges that (1986, p. 229). It was the only element of the program for which the President set limits. At a key meeting in early May, Reagan drew the line at 25 percent for his 3-year tax cut (Stockman 1986, p. 239). He would accept nothing less. He never set comparable targets for expenditure cuts. What is more, he ruled out, early on, significant cuts in entitlement and domestic programs that comprised some 40 percent of the domestic budget (Stockman 1986, pp. 130–131). Together with defense, which was always off-limits, the President sequestered more than two thirds of the entire budget from expenditure cuts.

Ronald Reagan believed in a substantial portion of the welfare state. He opposed the Great Society programs but the old New Deal Democrat in him, despite all his rhetoric, never seriously questioned the entitlement programs, especially Social Security. In spring 1981, with strong advice from Jim Baker, his chief of staff, and Dick Darman, Baker's aide, Reagan pulled the plug on multiyear cuts of some $150 billion in the Social Security program (Stockman 1986, pp. 181–192; Niskanen 1988, pp. 36–40). This decision, together with Stockman's mistake in February basing outyear defense growth on the new Reagan budget of $222 billion rather than on the old 1980 defense budget of $146 billion, accounted for almost the entire budget shortfall of subsequent years.

In September 1981, Reagan consigned the Social Security issue to a bipartisan commission chaired by Alan Greenspan. Eventually this commission announced a $160 billion fiscal package that provided three quarters of the financial requirements for Social Security from tax increases and general tax revenues and only one quarter from benefit cuts (Roberts 1984, p. 280). Together with tax increases that had been voted in 1977 to take effect in the 1980s, Social Security taxes went up continuously during the Reagan administration, offsetting in considerable part the reductions in income taxes. Stockman argues that Reagan was "more than willing to fight for" the cutbacks in Social Security benefits but that "his managers," presumably Baker and Darman, "ran up a white flag and kept him in the dark" (1986, pp. 191–192). The argument does not wash. Reagan got the same kind of advice from the same "managers" on tax cuts and rejected it (Stockman 1986, pp. 239, 313). The President simply did not fear the budget deficit as much as Stockman and the new as well as traditional fiscal conservatives did. As Roberts reports triumphantly, the President's program never set a balanced budget as a *goal*, merely as a hoped-for *result* (1984, p. 105).

If tax cutting was Reagan's highest priority, bringing inflation down and tight money were a close second. Unlike Carter, Reagan did not make employment at all costs the centerpiece of his program. Here he disappointed the one-dimensional supply-siders, who wanted expansion above all else and who believed that Volcker was deliberately sabotaging the tax program (Roberts 1984, p. 114). Reagan met regularly with Volcker, never asking him in 1981 to ease or tighten the money supply (Anderson 1988, pp. 249–253) and never criticizing the Fed publicly, even in the depth of the recession of 1982, when Republicans lost 26 seats in the House of Representatives. But Reagan did not put monetary policy before his tax cuts, and the large current and prospective budget deficits that resulted from the tax cuts may have put Volcker under pressure in the winter of 1981–1982 to tighten money further, even though the velocity of money began to drop during this period. Volcker may have concluded that the consensus on disinflation would not last very long, given the pressures on monetary policy of large fiscal deficits, and that inflation would have to be stilled as quickly as possible (Niskanen 1988, pp. 168–169). As it turned out, he had to loosen monetary policy significantly in late summer 1982 to accommodate financial strains both at home and abroad. Thus, Reagan's fiscal and monetary policies, taken together, undoubtedly made Volcker's job economically more difficult, but in 1982 especially, they also made Volcker's success politically possible. As Stockman concludes, "there is . . . little doubt that Volcker's feat would not have been possible without Ronald Reagan's unwavering support during the dark days of 1982" (1986, p. 378).

When all was said and done, therefore, the President got exactly what he wanted and he stuck with it. He accepted modest tax increases in 1982 and 1984 but won reelection in 1984 in good part on a firm pledge not to raise taxes further. As we have seen, he never had the same ardor for expenditure cuts and backed off even more after election defeats in 1982 and the booming economic recovery in 1983 and 1984. He became the biggest budget buster in history, nearly tripling the federal debt from $1 trillion in fiscal year 1981 to $2.6 trillion in fiscal year 1988 (*ERP* 1989, p. 399). He accepted the political consequences of tight money in 1981–1982 and reappointed Paul Volcker in 1983.

His program was constant—tax cuts, budget deficits, and low inflation. It was neither an exclusive Keynesian program that ruled out any role for monetary policy nor a conservative Keynesian program that placed limits on budget deficits. It was mostly conventional free market economics with an emotional but not financial acceptance of such New Deal elements as Social Security (see Chapter 3 for definitions of exclusive and conservative Keynesian, as well as conventional free market, economic policies). Its premises as well as its results, which we examine later, do not fall into easy descriptive categories. Under the flag of Reaganomics, the United States achieved considerable progress during the 1980s, both at home and abroad, but at the same time the program bore inevitable long-term contradictions. At some point, when foreign borrowing tailed off, budget and trade deficits would have to be paid for by the mobilization of domestic resources, either through taxes or inflation. But that is a subject for Chapter 9. Now we turn to the early international economic diplomacy of the Reagan administration, which made domestic policy reform once again the centerpiece of world economic discussions.

International Conflict

If Reaganomics offered little coherent explanation of domestic economics, it said practically nothing at all about international economics. In 1981 the administration was almost totally absorbed in domestic reform, both military and economic. Yet domestic reform, particularly price stability and flexible markets, which the Reagan administration sought to retrieve, was an essential element of restored international market stability and growth. The early critics of Reagan policies ignored this fact, as some of them had as policymakers in the 1970s (Bergsten 1981; Oye, Lieber, and Rothchild 1983). Moreover, the Reagan administration was right to play down the form of international cooperation that characterized the 1970s. As we noted in Chapter 6, this cooperation frequently sacrificed domestic reform, especially medium-term fiscal and microeconomic reform, for the sake of international coordination of short-term exchange rate and monetary policies. In 1981 such coordination would have endangered efforts to lower inflation. It is unlikely, for example, that the United States could have negotiated an agreement at the summit table, especially with expansionist-minded France, to tighten money and disinflate the world economy. In contrast to 1947, when the United States dominated at the bargaining table and large international markets did not exist, the United States in 1981 had less direct influence over international negotiations and had to rely more on indirect influence through the large and highly interdependent world marketplace that now existed. Thus, the decision in early 1981 to discontinue extensive intervention in international exchange markets was fully consistent with the administration's desire to protect its disinflation program, especially from expansionist programs in other countries, and its readiness to use market forces to pressure other countries to move in the same direction.

What was not necessary in 1981–1982, however, was for the Reagan administration to appear to reject the concept of intergovernmental cooperation altogether. There was a role for governments to play in international policy, particularly to stabilize financial markets and promote liberal trade policy (the role of the Bretton Woods institutions). The President did back the role of the IMF and World Bank in speeches to those institutions in the fall of 1981, and the administration mobilized a much more active international economic policy in the winter of 1981–1982 to prepare for the Versailles Summit. But this effort came only after the impression had been created that the United States did not care about the impact of its policies on others.

In one sense, the administration should have stepped up the rhetoric of international cooperation in early 1981, even as its domestic policies began to project powerful and necessarily (e.g., to bring down inflation) disruptive influences into the international marketplace. But it did not have the confidence to do so, especially when its policies produced an unexpectedly harsh recession in the fall of 1981. By the time the Reagan administration gained confidence in the recovery of 1983–1984, its earlier policies had left enough barnacles of bitterness within the international community that international cooperation never achieved the potential that it might have to support a different choice-oriented approach to international economic policy relations (see Chapter 2).

The Ottawa Summit

The Ottawa Summit featured a confrontation of diametrically opposite domestic policy approaches to growth and world economy. Ronald Reagan and Great Britain's Margaret Thatcher advocated disinflation, fundamental fiscal policy reforms, and a general rollback of government. François Mitterrand of France advocated comprehensive economic planning and stimulus. The confrontation, pitting the extreme ends of the policy spectrum against one another — free market, laissez-faire policies against deep Keynesian policies — marked the natural outcome of a process of international cooperation in the 1970s that either ignored domestic policy differences or sought to reconcile these differences at higher and higher levels of inflation. The Ottawa Summit would have had great difficulty addressing these differences, even if there had been adequate time to prepare for it. But there was no time. The new U.S. government had been in office for only 6 months, and the French government took power only 6 weeks before Ottawa. Moreover, the U.S. government opposed the type of summit process that had evolved in the 1970s and that had been symbolized in the Bonn summit of 1978 in which detailed negotiations were taken up and specific agreements reached. As Myer Rashish, the Under Secretary of State for Economic Affairs and Reagan's personal representative for Ottawa, explained:

> Consciously, this summit is going to be different. While the last few summits tended to produce agreement on specific undertakings — this summit will not. It's much more valuable to use the limited time available . . . to have a full discussion of major topics [NYT, July 5, 1981].

The issues were sharply drawn. United States interest rates and the high dollar were the immediate concerns. But the leading issue at Ottawa, on which differences were equally marked, was the relatively narrow topic of North–South relations and Global Negotiations, a carry-over subject from the Venice Summit in 1980, which had been held under the cloud of the second oil crisis. United States negotiators worked to widen the agenda. East–West and West–West trade issues were eventually added, but these issues were lower priority for other countries and not the highest priority for the United States (see later discussions of West–West trade in this chapter and East–West trade in Chapter 10). Finally, political topics held considerable interest. France had just brought communists into the new government, and it was unclear where it would go on critical alliance issues. United States–French differences on third world political issues (e.g., Nicaragua) were already well known.

Reagan fiscal and monetary policies, by whatever specific mechanism (i.e., large budget deficits, tight money, or higher after-tax returns on investment), had sent American interest rates and the American dollar soaring. From January to July 1981, U.S. long-term interest rates went up 171 basis points (although short-term interest rates increased only marginally), and, from the fourth quarter of 1980 to the third quarter of 1981, the trade-weighted, real value of the U.S. dollar appreciated by 25 percent (*ERP* 1982, pp. 311, 345). At Ottawa, European leaders complained bitterly about these developments. Helmut Schmidt of Germany protested vociferously that U.S. policies had precipitated the "highest rates of interest [in Germany] . . . since the birth of Christ, as far as real interest rates are concerned" (*NYT*, July 21, 1981). High real rates of interest, he contended, squeezed investment and growth and thus exacerbated serious unemployment problems in Europe. Mitterrand warned Reagan that "unemployment in France as well as in Germany and other countries in Europe was getting to a flash point, a point where it might cause social upheaval" (*NYT*, July 21, 1981).

Schmidt and other European officials recommended a different policy mix for the United States (*NYT*, July 22, 1981). They wanted the United States to pursue tighter fiscal policy and looser monetary policy. The major allies, excluding France, were pursuing restrictive fiscal policies, seeking to repair the budgetary damage of expansionist programs in the 1970s particularly after the first oil shock (Henning 1987, p. 15). They hoped to use monetary policy to stimulate their economies, but the U.S. mix of tight money and loose fiscal policies, driving up U.S. interest rates, prevented them from doing so unless they were willing to accept a depreciation of their currencies.

At the time, the United States had the stronger case, but Reagan officials failed to make it convincingly to the allies or the media. Had the U.S. eased monetary policy, allowing the allies to do so, it is conceivable that inflation would not have come down as rapidly and widely as it did in the United States and the industrial world in the early 1980s. Certainly the process of reversing expansionist policies in France and other countries (e.g., Italy) would have taken considerably longer. And if the United States had tightened fiscal policy along with all other countries, the world as a whole would have probably experienced the kind of anemic growth of domestic markets that Europe and Japan experienced during the first half of the 1980s. Everyone would have depended on foreign markets for growth, and there would have been no growing foreign markets. Europe, except for Great Britain, was looking to continue policies of loose money, export-led growth, and fiscal stabilization at high levels of taxes and expenditures, whereas the United States was gambling on a new approach.

The issue that reflected these differing economic priorities and that came to symbolize allied disagreement was exchange market intervention. The Reagan administration believed that divergences among underlying domestic economic policies, particularly high and volatile inflation differentials, exerted the primary influence on exchange rate movements (Sprinkel 1981). It did not believe that speculation played a significant role except in highly disorderly markets. Hence, it accepted intervention only to counter disorderly markets (which occurred three times in 1981–1982). European governments, on the other hand, favored more frequent intervention to influence speculation and stabilize exchange rates. They had larger international sectors and suffered more from disruptive exchange rate movements. They preferred more managed exchange rates, particularly within the European Monetary System (EMS) set up in 1978 to protect European currencies against the fluctuating dollar.

To slow or reverse the rise of the dollar on a sustained basis, the Reagan administration would have had to ease monetary policy (or, more accurately, pressure Volcker publicly to do so) or give up its tax cut. Unwilling to compromise with either inflation or high taxes, it refused to buffer the rise of the dollar by intervening in exchange markets. Although intervention to achieve short-term objectives would have been consistent with its policy, it feared that such intervention would dilute its long-term objective to encourage disinflation in Europe. (See comments to this effect by Under Secretary Beryl Sprinkel in *Business Week*, June 21, 1982, p. 37.)

European officials complained bitterly about the U.S. decision to discontinue exchange market intervention. But in retrospect, as Karl Otto Poehl, who later became President of the German Bundesbank, observed, with a "fixed exchange rate between the U.S. dollar and its major partner currencies, the double digit inflation of the 1970s very probably would have accelerated distinctly during the 1980s" (1987, p. 26). Moreover, European officials had a choice. They could have accepted a larger depreciation of their currencies and nurtured an expansion through some modest fiscal stimulus (Krugman 1985, pp. 44–45). France did so for a while but it was unable to convince Germany or Britain to go along with this expansionist effort. By not going this route, German officials opted instead for the U.S. priorities to reduce inflation, even while using U.S. policies as a convenient whipping boy, in part to disguise their own differences with France (which in 1981, like Germany, had a socialist government).

North–South issues at Ottawa were equally divisive. These issues were a top priority for the host, Prime Minister Pierre Trudeau. In contrast to macroeconomic issues, Trudeau hoped to accomplish something specific in the North–South area, namely, a commitment of the summit governments to endorse the immediate convening of Global Negotiations, a conference of all U.N. members (over 160) called in 1980 to deal comprehensively with international economic issues in the wake of the second oil crisis (*NYT*, July 16, 1981). The new U.S. government had already agreed in March 1981 to attend a smaller North–South conference (only 21 countries) in Cancun, Mexico, scheduled for October of that year. Trudeau hoped to secure a U.S. commitment to convene Global Negotiations as well. (Global Negotiations, also known as GNs, was always referred to in capital letters to designate a specific U.N. conference, rather than a general process of negotiations; see later discussion.)

The differences between the new U.S. government and the other summit countries on North–South issues, however, were enormous. These differences emerged starkly in a joint paper that the personal representatives of the summit leaders (known as sherpas, after the Nepalese tribesmen who assist climbers to the summits of the Himalayan Mountains) drafted in the preparations for Ottawa. This paper, as originally commissioned by the Venice Summit in 1980, was to focus narrowly on aid policies and the North–South political dialogue. For the new Reagan administration, however, aid and the U.N. political discussions were the least important aspects of the development problem. Development, the new administration believed, was much more a matter of real economic developments, specifically domestic and trade policies in the developing countries themselves. Consistent with its own approach at home, the United States wanted to put domestic policy reform at the top of the North–South agenda. It sought to urge developing countries to reconsider long-practiced policies of microeconomic government intervention and import substitution (trade protection) and to give greater emphasis to monetary discipline, fiscal and regulatory reform, and private sector capital formation and entrepreneurship. It also urged progressive liberalization of trade and investment policy and invited developing countries to join vigorously in the new round of trade negotiations that would be launched by a proposed GATT ministerial meeting in November of 1982.

Europe and Japan saw the priorities in exactly the reverse order. The political dialogue was too important to be ignored, they argued. Political unity for the South, in groups such as the G-77 (the developing country bloc in the United Nations), was a natural expression of the common anticolonial experience of the developing countries and their newly gained sovereignty and independence. Respect for this unity and independence required the North to negotiate directly with the South and precluded any international discussion of sensitive and sovereign domestic policies. Moreover, aid was key to this discussion because it expressed political commitment by the North to the South. Europe, of course, tied aid to the export of European manufactured goods to the South or to commodity agreements to stabilize the prices of raw material imports from the South. Thus aid made good economic sense, given Europe's trading patterns with the South (largely importing raw materials and exporting manufactured goods). As Europe saw it, liberal trade policies were primarily a Trojan horse for bringing American manufactured exports and multinational capital into developing country markets.

The preparation of the joint paper was a difficult but useful exercise, as it started the process of turning allied and world views away from the interventionist approaches of the 1970s to the market-oriented approach that increasingly characterized the 1980s. But at Ottawa, the focus remained on Global Negotiations and a new World Bank aid institution known as the Energy Affiliate. As soon as the American delegation arrived in Ottawa, Prime Minister Trudeau made an urgent appeal that the leaders endorse Global Negotiations publicly, even before the summit began. The issue dominated behind-the-scenes negotiations throughout the summit and was eventually papered over with artful wording in the communiqué, which stated that the summit countries "were ready to participate in preparations for a mutually acceptable process of global negotiations in circumstances offering the prospect of meaningful progress" (for text of the Ottawa Communiqué, see *NYT*, July 22, 1981). For the American delegation, this wording allowed for at least four stops between the Ottawa declaration and the actual convening of Global Negotiations: (1) The summit countries agreed to participate in preparations only, not the negotiations themselves; (2) Global Negotiations itself was referred to as a process (hence small caps), not as a specific U.N. forum, opening up the possibility that other, more manageable forums might be substituted for the U.N. forum; (3) whatever process emerged had to be "mutually acceptable"; and (4) this process had to hold prospects for "meaningful" progress. From the allies' point of view, despite these qualifiers, the statement could be portrayed as another step forward toward the inevitability of Global Negotiations, and they rushed off to reassure the developing countries that they would bring the Americans around to a full endorsement of GNs at the Cancun Summit in Mexico in October.

The Ottawa Summit did not register significant decisions, but it was not just a media event, as it was billed at the time. Behind the scenes, as one French official noted, the discussions were "sometimes brutal" (*NYT*, July 22, 1981; see also Putnam and Bayne 1984, p. 156; Regan 1988, pp. 258–259). To some extent, the United States obscured the serious character of the discussions by its own public relations blitz. Anxious to make the President's first diplomatic encounter a success, Reagan officials commandeered a cluster of vacation homes in the hills around the summit site at Montebello and briefed the press almost hourly on Reagan's contributions. This exercise defeated the purpose of holding the meeting in Montebello — a rustic hunting lodge about 50 miles east of Ottawa — while the principal media corps remained in Ottawa. It also antagonized the allies and created ill will that carried over to the Versailles Summit the next year. Nevertheless, the public relations confrontation at Ottawa was not grafted onto the situation. It happened in part because the participants took the issues too seriously, particularly the differences over domestic policy, not because they took them too lightly. That the world and the media expected more from Ottawa, given the serious differences over issues, reflected the extent to which the world had become hooked from the 1970s on a process of international cooperation which produced agreements without consensus on substantive domestic policy issues.

Cancun and the Caribbean Basin Initiative (CBI)

From the beginning, the new administration gave priority to regional relations with developing countries, particularly in North America. When he announced his candidacy in November 1979, President Reagan spoke of a North American Economic Area to include Mexico, Canada, and the Caribbean countries; and during the campaign in October 1980, he promised: "We will initiate a program of intensive economic development with cooperating countries in the Caribbean" (1980). This concept of regional cooperation, driven by security concerns in Central America and a desire to showcase a more market-oriented approach to development, led to early meetings between President Reagan and President Lopez-Portillo of Mexico (in January before the inauguration) and Prime Minister Edward Seaga of Jamaica. It also led President Reagan to accept, in March, President Lopez-Portillo's invitation to attend the North–South summit in Cancun, Mexico, in October 1981.

The Cancun Summit was notable as one of the few occasions in the contemporary world economic system when heads of state and government from both industrial and developing countries met to discuss common problems. The chance existed at Cancun to perpetuate this type of smaller North–South forum, but the Cancun forum was suspected as an attempt to confuse and circumvent the proposal for Global Negotiations. Thus, the October meeting focused on Global Negotiations, a dying proposition, while a new, potentially useful forum that might have brought industrial and developing countries closer together on a continuing basis was passed up.

The Cancun Summit gave the Reagan administration an opportunity to develop its case for a different, more market-oriented approach to development. At the IMF–World Bank meeting in September, President Reagan had stressed the importance of a "shared vision of growth and development through political freedom and economic opportunity" and praised the Bretton Woods institutions for working "tirelessly to preserve the framework for international economic cooperation and to generate confidence and competition in the world economy" (1981a). Two weeks later, in Philadelphia, he stressed domestic reforms. Speaking first about the United States, he noted how reducing U.S. inflation and interest rates by one percentage point benefited developing countries by $1 billion. "By getting our own economic house in order," he proclaimed, "we win, they win, we all win." Similarly, he said, we must

> examine cooperatively the roadblocks which developing countries policies pose to development, and how they can best be removed. For example: Is there any imbalance between public and private sector activities? Are tax rates smothering incentives and precluding growth in personal savings and investment capital [1981b]?

By posing such fundamental domestic issues delicately as questions, President Reagan suggested how far out of line the U.S. approach still was from that of other countries (Lubar 1981). Four years later, the development dialogue, under the impact of the Reagan agenda first articulated at Cancun, would embrace the once radical ideas of domestic monetary and tax policy reform, deregulation, privatization, and more liberal trade policies. But in 1981 the overriding issues were still ones of process not substance. What to do about Global Negotiations?

The U.S. delegation was not of one mind. Foreign policy officials, including Secretary of State Alexander Haig and U.N. Ambassador Jeane Kirkpatrick, worried about the political cost of repudiating Global Negotiations. The Secretary of the Treasury and key White House advisers, on the other hand, worried about the dilution of the President's domestic agenda and priorities. Above all, they could not see the President endorsing a U.N. process that represented the worst type of bureaucracy and political maneuvering, which the administration was committed to oppose at home as well as abroad. Neither group wanted to see the President isolated at Cancun.

At Cancun, the United States accepted a compromise to take Global Negotiations another step forward, but this compromise was ultimately not accepted. Worked out by Canada, Austria, and Mexico and backed by Yugoslavia, India, and Nigeria, the compromise called for the two cochairmen of the Cancun meeting (Austria and Mexico), together with the U.N. Secretary-General, to convene at the United Nations by the end of 1981 an informal group, consisting primarily of Cancun countries, to discuss "preparations for a mutually acceptable process of global negotiations in circumstances offering the prospect of meaningful progress" (picking up the Ottawa Communiqué language). This agreement recognized the role of the United Nations "while emphasizing the competence, functions and independence of the Specialized bodies as defined in the Association Agreements." It therefore safeguarded the IMF and World Bank, whose authority and weighted-voting procedures the developing countries were hoping to subordinate to Global Negotiations as part of their campaign to restructure and democratize international economic institutions.

The United States accepted the compromise reluctantly. It was looking for some way, as the President put it at Cancun, "to carry out the commitment in the Ottawa Summit Declaration to conduct a more formal dialogue — bilaterally, with regional groups, in the United Nations, and in the specialized international agencies" (1981c). It regarded the smaller Cancun-type meeting as more acceptable than Global Negotiations and even toyed with the idea of a full-blown but one-time U.N. Conference on Economic Growth (originally suggested by Under Secretary Myer Rashish) if the momentum to launch Global Negotiations continued to mount.

At the last minute in Cancun, however, the compromise agreement was rejected. Venezuela and Algeria, two long-standing advocates of the New International Economic Order, saw it as a diversion from Global Negotiations. Supported by Tanzania and Guyana, they preferred to return to the U.N. arena to work for the original idea of Global Negotiations. From November until the following spring, therefore, the issue returned to the United Nations; and the by then familiar and fruitless process of "wordsmithing" new resolutions resumed once again.

The G-77 presented a new draft resolution in New York in November. At a National Security Council meeting in early December, U.S. officials agreed to make a counterproposal. The key issue at this point was whether the United States should agree to convene the Global Negotiations first and then consider, in a preliminary phase, issues of procedure, agenda, and timetable (specifically, the key questions of whether Global Negotiations would have jurisdiction over the functional areas covered by the specialized agencies such as the World Bank and the IMF) or whether the United States should agree only to convene a preliminary conference to consider these issues (the U.S. commitment at Ottawa). Haig, Kirkpatrick, and Vice-president George Bush believed the United States should agree to convene Global Negotiations first. They felt the United States was isolated, and they doubted that the G-77 resolution would ultimately be adopted in any case. Ambassador Kirkpatrick believed it was time to throw some politics at the development problem. GNs had become a litmus test of good will, and the United States, she believed, could afford to do the "dance of the seven veils" without really conceding anything. Secretary of the Treasury Regan and senior White House aides opposed the idea of convening Global Negotiations before resolving controversial issues. They felt the United States was sliding into something it could not control, and White House aides read from portions of U.N. Resolution 34/138, which was the basis for Global Negotiations, to show how far out of line this forum was from the President's economic philosophy. The President too feared getting into Global Negotiations step by grudging step and decided to accept a preliminary conference only. He wanted to shift attention to substantive issues and urged that the U.S. Agency for International Development expedite the dispatch of high-level agricultural task forces to developing countries, which he had called for upon his departure from Cancun.

Predictably, the G-77 in New York rejected amendments calling only for a "preliminary" conference. The issue died until March, when a new chairman of the informal negotiating committee in New York offered another draft. At this point, to head off the process of progressive isolation of the United States in New York, a game the European allies enthusiastically played, the United States decided to use the preparations for the Versailles Summit in June 1982 to fashion a compromise with the allies that would tie them to the United States without giving away any of the essential U.S. conditions for launching the talks. If this strategy worked, the developing countries would then reject the allied compromise, sealing the fate of the GNs, but the United States would not be identified as the country that killed GNs. The idea was to avoid both the economic consequences of convening GNs and the political costs of not convening it. As we will see shortly, the strategy succeeded.

While the sterile deliberations on the development dialogue proceeded in New York, the United States turned its attention to the formulation of a substantive program for regional economic development. The Caribbean Basin Initiative (CBI), announced in February 1982, reflected the new priorities for trade, private investment, and domestic policy reform that the administration stressed at Ottawa and Cancun. Almost a year in preparation, the CBI represented a historical departure for postwar U.S. trade policy. For the first time, the United States advocated a regional preference zone, a one-way free trade area offering guaranteed duty-free access for Caribbean exports to the United States over a 12-year period. Although most Caribbean exports already entered U.S. markets duty-free, they did so under the Generalized System of Preferences, a global arrangement for most developing countries in which duty-free access had to be renewed year by year. More assured, longer-term access to the U.S. market was designed to stimulate private investment in the Caribbean and to encourage related market-oriented domestic policy reforms.

The plan originally called for American firms investing in the Caribbean to receive the 10 percent investment tax credit, which applied at the time only to investment in the United States. That provision had to be withdrawn under pressure from Congress worried about the loss of American jobs. The trade provisions were also significantly weakened by the exclusion of such sensitive products as textiles, leather goods, and sugar. In the case of sugar, the United States actually tightened import quotas in 1981 largely to win votes for its domestic budget program, a clear example of the priority that domestic objectives extracted from foreign policy initiatives in 1981 (Niskanen 1988, p. 45). As the CBI developed, aid became a bigger part of the package than was originally intended, sometimes doing little more than substituting for the trade benefits that were being denied.

The war in Central America further distorted the program. The administration debated in January 1982 whether emphasis should be given to security or economic aspects of the CBI. The foreign policy agencies urged a strong focus on geopolitical and security threats in the Caribbean, whereas White House aides preferred a softer, more economic-oriented approach. President Reagan's speech at the OAS in February 1982 put security issues up front but balanced the initiative with strong economic proposals. Over time, however, the security aspects dominated. With the report of the Kissinger Commission on Central America, the economic content of the program also shifted to a more traditional development aid approach (Kissinger 1984).

The Versailles Summit

The winter of 1981–1982 was one of stock-taking for the Reagan administration's foreign economic policy team. The recession was in full swing. Unemployment peaked at 10 percent in the United States and went as high as 16 percent in some European countries. After increasing by 25 percent over the previous year, the trade-weighted value of the dollar leveled off in nominal terms but continued to climb in real terms. It also moved up steeply against the French franc and steadily against the Japanese yen (*ERP* 1982, p. 345; *ERP* 1984, p. 331). United States macroeconomic policies were no longer abstractions; they were producing painful results. And they looked increasingly like the root canal economics of traditional conservatism rather than the promised land of supply-side economics.

Speaking in London in October 1981, George Shultz, then President of the Bechtel Group, challenged the United States and other Western countries to become more aggressive in international economic policy. "Can we aspire again to grand objectives," he asked, "to put in place a Bretton Woods II, in effect a renewed International Economic Constitution?" Applauding the emphasis on domestic market-oriented reforms, he nevertheless called for a new international "offensive" on trade, foreign investment, economic development, and economic policy issues (Shultz 1981).

A think piece, circulated among U.S. foreign economic policy officials in January 1982, called for doing more. "Is there something missing in U.S. foreign economic policy?," the think piece asked. The emphasis on market-oriented domestic policies, it noted, was long overdue and "provided a badly needed corrective to past emphasis on public sector actions and institutions." But even in domestic policy, the paper pointed out, the administration accepted an important role for government. "The social safety net reflects government's domestic responsibilities for education, health, agricultural and industrial infrastructure and maintenance of decent standards of living for the old, handicapped and unemployed." "Is there a comparable concept," it asked, "for government's international economic responsibilities, as understood by this administration?" "If there is," it concluded, "then we should develop and emphasize this concept more, since the international system is diverse, and the role of government is larger in practically all of the allied countries than in the United States."

The paper went on to critique prevailing concepts of international cooperation that emphasized interdependence and the new international economic order (NIEO). These concepts, it argued gave too much attention to either short-term responses to economic interdependence or long-term requirements of international institutional change. Fine tuning the world economy was both economically undesirable and politically unrealistic, and restructuring international institutions was economically unnecessary. Yet "some conceptual glue is needed," the paper argued, "to restrain the divergences in economic policy and outlook among countries, to increase the awareness of leaders about the effects of their policies on one another, and to maintain and enhance a sense of community and confidence among the industrialized and more broadly developing countries of the world."

The paper settled on the concepts of system to emphasize medium-term and enduring economic outcomes and of community to stress common political purposes. The idea of community, it explained, is less precise in economic terms but goes further to recognize common human and moral responsibilities at the international level (stressed by President Reagan at Philadelphia) and not just the need to create more wealth. At the same time, it allows for more pluralism and autonomy than either the *dirigiste* concept of interdependence or the authoritarian concept of a new international economic order.

Under the concepts of system and community, the think piece called for "common analysis of international economic problems . . . to move domestic economic policies in less diverging directions" and to study the "capital and exchange rate consequences of diverging macroeconomic policies [as well as] the domestic economic consequences of moving back to a less flexible exchange rate system." These ideas became the basis of Reagan administration initiatives at the Versailles Summit in June 1982 to establish a new process of convergence or multilateral surveillance among the principal industrial countries and to undertake a joint study of the consequences of exchange market intervention.

At a preparatory meeting in Celle St. Cloud outside Paris in February 1982, the U.S. summit team presented proposals for a new forum of G-5 finance ministers that would include the Managing Director of the IMF. This forum would conduct "multilateral surveillance" of the interrelationship of domestic economic policies among the principal industrial countries. The idea, as Beryl

Sprinkel, Under Secretary of Treasury for International Monetary Affairs and one of the chief architects of the proposal, later explained, was to create a "small, intensive and confidential mechanism through which the finance ministers and central bank governors, with the assistance of the Managing Director of the IMF, [could] explore their differences and try to reach common views that promote a convergence of economic performance" (1984). Since the first summit at Rambouillet, France, in 1975 and the subsequent Jamaica Accords in 1976, the IMF had conducted only bilateral surveillance of economic and exchange rate policies with each member country separately. The convergence proposal now added a complementary multilateral process to this bilateral one, and although not formally under the auspices of the IMF, the multilateral process involved a leading informal role for the IMF director (see the following discussion) and, as a summit mechanism, operated directly under the auspices of the heads of state and government.

The proposal to study exchange market intervention was linked to the convergence concept. At Ottawa, Reagan officials had spurned exchange market intervention because domestic policy priorities among the summit countries diverged radically and officials feared that intervention would compromise their longer-term goal of domestic policy convergence around low inflation and more flexible markets. They argued that policy convergence, not exchange market intervention, would stabilize exchange rates (see Regan's comments, *IHT*, May 18, 1982). Volatile exchange rates came about in the early 1970s from the loss of common domestic policy commitments, not from large and integrated capital markets operating primarily on speculative motives. Intervention might be able to counter short-term disturbances in the market, but it could have no effect on long-term exchange rate stability without convergence of fundamental domestic policy priorities around common performance indicators. American officials were confident the intervention study would confirm this view. Meanwhile multilateral surveillance would put common domestic policy disciplines back into the center of international economic diplomacy.

The allies welcomed the American proposals, particularly the flexibility they detected on U.S. policy toward exchange market intervention. They would subsequently play up the extent to which U.S. attitudes had changed on exchange market intervention, precipitating a nasty exchange between U.S. and French officials after the summit (see Chapter 10). But at Celle St. Cloud the finance sherpas quietly discussed the medium-term economic and monetary policy objectives of convergence that would appear subsequently in the annex of the Versailles Communiqué. (For a copy of the communiqué and annex, see *The Times*, London, June 7, 1982.) The stickiest issue was a procedural one: Who should participate in the multilateral surveillance exercise? That issue was not resolved until the next preparatory meeting at Rambouillet in April (Putnam and Bayne 1984, p. 160). The British, with American support, proposed that the exercise be limited to the countries whose currencies made up the basket for the IMF's special drawing rights (SDRs), essentially the G-5 countries. Technically, this suggestion made sense, and eventually it prevailed. Politically, however, Canada and Italy objected, but because the convergence exercise was to be confidential, they finally relented. Several years later, when the United States launched a much more dramatic surveillance exercise at the Plaza Hotel in New York in 1985 and then elaborated this exercise at the Tokyo Summit in 1986, political considerations forced the inclusion of Canada and Italy (see Chapter 9).

Multilateral surveillance differed from earlier international economic policy coordination in three respects (Nau 1984, pp. 18–26). It focused on the consequences or performance of domestic economic policies, not on the direct, negotiated adjustment of these policies themselves, such as the Bonn Summit had emphasized. It minimized formal, institutional arrangements, preferring open-ended private discussions among high-level officials to binding public agreements that often raised expectations and could not be enforced. And it focused on medium-term, structural (i.e., microeconomic) policy reforms to bring down inflation and free up labor, capital, and product markets. Thus, multilateral surveillance expanded the scope of international economic policy coordination as practiced in the 1970s—to include medium-term macroeconomic and structural policy objectives, not just short-term exchange rate and monetary policy coordination—even as it moderated the intensity of that coordination, putting more emphasis on steady, consistent objectives than on frequently negotiated adjustment of economic policies.

North–South issues were also intensively prepared for the Versailles Summit. A new G-77 draft for Global Negotiations, presented in New York in March, contained four areas of continuing disagreement: (1) how to reference the original U.N. resolution 34/138 calling for GNs, which Reagan's advisers found alien to the President's economic philosophy; (2) the old issue of whether to convene Global Negotiations first or settle key issues before convening it; (3) the jurisdiction of GNs over the specialized agencies; and (4) the authority of GNs to create parallel, ad hoc groups in areas of competence of existing institutions, such as monetary affairs in the case of the IMF. The G-77 document called for Global Negotiations to be convened first "in accordance with" U.N. resolution 34/138 and to respect the jurisdiction of the specialized agencies, but it then went on to authorize Global Negotiations to create ad hoc groups and to issue "appropriate objectives and guidelines" to these groups as well as to the specialized agencies. The United States and to some extent the British feared that these ad hoc groups might compromise the independence of the specialized agencies. The allies felt that the language was ambiguous enough to protect the specialized agencies.

A member of the U.S. summit team met with EC leaders in Paris 1 month before the summit to propose a compromise. The United States hoped to draw the allies into an agreement that would stop the endless maneuvering in New York. The United States said it would agree to language leading to the convening of Global Negotiations first if the summit allies would agree with the United States on iron-clad language in the rest of the resolution to protect the specialized agencies. The proposal was accepted. At the summit, therefore, a working group was set up to negotiate the detailed language of the agreement.

The working group operated on several critical premises, which were specified in writing: (1) that the group would propose amendments to the G-77 draft that had a "fair chance to be acceptable to the 77 Group," (2) that the Canadian government would put the amendments forward in New York on behalf of the summit countries, and (3) that the summit leaders would not depart from the common language they agreed to at Versailles except by "unanimous consent." The first premise was the allies' assurance that the United States would negotiate seriously. The third premise was the American assurance that the allies would no longer break ranks with American negotiators in New York without having to take the issue back to their heads of state. The U.S. delegation was convinced that the heads of state would not want to deal with this issue again, once a formal agreement had been reached at Versailles.

On the basis of these premises, the working group agreed to specific amendments in the G-77 draft. The first amendment accepted the convening of Global Negotiations first but obliged this conference to reach a "consensus" on procedural issues before it could proceed to the substantive phase of its deliberations. A second amendment added specific wording to the G-77 draft that the specialized agencies would be respected "by the conference," and a third amendment precluded ad hoc groups that duplicated "existing appropriate fora." The latter wording was artful to say the least. For the Americans it meant that, for such topics as money, the specialized agencies (in the case of money, the IMF) would always be considered the "appropriate" existing fora and therefore could not be duplicated. For the allies (and the LDCs they were in effect negotiating for), it meant that the specialized agencies, though existing, might not be considered appropriate and therefore ad hoc groups might be created that could duplicate the work of the specialized agencies. Finally, the summit agreement had the G-77 draft "recall" U.N. Resolution 34/138, rather than suggest action "in accordance with" that resolution.

On the basis of this agreement, the United States accepted communiqué language at Versailles that declared "there is now a good prospect for the early launching and success of global negotiations, provided that the independence of the specialized agencies is guaranteed." Briefly, in the final discussions at the summit on Sunday, June 6, the North–South issue became hostage to the East–West dispute. With the East–West credit issue still to come, President Reagan deferred final approval of the North–South communiqué language. What followed then was a spirited and at times confusing discussion of the communiqué language on East–West credits (see Chapter 10). Eventually, after more than 3 hours, that issue was settled — so it seemed — and the summit leaders returned to the North–South issue. They explicitly noted the detailed amendments to the G-77 draft that had been negotiated among their representatives in the working group and, in so doing, gave the unpublished working group document the same force of agreement as the North–South language in the communiqué. This was confirmed in subsequent communications among the allies immediately after the summit. It was an important point for the U.S. delegation because it meant that the allied agreement on the specific amendments to the

G-77 draft could not be changed without the explicit consent of the summit leaders. It was hoped that this would be the last round of allied maneuvering on Global Negotiations.

As it turned out, the summit agreement was the last round of Global Negotiations. The Canadians offered the amendments in New York in late June. The United States continued to insist, as its interpretation of the working group document allowed, that no ad hoc groups could be created except in areas such as energy, where no appropriate specialized agencies existed. OPEC countries did not want an ad hoc group in energy, and ultimately countries like Venezuela and Algeria prevailed upon the G-77 to reject the allied proposals. Global Negotiations quietly faded from the scene. After Versailles, even Prime Minister Trudeau washed his hands of the whole matter.

The Versailles Summit, although different in approach and substance from previous summits such as the Bonn Summit in 1978, was every bit as serious and ambitious. The "crown jewel" of the meeting was the multilateral surveillance accord. As practiced over the next several years it involved discussions of the G-5 countries prior to regular G-5 sessions (usually held before IMF Interim Committee meetings) in which the Managing Director of the IMF presented candid and hardhitting written critiques of the policies of the United States and the other countries and prompted serious discussion among the finance ministers concerning the relationships between their policies (Henning 1987, pp. 19–20). In this form, multilateral surveillance epitomized the new choice-oriented approach to international economic diplomacy that the administration pursued from 1982 to 1985. The confidential character of this process, however, together with the more publicized disputes at Versailles over exchange market intervention and East–West trade issues (see Chapter 10), obscured its significance. Nevertheless, the convergence agreement at Versailles was the most significant step in international economic policy cooperation since the Rambouillet Summit in 1975, and as two students of the summit process noted, "Versailles would prove to be a substantive success, if a procedural failure" (Putnam and Bayne 1984, p. 170).

Trade and Debt Crises

The Versailles Summit was barely over when the alliance and the world economy, for that matter, began to fall apart. In June and July 1982, the alliance was hit with a triple trade shock—the East–West pipeline imbroglio, a steel trade dispute, and the continuing high-interest-rate and high-dollar feud. Soon thereafter, the debt crisis struck. In August, Mexico let it be known that it would have to suspend interest payments on its $80 billion debt. Brazil and Argentina followed suit in the fall. The GATT Ministerial Meeting in November 1982 added a fourth point of serious international contention—the explosive agricultural dispute between the United States and Europe. In a sense, the breakdown of international economic relations was not surprising. Domestic economic policies in the United States and some European and third world countries had been moving in opposite directions for several years. The widening fault in domestic policy priorities inevitably sent shock waves through the trade and financial systems. The Versailles Summit had addressed the need to initiate convergence of domestic policy priorities, but it could not arrest the consequences of diverging priorities from previous years.

Early Reagan administration's priorities addressed domestic policy reforms—price stability and flexible markets—and relatively neglected trade and financial initiatives. Thus, the financial crisis of August 1982 came as a surprise, not only to the administration but also to the Federal Reserve Board. Indeed, the Fed may have had the major responsibility for unintentionally precipitating the crisis. The world was experimenting for the first time with disinflation under dramatically new conditions of tightly integrated world trade and financial markets. The Fed's haste to disinflate the U.S. economy and the administrations's gamble with bold new fiscal initiatives clearly overloaded world markets and in the fall and winter of 1981–1982 nearly brought about their collapse.

Nevertheless, U.S. policies did not cause the debt crisis. The causes lay in the excessive domestic and foreign borrowing of the 1970s unaccompanied by sufficient adjustments in the real economy (see Chapter 6). Administration and Fed policies brought the debt problem to the surface sooner than might have otherwise been the case, and until summer 1982 the United States provided precious little conceptual understanding or compassionate leadership to "hold the hand" of the world community while the latter was going through a painful and difficult process of disinflation. But disinflation itself was inevitable. As the Bank for International Settlements pointed out at the time, the accumulation of debt in the late 1970s "would have been unsustainable even if world demand . . . had continued to grow at a fast pace and interest rates had remained at low levels" (Bank for International Settlements 1983, p. 6).

When the debt crisis struck, the administration and Federal Reserve Board evolved a case-by-case approach to the problems of individual indebted countries that relied heavily on the IMF to promote domestic and trade policy reform in these countries. The administration outlined five steps to cope with the crisis:

1. Bridge financing to prevent immediate collapse.
2. An IMF-negotiated stabilization package to facilitate macroeconomic policy reform.
3. Rescheduling of and perhaps some new commercial financing.
4. Longer-term structural reforms in developing countries.
5. Renewed economic growth in the industrial world.

The administration emphasized transitional finance but not large new resource transfers to the indebted countries. In 1981 it had opposed an increase in IMF quotas and, at the Helsinki IMF meetings in May 1982, before the debt crisis, it signaled a willingness to approve a "slight increase." When the debt problem spread in the fall of 1982, the administration eventually approved a 47.5 percent increase in IMF quotas. But this amount was still less than half that advocated by other countries, some of which were calling for increases of

200–300 percent. Consistent with its development approach outlined at Ottawa and Cancun, the administration gave priority to domestic economic and trade policy reforms, not to generous financial transfers.

Where the administration's debt policy fell short, and would continue to do so until 1985 (Baker Plan; see Chapter 9), was in outlining a longer-term, more comprehensive approach to structural reforms in the developing countries. Although this was the fourth point of the administration's five-point approach, it was never fleshed out. The assumption was that the financial squeeze alone would move developing countries toward the necessary overhaul of their highly protected, public sector–dominated economies. As in the case of the Marshall Plan in the late 1940s, however, international leadership was needed to establish the broad premises of reform and provide longer-term development assistance tied directly to such reform (see Chapter 4). The administration talked wistfully about closer integration of the developing countries in the world economy, and President Reagan visited Brazil and other Latin American countries in November 1982 to demonstrate his personal concern for the heavily indebted countries. But the administration failed to mobilize the World Bank or the GATT to address domestic microeconomic and trade policy reforms. Too much of the burden fell on the IMF, which had only short-term macroeconomic stabilization objectives and could not really monitor trade policy reforms.

As a result, the adjustment that occurred in developing countries after 1982, especially in Latin America, came about largely through import restrictions and export subsidies (see Chapter 9 for data and details). United States markets in Latin America disappeared, and U.S. import-competing industries, already reeling from the high dollar, felt the added pressures in U.S. markets of developing country exports heavily subsidized by foreign governments. The debt crisis thus compounded the already serious problems in the U.S. and world trade system. World trade in 1981 had declined in value for the first time since 1945, and in 1982 it actually declined in volume by 2.5 percent (GATT 1985–1986).

Trade policy was always the Achilles heel of the administration's recovery program. By logic, trade should have been the leading edge of the administration's international economic strategy, being perceived in the choice-oriented approach as far more important than aid or exchange rate coordination. Yet by neglect as well as by consequence of macroeconomic policies, trade policy languished. As noted earlier in this chapter, it was not even mentioned in the administration's first economic program. In April 1981 the administration convinced Japan to impose voluntary export restraints on automobiles, even though the International Trade Commission had earlier rejected economic arguments for such relief. The administration issued a trade policy statement in July 1981, but this statement was a study in ambivalence, calling for both free and fair trade and bilateral reciprocity as well as regional and multilateral initiatives. Most important, the administration's macroeconomic policies and the resulting high dollar (encouraging imports and discouraging exports) shifted enormous pressures to the trading sector, increasing calls for protectionism, exchange rate manipulation, and export subsidies.

The think piece circulated within the Reagan administration in January 1982 noted the ambivalence. The reciprocity approach, it pointed out, "implies the goal of open markets but is ready to impose the opposite." Moreover, if it is a "bargaining tool" to open markets, "where do we want to move the multilateral trading system through hard bargaining on the basis of reciprocity?" Indeed, "what do we mean by opening markets," it queried, "when we interpret subsidies or trade distorting measures so broadly as to strike at the core of domestic economic policies (if not cultural values) in some countries, particularly those [Japan] which are sectorally centralized and feature strong government roles?"

If U.S. trade policy was ambivalent, it was no less ambitious — but in several directions at once and without the highest level of support within the administration. While the administration worked to gain congressional passage of the CBI, it prepared far-reaching objectives for the GATT Ministerial Meeting in November 1982. These included resisting protectionism, strengthening GATT safeguards and dispute settlement mechanisms, liberalizing markets in agriculture, services, high technology sectors, and trade-related investment areas, and integrating developing countries more effectively through a North–South GATT trade round. For a world deep in recession and debt, the U.S. program was too much. Even for the U.S. administration, it was too much. Asked to give a speech on trade before the GATT Ministerial Meeting, the President settled for a 5-minute radio address. Substantively overloaded and politically undersupported, the GATT Meeting almost blew up. United States–European relations soured further over the agricultural issue, and the United States clashed bitterly with India and other developing countries over trade in services and other issues.

The GATT Ministerial Meeting would influence U.S. trade policy significantly for the next 3 years. In retrospect, it was disastrously ill timed. Had it occurred 1 year later, as world recovery gathered steam, it might have sparked new movement in the multilateral trading system. As it was, it had the opposite effect. It took the steam out of the multilateralists in the U.S. administration and tilted the balance in favor of the unilateralists, who sought tougher trade laws, and the "reciprocity advocates," who demanded retaliation against unfair trade practices. Even as the economy recovered in early 1983, trade policy never recovered. The administration continued to press for international consensus on a multilateral trade round, especially at the annual economic summits. But it now increasingly applied its market power to initiate bilateral and discriminatory free trade discussions with various countries, such as Canada, Israel, and the Association of Southeast Asian Nations (ASEAN).

United States use of market power in trade policy was not unlike its use of market power in macroeconomic policy to lower inflation and revive investment incentives in the world economy. The use of such power was appropriate as long as the multilateral objective, which this tactic was designed to serve, was kept in view. In fact, however, bilateralism began to compete with rather than complement multilateralism. The administration did not make another determined effort to launch a multilateral trade round until late 1985 under a new U.S. trade representative. By that time a soaring trade deficit made it politically impossible to pursue multilateral objectives without increasingly tough bilateral actions (e.g., 301 actions) that seriously undercut the credibility of U.S. leadership in the multilateral arena.

Choice-Oriented Economic Diplomacy Reassessed

In sum, 1982 was a tumultuous year for the new U.S. administration and its revolutionary program for U.S. and world economic revival. Instead of reviving American economic power, its policies plunged the U.S. and world economy into the worst recession of the postwar period. What is notable is that the Reagan administration, despite these setbacks, stuck with its program. Unlike administrations in the 1970s, it did not succumb to quarterly revisions of its economic plans. In retrospect, this stubbornness served it well. By holding course, the Reagan administration created the opportunity for genuine structural shifts to take place in international economic conditions and policies. One year later, prices would be falling in practically all the OECD countries, and market reforms to lower tax burdens and privatize public sector activities would be taking hold in industrial and developing countries. Where the administration ultimately went off track was in failing to exploit the strong recovery of 1983–1985 to complete its program and reduce the budget and trade deficits. Eventually, continuous stubbornness became counterproductive, no less than continuous stop-and-go policies had been in the 1970s. The administration was right not to change course in 1982 after only 6 months of implementation of a new, albeit incomplete, program; but it was wrong not to do so for another 6 years.

As a result, U.S. diplomacy remained compromised and weak. An ambitious diplomacy that sought in 1982 to create a new process of multilateral surveillance, manage the debt crisis, and launch new multilateral negotiations to liberalize trade stalled because American power and credibility remained too weak. Ironically, when American power and prestige returned in 1983–1984 on the basis of a rapid investment-driven recovery, American diplomacy flagged, as prosperity seemed to remove the need to nurture and sustain the new style of choice-oriented international economic diplomacy.

8

Rebound 1983–1985:
Missed Opportunity for
International Cooperation

A choice-oriented perspective reveals both the potential and problems of U.S. international economic diplomacy from 1983 to 1985. Acting decisively at home, the United States reversed decades-old domestic policies of inflation and intervention and launched a recovery in which investment played a larger role than in any previous postwar recovery. Abroad, the United States attempted a new medium-term international diplomacy based on convergence of economic performance among major industrial countries around low inflation and renewed market incentives and a revitalization of world trade and financial links between industrial and developing countries. But in the end the new choice-oriented diplomacy flagged because the Reagan administration lacked a full, integrated, intellectual understanding of its programs and ultimately failed to generate the political will and support to follow through on its programs, especially the programs to reduce government spending and liberalize world trade.

Reagan administration policies paid off in 1983 and 1984. The decision to resist the stop-and-go tendencies of the 1970s and stick with the basic program in 1981–1982, despite imbalances generated by relatively tight and volatile money growth and massive budget deficits, produced significant benefits. Real GNP in the United States rebounded in 1983 and 1984 by 3.6 and 6.8 percent, respectively. Inflation (CPI) descended to 3.2 percent and 4.3 percent, respectively, and civilian unemployment dropped from 9.7 percent in 1982 to 7.2 percent in 1985 (*ERP* 1988, pp. 251, 293, 317).

When President Reagan hosted the other industrial countries at the Annual Economic Summit in Williamsburg, Virginia, in June 1983, the American economy was on its way back. The summit declaration read like a testimonial to the Reagan economic philosophy. What is more, the influence of Reaganomics spread. Inflation dropped throughout the industrial world, standing by mid-1985, at 4.5 percent in these countries, compared to 13 percent in mid-1980. Growth abroad expanded, boosted by American imports. From

1983 to 1985, real GNP growth in the OECD countries averaged 3.6 percent per year (OECD 1987b). Even more impressive, Reagan's magic of the marketplace began to be imitated around the world. Making markets more flexible and competitive caught hold. Slowly, the philosophy of statism that had dominated the previous decade began to retreat. Structural adjustment, deregulating labor, capital, transportation, and communications markets became the new common wisdom for achieving greater competition, higher growth, and accelerated productivity.

Regrettably, the administration failed to exploit vigorously the opportunity offered by the new popularity of its policies. It undertook no new initiatives to press its economic philosophy in the international economy—at the IMF, the World Bank, the GATT, the United Nations, or any other international forum. It moved forward incrementally at the Williamsburg, London, and Bonn economic summits from 1983 to 1985 but was still not completely sure of itself at the Williamsburg Summit, was uninterested in rocking the boat at London before a presidential election, and was in the early stages of a transition of key foreign economic personnel and policies at Bonn. Perhaps the administration refrained from pursuing international initiatives more vigorously because it was never sure how its domestic program meshed with international needs or because it sensed a certain vulnerability as long as it refused to complete the reform of its own domestic economic policies. Whatever the reason, it settled largely for rhetoric.

Domestic imbalances inherited from 1981 to 1982 persisted. Although the administration accepted some tax increases in 1982 and 1984 (and 1983 in the case of Social Security) that were inconsistent with its program, it failed to implement vigorously budget cuts that were consistent with its program. It addressed neither of the two big ticket items in the budget—entitlements and more efficient use of soaring defense expenditures. Meanwhile, monetary policy remained volatile; it was looser after summer 1982 but contracted once again from August 1983 until late 1984. Real interest rates remained high and to some extent delayed the spread of recovery to the heavily indebted developing world. The dollar rose from early 1983 to the end of 1984 by another 20 percent in real trade-weighted terms. And the U.S. current account deficit soared from $2.3 billion in the first quarter of 1983 to $31.8 billion in the fourth quarter of 1984 (*ERP* 1985, p. 351; *ERP* 1986, p. 366).

Economic Recovery

The performance of the U.S. economy in 1983–1984 was nothing less than spectacular. Administration forecasts in early 1983, deliberately cautious as an antidote to the "rosy scenarios" of 1981 and 1982, were spectacularly wrong again, this time on the low side. In fact, economic officials in the administration had no agreed explanation for the sudden and dramatic reversal. The supply-siders attributed growth to the tax cuts finally implemented in fall 1981. The monetarists saw it as the result of the relaxation of monetary policy in late summer 1982. The new chairman of the Council of Economic Advisers, Martin Feldstein, was skeptical of both the supply-side and monetarist explanations but had no clear explanation of his own. He was most concerned with the budget deficit and pushed for a contingency tax in the 1983 budget. For him, deficits were not the stimulus for recovery that Keynesians believed but the primary cause of high interest rates, which in turn drove up the dollar. High interest rates and the high dollar then crowded out domestic investment and squeezed the trading sector, both export and import-competing manufacturers (*ERP* 1984, pp. 38–40).

Thus, there were three broad, competing explanations for the economic recovery after 1982 – the Keynesian explanation based on large budget deficits, the monetarist explanation based on looser money, and the supply-side explanation based on higher after-tax returns on investment from the 1981 tax cuts. These different explanations corresponded roughly to the different schools of economic thought represented in the Reagan administration. Which of these explanations is correct continues to be disputed today, not only within the economics profession but also among the broader public. (For views that attribute recovery variously to money policy, tax policy, and budget deficits, see, respectively, Stein 1984, Chapter 8; Tatom 1986, 1987; and Stockman 1986).

Strong Investment-Led Recovery

My purpose here is not to sort out these various claims but to salvage from the debate what appears to be an important difference of performance between the Reagan recovery and previous ones. The Reagan recovery in 1983–1984 was a strong investment-led recovery, more so than any of the previous recoveries in the postwar period (excluding the recovery in 1949). This difference escaped most media attention at the time and remains shrouded today in the puzzling debate about the decline of American competitiveness and power. Amidst all the concern about America's domestic and foreign debt, few commentators have noted the surge in American manufacturing investment and productivity that has taken place since 1981. Unless this surge is an accident or pure luck (explanations sometimes favored by structuralist arguments; see Chapters 1 and 2), it must have had something to do with American policy choices in the early 1980s, even if economists from conflicting priesthoods may never agree on the precise causal link between their policy choices and economic outcomes.

Figure 8-1 shows the various sector contributions to real GNP growth for the first eight quarters of the expansion in 1983–1984. Whereas the contribution of personal consumption expenditures was roughly the same in the current recovery, compared to the typical postwar recovery (3.3 versus 3.2 percent), the contribution of nonresidential fixed investment (plant and equipment) and particularly producers' durable equipment was three times as large (1.8 and 1.5 percent, respectively, compared with 0.6 and 0.5 percent). These data from government sources are confirmed by private studies. Figure 8-2 shows, over the same eight-quarter period, that personal consumption rose 5.0 percent per

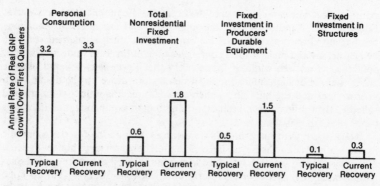

Figure 8-1 Sector contribution to real GNP growth: typical vs. current recovery. (Source: *ERP* 1985, p. 30. See Appendix Table A-6 for details.)

year in the 1983–1984 recovery—compared to 5.1 percent per year in previous recoveries and 6.0 percent per year in the strongest previous recovery in 1970—and that other fixed investment and producer durables rose at rates of 15.0 and 18.6 percent per year, respectively, in the current recovery compared to 6.9 and 9.4 percent per year in previous recoveries and 7.4 and 10.7 percent per year in the strongest previous recovery in 1970.

The current recovery slowed perceptibly after mid-1984. Yet even if we extend the data to cover an additional four quarters (i.e., the first 3 years of the recovery through 1985), the current recovery continues to outperform previous

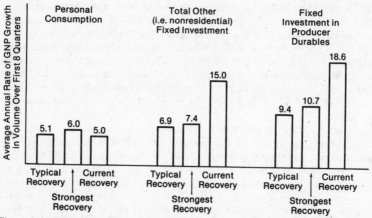

Figure 8-2 1983–1984 recovery vs. past recoveries. (Source: Marris 1985, p. 43. See Appendix Table A-7 for details.)

recoveries in terms of nonresidential fixed investment, particularly producers' durable equipment. Figure 8-3 shows that, over this period, personal consumption actually contributed slightly less to the current than to previous recoveries, whereas nonresidential fixed investment, and specifically producers' durable equipment, contributed nearly twice as much. These data exclude the 1958 recovery, which did not last 3 years. And if the 1954 recovery is also excluded, investment performance in the current recovery assumes a more typical pattern (Blanchard 1987, p. 32). Apparently, investment performance in the 1954 and 1958 recoveries was particularly weak, pulling down the averages for all previous recoveries. Nevertheless, even excluding the recoveries in the 1950s, which clearly biases the calculation, studies conclude that investment in the current recovery was not inhibited by the economic policies of the first term Reagan administration (Blanchard 1987, p. 33).

Two other features of the 1983–1984 expansion stand out by comparison. Over the first eight quarters, the role of exports was significantly lower in this recovery than in previous ones, and the role of imports was considerably higher (see Tables 6, 7 and 8 in Appendix). Similarly, the role of public expenditures was higher in this recovery. Both of these patterns remain strong even when the four quarters of 1985 are added to the data. The two features reflect the much larger federal budget deficit and higher dollar that accompanied the 1983–1985 recovery.

Although the budget and trade deficits, therefore, are a distinguishing feature of this recovery compared with previous recoveries, they have not been associated exclusively or even primarily with a personal or public sector consumption binge in the United States, at least not through the end of 1985. During this period it could be said that the United States was behaving like a smart borrower. It was borrowing from abroad to finance large budget and trade deficits. But relative to previous recoveries, the private sector was spend-

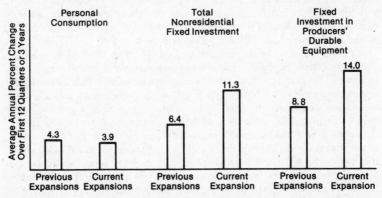

Figure 8-3 Growth rates of real GNP components: current vs. average previous expansions. (Source: *ERP* 1986, p. 39. See Appendix Table A-8 for details.)

ing this money, as well as domestic savings, more for investments and less for consumption. It was refurbishing America's industrial plants and outfitting its management and inventory systems with new computer and communications equipment. And although public sector spending was higher, this spending was now going, proportionately, more into defense and defense-related industries and less into transfers and consumption (although some subsidies, especially to agriculture, were rising). Because defense expenditures tend to contribute more to output than to consumption, public sector spending was also contributing more to higher investment. America was putting in place the productive capacity with which to repay its foreign debt.

All of this investment substantially boosted American manufacturing productivity and increased overall productivity as well, though not to its pre-1973 levels. Productivity in the business sector as a whole increased 1.4 percent per year from 1979 to 1986, compared to 0.6 percent per year from 1973 to 1979 and 2.8 percent from 1948 to 1973. Manufacturing productivity, however, rose 3.5 percent per year from 1979 to 1986 compared to 1.4 percent from 1973 to 1979 and 2.8 percent from 1948 to 1973. From 1981 to 1987, it grew at an even stronger pace of 4.1 percent per year, outstripping annual productivity growth in the halcyon years of 1948–1973 by 50 percent. The improvements were strongest in electrical and non-electrical machinery, instruments, food, rubber, textiles, and miscellaneous manufactures (for data here, see Kendrick 1989, Dertouzos et al. 1989, p. 28, and *ERP* 1989, p. 67).

Some studies spotted these developments at the time. An analysis in 1984 showed that American manufacturing difficulties from 1980 to 1982, as reflected in the trade balance, were almost entirely related to macroeconomic circumstances, not competitiveness factors. Between 1979 and 1983, according to this study, "performance in the manufacturing sector suggests that the rise in defense procurement and the strong growth in domestic investment have offset the negative effect of foreign trade" (Lawrence 1984, p. 8). "If over the next few years," this study concluded, "U.S. policy could reduce the government deficit without curtailing investment incentives, the combination of strong domestic demand and a declining foreign exchange rate could result in extraordinary prosperity for American industry" (Lawrence 1984, p. 145).

Whether the specific Reagan tax cuts of 1981 had most to do with this strong investment performance is another question. Some studies suggest that, in terms of the cost of capital, the tax cuts were probably offset by increase in real interest rates (Blanchard 1987, p. 23) or that investment was highest in sectors not particularly affected by tax reductions (Bosworth 1984) and may have been influenced more by a decline in real prices for products such as automobiles and computers (Niskanen 1988, p. 234). One study concluded, however, that the tax changes "may well have played a role in 1982–84 because they involved a rather *sudden* change in U.S. company taxation, and because they occurred at a time when there was an ample supply of surplus savings in the rest of the world" (Marris 1985, p. 50; see also Boskin 1987, p. 60). What was unusual about the recovery in 1983–1984, according to this same study, "was the *speed* of the pickup in investment rather than the level it reached" (Marris 1985, p. 43).

But the level of investment in the United States in the 1980s may have also been unusual, which provides even stronger evidence that the role of the tax cuts may have been significant. Although gross private investment after 1982 was no higher in nominal terms than previous levels, it was considerably higher in real terms. Figure 8-4 shows real nonresidential fixed investment as a percent of real GNP for every year since 1929. This ratio reached a postwar high in 1985, exceeding the levels of the late 1970s and suggesting that the strong investment activity in the early 1980s was not just a consequence of the postwar period's deepest recession in 1981-1982.[1] The ratio declined after 1985, coincident with changes in U.S. tax law that rescinded the tax benefits of 1981 for American industry, but surged ahead once again in late 1987 through early 1989 when the long-term benefits of the 1986 tax reforms took effect (see Chapter 9).

The specific positive effect of tax changes on investment does not have to be decided here, however, or confused with the effect of tax changes on national savings or tax revenues, which was negative, to conclude that the Reagan program, as it played out in 1983-1984, had many positive consequences and was nothing like the calamity that critics proclaimed in 1981-1982 and would reassert in 1986-1987 after growth slowed and foreign debt mounted (Friedman

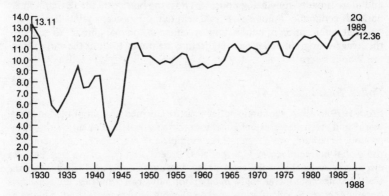

Figure 8-4 Real nonresidential fixed investment (percentage of real GNP). [Source: Federal Reserve Bank of St. Louis (obtained from John Tatom, Assistant Vice-president).]

[1]Real *net* investment was also higher than critics of the tax cuts acknowledged (e.g., Altman 1989). This measure takes into account the shorter life span of investment in equipment as opposed to plant or structures. Low capacity utilization in the early 1980s increased the incentive to invest in equipment compared to structures. When adjusted for the lower rates of capacity utilization, average real net investment as a percentage of real GNP was sharply higher in 1981 and 1984-1985 than the average for the previous peak years from 1956 to 1979 (Tatom 1989).

1988). The tax cuts had clearly made some contribution to the higher levels of investment in this recovery compared to previous ones. And budget cuts and deregulation had reduced waste and inefficiency in the American economy. In addition, what government expenditures remained were arguably more productive, involving reduced transfer payments, which tend to shift income from investment to consumption, and increased defense investments, which tend to boost output (even if direct government investment in civilian industry, assuming the government makes the right market decisions, might have been more productive still).

Why few of these positive features were adequately emphasized by the administration or reported by the news media is genuinely puzzling. How the country and the administration could wind up in the late 1980s in a competitiveness debate that pictured American manufacturing on the brink of disaster is even more puzzling (Cohen and Zysman 1987). Partisanship and continuing disarray in economic theory played a role. But, in the end, the administration had itself to blame. It did not engage the intellectual debate. It did not encourage conceptualization of its policies and performance, even post facto. It nurtured few academics in its Cabinet or inner circle, and the few who were there seemed partly embarrassed by the policies and entirely at a loss to explain them. The administration continued to settle for public relations and missed a golden opportunity to couple renewed American power with clearer purposes and more broadly convincing policies. Thus, the best years for Reagan's international economic diplomacy passed without the necessary intellectual and political mobilization of public support, either at home or abroad, to complete the Reagan program and effectively restore the policy triad of the earlier Bretton Woods system.

Policy Triad Flickers Abroad

From 1979 to 1983, economic policy priorities turned around in the industrial world and, to a lesser extent, in the developing world. Not all of this turnaround, to be sure, was a result of the example and pressures being generated in the global marketplace by U.S. policies. Indeed, the mix of policies pursued elsewhere, especially in the early going, was quite different from that of the United States. But increasingly, in all major industrial countries, the objectives became the same—disinflation, greater flexibility of markets, and, as long as there was a large demand for imports in the United States, freer trade. United States leadership in 1981–1982, resisting compromises with inflation and tax cuts, was clearly important and perhaps even critical in producing this outcome.

Margaret Thatcher in Great Britain pioneered the conservative economic "renaissance" 2 years before Ronald Reagan (Walters 1985). And she did it with a quite different policy mix. From 1979 through 1982, Thatcher gave priority to monetary policy (which in Britain is controlled by the Treasury) and the fight against inflation. In this period, despite a recession, Britain raised taxes to cut the budget deficit and reinforce the credibility of monetary policy. The supply-side program — curbing union powers, privatization and deregulation, and reduction and reform of taxes and social security benefits — did not follow until the second half of 1982. Even then tax changes in Britain have been more modest than in the United States. Top rates on personal taxes were reduced from 83 and 98 percent for earned and unearned income, respectively, to 60 percent and on corporate taxes from 52 to 35 percent (Matthews and Minford 1987, pp. 70–73). The result is that Britain has had no significant fiscal problems, compared to the United States, but it has had an unemployment rate almost double that in the United States (Krieger 1986).

Economic policy in Germany turned around somewhat later in the fall of 1982. From 1980 to 1982 a debate raged in Germany between a revitalized supply-side school represented by the economic institutes and the Bundesbank and a Keynesian school of demand-side advocates in the trade unions. After 1980 the socialist government of Helmut Schmidt sided with the supply school, calling for wage moderation, deregulation, and fiscal and monetary consolidation. But without adequate political support in the socialist party and with the economy in the throws of a recession, Schmidt's policy was not credible. In the fall of 1982 the small Free Democratic Party (FDP) left the socialist-led coalition and formed a new government with the Christian Democrats (CDU). Confirmed in elections in March 1983, the CDU–FDP coalition under Helmut Kohl proceeded more systematically to halve the budget deficit from 4.1 percent of GNP in 1982 to 1.9 percent in 1986 and to initiate modest tax and deregulation reforms (Hellwig and Neumann 1987, pp. 132–135). German policy, however, moved much more slowly on structural changes (e.g., confronting labor market rigidities) than either U.S. or British policy.

The turnaround was perhaps most dramatic and divisive in France. In 1981 the new socialist government of François Mitterrand had "embarked on an ambitious program of 'redistributive Keynesianism'" (Hall 1986, p. 194). The budget deficit jumped from zero in 1980 to 3.2 percent of GNP in 1983, even though taxes were increased on employers and upper-income groups. The public sector, after a wave of nationalizations, increased from 46.4 percent of GNP in 1980 to 52.0 percent in 1983. Inflation hung at 13.4 percent in 1981 and 11.8 and 9.6 percent in 1982 and 1983. The trade deficit soared and the franc plunged. Despite all the stimuli, unemployment continued to rise from 7.4 percent of the labor force in 1981 to 8.3 percent in 1983 (data from OECD 1987b, reference tables).

French policies were totally out of line with more conservative, production-oriented policies in other countries. Eventually, the Mitterrand government had to adjust, tentatively in June 1982 and then decisively in March 1983. By the latter date, it had only two policy options for salvaging the franc: quit the European Monetary System (EMS) and take France behind protectionist walls or adopt more conservative policies to reduce consumption, inflation, and the disincentives crippling investments and encouraging capital flight. The external constraints imposed on the French situation by policy shifts in the United States, Great Britain, and Germany were significant. Fortunately for France and the industrial world, Mitterrand decided to stay in the open world economy and alter his domestic policy course. Thereafter, French policies mirrored the priorities of disinflation, deregulation, and in 1985–1986, tax reductions and reforms existing in other major countries (Hall 1987, pp. 55–58).

The economic turnaround of the early 1980s is evident to varying degrees in other countries in this period, both industrial (e.g., Italy, Japan) and, after the debt crisis of 1982, some developing countries. As the OECD reported in 1988, "almost all OECD member countries have begun to retreat on a number of intervention fronts," including subsidies and institutional and regulatory frameworks for government control of industry (1988, p. 232). World Bank studies documented similar rethinking in developing countries (World Bank 1986b, 1987b, 1988b). Whether all this change was deep enough or permanent enough to make a difference could be questioned and the changes "have not yet proved sufficient to justify or facilitate a real reversal of trade protectionism," in either developed or developing countries (OECD 1988, p. 232). But the patterns were real.

Figure 8-5 offers some aggregate indicators of the policy shifts in this period. In all developed country groups (G-5, G-7, and all industrial countries), growth of monetary aggregates (M_1 and M_2) receded decisively after 1980 compared to the period 1977–1979 and before, especially for the first 3 years of the 1980s. Expansion of the public sector also declined after 1980, and the public sector actually retreated after 1983 in all countries, shrinking at a rate of three to four percentage points per decade. Budget deficits are the one policy indicator showing a continuation of trends from the 1970s. But these numbers are distorted by the United States. If the United States is excluded, budget deficits in the other G-5 countries did not grow over this period (averaging 3.5 percent of GDP from 1974 to 1980 and 3.6 percent from 1981 to 1987). In Germany, Great Britain, and Japan alone, fiscal deficits actually contracted from 1980 to 1985 by a total of 4 percent of GNP, whereas the U.S. deficit ballooned by an amount equal to 3 percent of GNP (Henning 1987, p. 10).

It is tempting again to attribute these cross-country shifts to common circumstances or to external developments such as the establishment of the EMS in Europe or the softening of oil prices after 1981. But close examination does not warrant such conclusions. The EMS was hardly a zone of monetary stability in the early 1980s, experiencing seven realignments in its first 4 years (1979–1983). Exchange rates in Europe became more stable only after domestic policies converged and remain stable today only because monetary officials in the EMS countries are willing to align their policies for the moment with those of the inflation-sensitive Bundesbank in Germany. Similarly, end-user prices for oil declined after 1981 but did not return to their pre-1979 levels until after 1985 (OECD 1988, p. 156). Moreover, currency depreciations in Europe actually raised the price of imported oil in most countries during the early 1980s (the so-called third oil shock).

The decisive factor in this period, therefore, was not favorable circumstances or good luck but policy changes brought about by new policy ideas (e.g., deregulation of energy prices in the United States, which helped to bring oil consumption down and subsequently to lower oil prices). The competition

I. G-5 (France, Germany, Japan, United Kingdom, and United States)

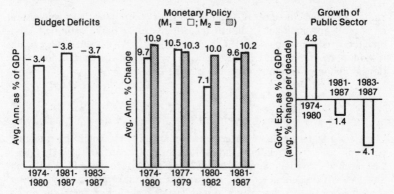

II. G-7 (G-5 plus Canada and Italy)

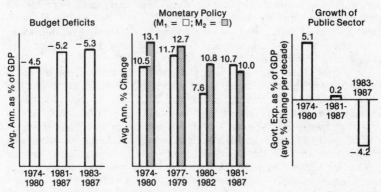

III. All Industrial Countries

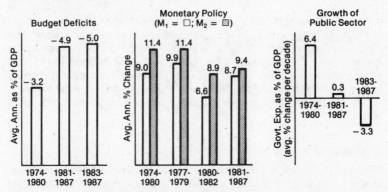

Figure 8-5 Economic policy shifts after 1979–1980. (Source: See Appendix Table A-9.)

among ideas had cut across political persuasions (e.g., the socialist conversion in France) and national traditions (as differences in basic institutions and practices persisted among the OECD countries) to tilt the advantage once again toward moderate, market-oriented policies, reversing at the margins the inflationary and interventionist tendencies of the 1970s.

Hesitant Diplomacy

The United States had an unprecedented opportunity to exploit the emerging economic recovery to reinforce and confirm the policy shifts of the early 1980s. The Annual Economic Summit in 1983 was set to take place in the United States at Williamsburg, Virginia, in May. The summit offered an ideal setting to strengthen further the process of multilateral surveillance, signal the possibility that exchange rates might become more stable in the future as domestic policies further converged, address vigorously and interrelatedly the trade and debt crises, and heal old wounds from Versailles on East-West trade issues. Indeed, some of these objectives were achieved at Williamsburg. But, somewhat surprisingly, Williamsburg was far less ambitious than Versailles, even though economic conditions had improved.

It was as if the allies and particularly the United States had expended themselves over the previous year in the contentious Versailles meeting and its aftermath. There was a tendency to pull back. The summit process itself came in for severe criticism. President Reagan would attempt a far less structured and more spontaneous summit at Williamsburg. It was not clear how such a summit would go, and there was a desire to be cautious, to resolve outstanding problems, particularly the noisy East-West trade confrontation of summer 1982 (see Chapter 10), rather than to attack new problems decisively. The mood, although hopeful, was hesitant.

In retrospect, 1983 was probably the best year for the administration to pursue a more aggressive international economic diplomacy. The following year was an election year; and 1985, the year after reelection, was a good year to focus again on outstanding domestic imbalances (as had been done initially in 1981). Although moving in the right direction, however, U.S. international diplomacy in 1983 came up short. Subsequent opportunities to press the domestic policy-oriented agenda at the London and Bonn Summits in 1984 and 1985 were less propitious. By late 1985, U.S. diplomacy had turned once again in directions reminiscent of the 1970s, encouraging exchange rate and monetary policy coordination while relatively ignoring domestic budget and trade policy reforms.

Scaling Back the Summit Process

The trade and financial crises that struck the industrial and developing countries in fall 1982 tested the very existence of cooperation in the Western world economy. In October 1982 President Mitterrand called the industrial country summit meetings "nearly worthless" (*NYT*, October 12, 1982), and in spring 1983 France actually considered leaving the open world economy altogether. Major developing countries — Mexico, Brazil, and Argentina — encountered unprecedented international debt problems, even as the North–South dialogue — at least, as represented by Global Negotiations and 1970s-style diplomacy — seemed to have fallen apart. The souring mood prompted U.S. Secretary of the Treasury Donald Regan to propose in December 1982 a new Bretton Woods Conference. Although the idea was off the cuff and had not been staffed out by the U.S. administration, it reflected the concern that the process, let alone the substance, of international cooperation was about to disintegrate.

Procedural issues, therefore, dominated early U.S. planning for the Williamsburg Summit (apart from East–West trade issues, which were being intensively negotiated in fall 1982; see Chapter 10). The summit date itself, though announced by the United States in October, was not agreed to by the French until the first summit preparatory meeting in Celle St. Cloud, France, in December.

In an October communication to other world leaders, President Reagan raised two additional procedural issues for Williamsburg — how to make the summit less formal and how to convey publicly the true nature of these meetings as forums for discussion rather than decision meetings without making them appear pointless. The two issues were interrelated. A less formal summit could not aspire to deal with specific issues and negotiations. Yet the press, expecting summit meetings to deal with such issues, would report the sessions as pointless. The leaders were caught on the horns of a dilemma. Both the "structuralist" summit at Bonn and the "choice-oriented" summit at Versailles had involved detailed preparations and negotiations. By forcing agreements, however, both summits had created misunderstandings and ill will — the Versailles Summit more immediately than the Bonn Summit. Moreover, the leaders preferred relatively private, informal meetings as the summits were originally intended to be. Thus, President Reagan and the other leaders aspired to bring the summit back to its origins, as difficult and temporary as that effort would turn out to be.

The challenge of the task was soon evident. In initial bilateral visits with European sherpas in late October, the U.S. sherpas, headed by the new Under Secretary of State for Economic Affairs Allen Wallis (Reagan's third personal representative in three summits), encountered general agreement on the need for a smaller, scaled-back summit but overwhelming skepticism that it could be achieved. Scaling back involved a number of possible modifications of the meetings: reducing the number of ministers and staff attending the summit, reducing press coverage and expectations, eliminating a formal communiqué, and holding fewer and later preparatory meetings so the summit discussion would be more spontaneous and relevant.

Reducing the number of ministers was difficult. Countries with coalition governments frequently wanted to increase the number. At Williamsburg the Germans wanted three ministers to attend instead of the usual two (foreign and finance ministers), because the finance and economic ministers of the new

German governing coalition came from different parties. One way around this problem was to arrange more opportunities for the heads of state and government to meet alone without ministers. In the previous eight summits the heads had met alone only for lunch and dinner. Although Prime Minister Trudeau tried and failed to arrange additional sessions at Ottawa, President Reagan succeeded in organizing such a session for the first morning at Williamsburg. Even interpreters were out of the room, translating through audio-visual equipment. The sessions proved extremely useful and were copied at future summits.

Reducing press coverage and expectations was equally difficult. Some governments with greater direct control over their press wanted to prohibit all press contacts during the summit until the final press conference. This was completely impractical for the United States. Another possibility was to play down the summit. In late October 1982, officials had in mind the model of the successful, informal NATO meeting in La Sapinière, Canada, which had started the process of narrowing differences on East–West trade issues (see Chapter 10).

La Sapinière had also issued no formal communiqué but worked with a so-called nonpaper, suggesting the possibility that the summits too might dispense with communiqués. President Reagan felt strongly that prenegotiated communiqués limited rather than facilitated the discussion among leaders. Other leaders agreed, but as the summit approached, political pressures began to build. Without a prenegotiated communiqué leaders could not anticipate surprises and possible embarrassments at the summit.

At a tense preparatory meeting in Williamsburg in April 1983, the U.S. delegation came up with an alternative to the prenegotiated communiqué. At the first preparatory session in San Diego, the U.S. team had initiated a "paper trail" of the sherpa discussions in the form of summary outlines. The U.S. team now proposed that these outlines be developed into a "thematic paper." This paper would formulate broader positions but stop short of formalized communiqué language. The U.S. team suggested that the paper be discussed individually with each country during visits to Europe in early May and then be approved at the last preparatory meeting in La Celle St. Cloud in mid-May.

The thematic paper told each summit leader what to expect at the summit but contained no bracketed language (as was the practice with prenegotiated communiqués). It was thought of as a common briefing paper for each leader providing an inventory of the issues and a point of departure for the summit discussion itself. In the end, some key passages were lifted from the thematic paper for the final communiqué (e.g., the hotly contested language on an international monetary conference; see further discussion later). But overall the Williamsburg Communiqué was only one fifth the length of the thematic paper, the communiqué's tone and emphasis were drawn directly from the discussions at the summit, and no draft of the communiqué was discussed among the sherpas at all until the end of the first full day of summit discussions among the leaders.

Thus, the Williamsburg summit probably went as far as any summit could to give the leaders an opportunity for both serious and spontaneous discussions. The heads themselves negotiated both the economic communiqué and the political declaration on Intermediate Nuclear Forces (released the first day). On several occasions, they delayed or interrupted formal sessions to stand or sit around in small groups to craft the crisp and concise language that distinguished the Williamsburg declaration from the more lengthy and tortured verbiage of previous communiqués.

The format for Williamsburg was useful, of course, primarily for dealing with large themes. It could not have accommodated the intricate package-deal type of negotiations of a Bonn or Venice (1980) summit. In part perhaps the procedure accounts for the failure of Williamsburg to enact a more aggressive international economic program. As one who participated in all the preparations, however, I do not believe the format was at fault. The types of decisions that would have given Williamsburg a genuine "Bretton Woods II" character (see further discussion later) were big ones. They could have been made easily within the Williamsburg format and indeed in some cases were actually initiated in the preparations but not followed through at the summit (such as the initiative for a trade and finance ministers' meeting; see later discussion).

Although scaled back, Williamsburg was not "laid back" (*WP*, March 3, 1983). The preparations were no less intense than those for the 1978 Bonn or 1982 Versailles summits. But these preparations were different. They were conducted primarily at high levels and concentrated on broad interrelationships rather than on fragmented issues. President Reagan, for example, conducted four rounds of personal correspondence with all his counterparts and several additional rounds with the French president on East–West trade and credit issues (see Chapter 10). Reagan was deeply and personally involved in all phases of the summit's preparations, once sternly admonishing his summit team not to be drawn into the negotiation of communiqué language (on Reagan's involvement, see *NYT*, May 26, 1983). After the summit, Prime Minister Trudeau paid President Reagan a public tribute for organizing a substantive yet spontaneous meeting:

> The President took a very big gamble that we would have an unstructured summit and still produce results. I must say I had to congratulate him for having won that gamble [*NYT*, May 31, 1983].

Williamsburg stands, therefore, as a model of a choice-oriented approach to summitry that looks for some outcome between the public relations exercise of an Ottawa Summit, on the one hand, and the intense and often bitter, detailed negotiations of a Bonn (1978) or Versailles summit, on the other. Williamsburg combined thematic directions and technical detail (e.g., the monetary annex), producing neither empty political results nor tortured, complex agreements subject to misunderstanding. It fell short on ambition and probably follow-up. Regrettably, agreements at Williamsburg were followed up less intensively than disagreements at Versailles. Why that was so becomes clearer as we look at the substantive issues and outcomes at Williamsburg.

Strategy for Williamsburg

The desire to focus on broad topics placed the emphasis in the Williamsburg preparations on interrelationships of economic issues. Apart from the relationship between economic relations and Western security (see Chapter 10), the key interrelationships were twofold: (1) that between domestic recovery and international relationships, particularly exchange rate alignments, and (2) that between trade and finance relationships.

At the first substantive preparatory session in San Diego in mid-March, the U.S. sherpas tabled two proposals to address these interrelationships. The first one built on the multilateral surveillance process established at Versailles and sought to develop more specific policy guidelines to implement the performance standards, particularly lower inflation, cited in the Versailles monetary declaration. The United States viewed a deepening of the multilateral surveillance process as particularly important because the emphasis was now shifting from the fight against inflation, which had dominated the Ottawa and Versailles meetings, to the fight against unemployment, which had soared in the 15 major OECD countries from 5.8 percent in 1980 to 8.7 percent in 1983 (OECD 1984, p. 163). The United States did not resist this shift to reducing unemployment but sought to discipline it (*NYT*, May 15, 1983). Moreover, it sought to keep the multilateral surveillance initiative roughly in balance with the discussion going on concerning exchange market intervention.

The intervention study commissioned at Versailles was moving ahead. By March it comprised some 6000 pages, including detailed case studies of previous experiences with intervention. The French had always made clear that the intervention subject was a priority for them. The United States, on the other hand, feared that the reflex to intervene, the most developed of all the international economic policy reflexes (a legacy of the 1970s), would get out ahead of the pace of policy convergence that the United States emphasized. If it did, such intervention would weaken the incentives for policy convergence.

The second U.S. initiative was a novel one. It called for a joint meeting of trade and finance ministers to address related issues of protectionism and debt. The hope was to use the debt crisis to spark new initiatives to strengthen the trading system and to put the developing countries, especially the middle-income, heavily indebted ones, at the center of a new multilateral trade round. Potentially, the trade–finance link was the initiative that could have sparked a revival of the full Bretton Woods policy triad. At Ottawa and Versailles, American diplomacy had stressed price stability and flexible markets. Now American proposals stressed trade, both to counterbalance the growing emphasis on debt and finance and to spur exports as the only substantive way to repay debt.

Ultimately, the U.S. initiatives on multilateral surveillance and the trade–finance link yielded only partial results. The primary responsibility for lack of progress has to rest with the United States, but the United States got a lot of help from the allies.

An early strategy memo for Williamsburg raised three options for U.S. policy: (1) continue basic domestic-oriented strategies pursued at Ottawa and

Versailles, (2) seek international solutions and call for a new Bretton Woods to reform the international trade and financial systems, and (3) pursue a series of incremental international initiatives consistent with the themes of Ottawa and Versailles. The second option, it was felt, posed too many risks, especially after the misunderstandings at Versailles and because it might revive the prospect of an international monetary conference and stir up interest once again in Global Negotiations.

An ambitious strategy might also be misinterpreted, it was feared, as a reversal of U.S. international economic policy. Old habits of international cooperation died slowly. A distinguished group of international economists meeting in Washington in December 1982, for example, issued an appeal for reflationary measures, noting that the decline in inflation provided "scope for policy to be shifted in an expansionary direction so as to promote recovery" (IIE 1982, p. 1). Similarly, the OECD staff called in early 1983 for a differentiated approach to economic recovery, a new name for the old approach of some countries acting as locomotives for economic growth. A more aggressive strategy, therefore, might be exploited by those who always had a greater tolerance for inflation. Conceivably, if the summit had been held in the second half of 1983, as the French had originally advocated in the early dispute over summit dates, the recovery might have been further advanced and the United States, as well as Europe and Japan, might have been able to act with greater confidence and risk.

In early 1983, however, the allies too were in no position to act boldly. President Mitterrand, who picked up the idea of a new Bretton Woods conference and pressed it at the OECD meeting in May 1983 (*NYT*, May 10, 1983), was in the midst of a wrenching domestic economic crisis. The issue was being settled in mid-March, just as the first substantive meeting for Williamsburg convened in San Diego. Because of the crisis, neither the French nor the German sherpa attended the San Diego meeting. It was still unclear how aggressive the new German government would be, especially in dealing with France, and the British were exceedingly cautious, given their elections in June.

So neither the politics nor the economics was right for bold action in early 1983. That is not an excuse for Williamsburg; it is an explanation. The United States went ahead with its convergence and trade–debt initiatives, but the broad vision was missing and the Europeans, especially the French, sulked — in what had become by now characteristic fashion — about any new initiatives until interest rates and the dollar had been brought down.

Strengthening Multilateral Surveillance

The multilateral surveillance meetings initiated at Versailles were confidential and received no publicity. By contrast, the intervention study, also commissioned at Versailles, was getting heavy press attention. The United States worried that the focus of international economic diplomacy would slip once again into exchange rate and monetary policy coordination. It set out therefore to strengthen the multilateral surveillance process without jeopardizing the confidentiality of this process. The risks were considerable, as the U.S. experience after 1985 showed when a much more political process of policy coordination led to high public drama and squabbling among finance ministers that contributed, in the end, to a worldwide stock market and financial crisis in the fall of 1987.

At San Diego in March, the U.S. delegation identified four specific policy criteria to guide the multilateral surveillance process: (1) stable monetary growth at moderate rates, (2) discipline over government expenditures, particularly transfer payments, (3) tax and regulatory reform to enhance flexibility and the role of market signals in growth, and (4) policies of openness and freer trade to enhance competition. These were not policies of monetarist austerity (e.g., see Putnum and Bayne 1984, p. 185). They were, in fact, the essential policies of the original Bretton Woods system as it had operated after 1947. They were longer-term policies, to be sure, and they were undoubtedly more difficult to achieve politically, because they involved basic structural (i.e., microeconomic) reforms rather than quick macroeconomic fixes. But, no less than proposals for reflation, they constituted a program for stronger growth.

Moreover, the U.S. criteria were not insensitive to time limits. Specific measures were proposed to assess progress on an incremental basis. These measures included interest and inflation rates, growth of monetary aggregates and base, relative size and direction of change in government budget expenditures and deficit, unemployment rates, productivity growth, real GNP performance, balance-of-payments developments, regulatory changes, and measures of the extent and direction of change in trade and capital restrictions. Such measures added short-term policy and performance indicators to go along with the long-term policy goals—low inflation, greater employment, and renewed growth—enumerated in the Versailles monetary undertaking.

The Reagan administration at this point believed that the more progress that could be made on convergence, the less need there would be for exchange market intervention, and, more important, the less harm such intervention might do to the basic commitment to discipline domestic economic policies to achieve a few essential common goals such as low inflation. The objective was to keep convergence and intervention commitments in parallel with one another. At Williamsburg, therefore, the United States was prepared to accept some minimal guidelines for intervention *if* the allies accepted some comparable guidelines for policy coordination in the areas of inflation and other performance indicators. Then, if others failed in their inflation performance, the United States would have no obligations to intervene in exchange markets. (For comment on a U.S. strategy memo at the time, see *NYT*, May 15, 1983.)

The problem for Reagan officials was never the issue of intervention per se but the widespread belief that intervention was the panacea for exchange rate and macroeconomic problems. The U.S. team hoped that this belief would be convincingly expunged by the intervention study that was under way. After exhaustive investigations, that study concluded that "intervention has often been effective in attaining short-term objectives . . . in response to such exchange market disorders as unsettled trading conditions or unexpected events of an essentially non-economic nature" and could also be "useful," although on occasion "counterproductive," for other short-term purposes, such as reassuring market participants or sending a signal of determination to markets. But for longer-term adjustments of exchange rates, the study found, intervention "did not appear to have constituted an effective instrument in the face of persistent market pressures" unless intervention "was permitted to have direct impact on the authorities' monetary liabilities," that is, unless authorities allowed the purchases or sales of local currency for stabilizing the exchange rate to alter the money supply and did not immediately offset (i.e., sterilize) these purchases or sales through open-market operations. Effective intervention over the longer-run, in brief, required "supportive domestic adjustments, especially in the field of monetary policy." Intervention might "buy time . . . if followed by appropriate policy changes; but buying time had occasionally been useless or even counterproductive in the absence of appropriate policy changes." (See the study of the Working Group on Exchange Market Intervention released April 29, 1983.)

The report confirmed both the U.S. view of the long-term effects of intervention and the European view of the short-term effects. It offered a solid analytical basis for strengthening the convergence process over the medium term and for somewhat more frequent intervention *toward that end* in the short term. The finance ministers, when they approved the report in late April, appeared to agree to such a linkage. The communiqué they released read in key parts as follows:

> Intervention will normally be useful only when complementing and supporting other policies. We are agreed on the need for closer consultations on policies and market conditions . . . and . . . are willing to undertake coordinated intervention in instances where it is agreed that such intervention would be helpful. [Statement released with the intervention study, April 29, 1983.]

The communiqué specifically identified when intervention might be helpful:

> Intervention can be useful to counter disorderly market conditions and to reduce short-term volatility. Intervention may also express an attitude toward exchange markets. [Statement released with the intervention study, April 29, 1983.]

Parallel agreements on guidelines for intervention and convergence were not to be achieved, however. At the summit preparatory meeting in Williamsburg in mid-April, the French, supported by the Italians and Canadians, refused to elaborate policy and performance indicators for the convergence exercise. Although Italy and Canada had political reasons to object because they were not included in the convergence exercise, all three countries had substantive reasons as well. They were the high-inflation countries at this point. Their currencies were under downward pressure in exchange markets (the Canadian currency less so because it was linked to the U.S. dollar). They wanted coordinated intervention to support their currencies. As they saw it, no doubt, intervention was justified to buy time. France had undertaken fundamental policy adjustments in March; Italy and Canada had done so in mid-1982. But they were unwilling to commit to more specific policy and performance targets in the convergence exercise. The French, in particular, feared outcomes at Williamsburg that would identify their new policies politically with those of the conservative governments dominating the Williamsburg scene. The substance of convergence was present in French policies, but it could not be acknowledged in the convergence process at the summit. Because the changes in French policy were relatively recent, the United States was not yet sure that the substance was actually there.

Accordingly, the United States decided not to alter its intervention policy. It was convinced that U.S. policy was on the right track and that greater convergence would simply require more time. Inflation rates had come down dramatically in the other summit countries and interest rates with them. United States short-term rates had been cut in half since January 1981, from 18 to 9 percent. Similarly, U.K., German, and Japanese rates had fallen by two to six percentage points (Treasury data, March 31, 1983). The argument that high U.S. interest rates were at the core of Europe's problems seemed less convincing. Real U.S. interest rates remained high if measured in terms of current inflation, but they were not abnormally high in terms of anticipated inflation. Markets still needed to be convinced that the improvements in inflation performance would be sustained in the future. More active intervention in exchange markets would have sent exactly the wrong signal to the markets, particularly if such intervention was unaccompanied by a convincing demonstration at the summit that firm domestic policy adjustments were being made.

The Reagan administration also had no significant intention or plans at this point to reduce its own budget deficit. Presumably, recovery would help to do that, but the administration had been weakened politically in the 1982 congressional elections and felt less confident than before that it would be able to achieve further budget cuts. Undoubtedly, these domestic constraints also reduced the administration's enthusiasm for its own convergence proposals. Thus, the opportunity to make significant parallel progress on convergence and intervention was not fully exploited.

The Williamsburg Summit did go on to approve an annex to the communiqué entitled "Strengthening Economic Cooperation for Growth and Stability." This annex spelled out more specific, near-term policy actions to guide the convergence process over the medium term. It read in many respects like a catalogue of the new conservative economic policy consensus for noninflationary growth:

(1) *Monetary Policy.* Disciplined, non-inflationary growth of monetary aggregates, and appropriate interest rates, to avoid subsequent resurgence of inflation and rebound in interest rates, thus allowing room for sustainable growth.

(2) *Fiscal Policy.* We will aim, preferably through discipline over government expenditures, to reduce structural budget deficits and bear in mind the consequences of fiscal policy for interest rates and growth.

(3) *Exchange Rate Policy*. We will improve consultations, policy convergence and international cooperation to help stabilize exchange markets, bearing in mind our conclusions on the Exchange Market Intervention Study.

(4) *Policies Toward Productivity and Employment*. While relying on market signals as a guide to efficient economic decisions, we will take measures to improve training and mobility of our labor forces, with particular concern for the problems of youth unemployment, and promote continued structural adjustment, especially by:

- Enhancing flexibility and openness of economies and financial markets.
- Encouraging research and development as well as profitability and productive investment.
- Continued efforts in each country, and improved international cooperation, where appropriate, on structural adjustment measures (e.g., regional, sectoral, energy policies). [For text of Williamsburg Communiqué, see *NYT*, May 31, 1983.]

These commitments countered the short-term, fine-tuning prescriptions urged on summit leaders by other experts during the spring of 1983. But they did not suffice to raise the convergence exercise to new levels of effectiveness, in terms of either political commitment or public support. The exercise continued, and convergence in terms of policy outcomes actually accelerated from 1983 to 1985, as even critics conceded (Bergsten 1985). But the exercise could not withstand an indefinite postponement of America's commitment to reduce its budget deficit. Eventually, after 1985, the exercise was transformed into a more conventional exchange rate and monetary policy coordination exercise (see Chapter 9).

The Trade-Finance Initiative

The U.S. initiative to link trade and debt and reinvigorate international institutions came up with even less results at Williamsburg than the convergence initiative. The GATT Ministerial had dealt a serious blow to U.S. proposals for an early trade round. But in early 1983, U.S. officials saw a slight opening to exploit the debt crisis and emerging recovery to attack the trade problem from another angle. Rather than address bitter agricultural and other disputes among the industrial countries head on, the United States would appeal to the summit countries to halt protectionism and to initiate new trade liberalization *for the sake of developing countries*, particularly the heavily indebted ones. To this end, U.S. officials proposed at San Diego a series of joint meetings of trade and finance ministers, initially among the industrial countries but eventually to include selected developing countries as well. These meetings would address both substantive interactions of trade and finance and procedural initiatives to revive cooperation among the Bretton Woods institutions, particularly the GATT, IMF, and World Bank. The U.S. effort was sincere and, had it succeeded, the results would have been good for the world. But it foundered on both hard differences of interest and lack of good will among the allies.

The North–South trade initiative addressed U.S. trade interests more than those of the allies. While manufactured exports from developing countries had doubled in the 1970s as a percentage of consumption of manufactured goods in all industrial countries, this ratio had tripled in the United States from 1.1 percent in 1973 to 3.0 percent in 1983 (Brock 1984, p. 1039; Balassa et al. 1986, p. 160). In 1981 the United States absorbed 52 percent of all the manufactured exports of the developing countries, a percentage that grew by 1986 to 68 percent (Nau 1988a). Most of these exports came from a few countries, primarily the newly industrializing countries (NICs) in Latin America (Brazil, Argentina, and Mexico) and the Far East (Korea, Taiwan, Singapore, and Hong Kong). Similarly, the United States shipped a disproportionate share of its manufactured exports to developing countries. Exports from all industrial countries to the third world rose from 23 percent of total exports in 1973 to 28 percent in 1980. But in the case of the United States, manufactured exports to developing countries rose in 1980 to 40 percent of total exports and exceeded amounts going to traditional markets such as Europe and Japan (Brock 1984, p. 1039).

Europe's and Japan's trade with developing countries not only grew more slowly, but was of a different character. In 1980 the European Community imported about the same value of goods as the United States from non-oil-producing developing countries — around $45–50 billion — but the composition and sources of these imports were radically different. Europe imported twice as much foodstuffs as the United States and three times as much raw materials. And although Europe and the United States imported similar amounts from the ASEAN countries in Asia and until 1980 from Latin America, Europe imported six times as much as the United States from Africa. Proportionately, Japan imported even more foodstuffs and raw materials than Europe, and its trade was the most concentrated of all the industrial countries, primarily in Asia. From that region, Japan imported about the same absolute amounts of non–energy products as the United States, but from Latin America it imported less than a fourth of the amounts that the United States and Europe imported and from Africa about the same as the meager amounts the United States imported. European and Japanese exports to the developing world were also less significant than those of the United States. For Europe, exports to developing countries were less important than trade within the Community, and for Japan exports to the United States dominated.

The debt crisis, therefore, had a very different impact on the trade interests of the summit countries. The United States, trading primarily with the Latin American and Asian NICs, the most heavily indebted countries, was hit the hardest. United States exports to those countries dropped dramatically. From 1981 to 1984 the heavily indebted countries cut imports by a total of $43 billion, 40 percent of this, or $16 billion, coming from the United States alone. Meanwhile debtor countries stepped up exports to the United States. Brazil's exports grew by $3.5 billion, or 85 percent; Korea's, by $5 billion, or 90 percent; and Chile's, by $450 million, or 75 percent (Nau 1988a).

Recovery failed to revive U.S. exports to the debtor countries. Although IMF agreements generally tried to encourage debtor countries to liberalize imports, these countries often found they could meet IMF balance-of-payment targets most readily by restricting imports and subsidizing exports (see Chapter 9). What is more, when developing countries did liberalize under IMF arrangements, as, for example, South Korea and Mexico did, they got nothing in return from industrial countries. Reciprocal trade concessions could be achieved only in the GATT, but few developing countries participated in the multilateral trade negotiating process.

For the United States, therefore, a North–South round in the GATT offered a way to enable debtor countries to liberalize while also keeping U.S. markets open. In 1983 the Reagan administration was wrestling with renewal of the Generalized System of Preferences (GSP). A preference program from the 1970s that offered duty-free access to imports from developing countries, this arrangement was increasingly unpopular on the Hill, particularly with respect to the NICs. Congress felt these countries should graduate from aidlike trade arrangements such as GSP. Under pressure from Congress, the administration was inserting a negotiating element into GSP, such that the NICs would be able to retain certain GSP benefits only if they began to reciprocate with greater access for U.S. and other industrial country exports to their own markets. Why not generalize this process of reciprocal negotiations in the GATT? A North–South round would accomplish this. Moreover, from the U.S. perspective, expanding trade with developing countries was, as U.S. Trade Representative William Brock explained, "the only sure way out of the debt situation in the long term" (1984, p. 1041).

The U.S. debt strategy up to this point emphasized short-term IMF adjustment programs, commercial bank loan reschedulings, and modest amounts of new bank money (presumably encouraged by the good housekeeping seal of IMF agreements). The strategy needed a longer-term adjustment incentive and mechanism. In early 1983 the United States saw trade as such a mechanism. Trade would lead, and finance would accommodate, an expansion of flows of real goods and services between the industrial and developing countries. The initiative to convene joint meetings of trade and finance ministers would nurture this linkage. It was a sound strategy. In 1985, under the Baker Plan, the emphasis would shift to finance, and trade would fall victim to an exploding trade deficit and protectionism on Capitol Hill. Unfortunately, in 1983, the proposals for a trade round and trade–finance linkage would prove to be too much both for Europe and Japan, which continued to oppose the new round, and for the United States, which failed to mobilize adequate domestic support for the trade round or to come up with new financial incentives to make trade liberalization possible, especially in already heavily indebted developing countries.

The European and Japanese allies saw it all differently. They were less dependent on developing countries for export markets, and their banks moved much earlier than U.S. banks to provision reserves against losses that might result from the debt crisis. For them, therefore, it was less urgent to expand trade with developing countries. Moreover, they felt that continuing high interest rates and the high dollar in the United States and the uncertain prospects for economic recovery in other industrial countries were more harmful to developing countries than the lack of new trade and finance initiatives. The French, in particular, saw no change in developing country interest in a trade round since the GATT Ministerial Meeting in 1982, and they as well as the conservative German government (whose coalition partner, the FDP, shared some of the same views toward the third world as the French socialists) wanted more emphasis on the political dimensions of North–South relations, not just the interrelated economic aspects. Along with the Canadians and the Italians, they wanted the Williamsburg Summit to respond positively to the UNCTAD (U.N. Conference on Trade and Development) VI meeting scheduled for Belgrade in June 1983. Importing more raw materials and foodstuffs from developing countries, particularly from poor ones in Africa, the European allies favored commodity agreements and aid, which UNCTAD programs emphasized.

Table 8-1, which was prepared for the U.S. delegation at the Williamsburg Summit, shows the differences between Europe and the United States on the trade and aid aspects of development. In 1981 the EC provided two and one half times more aid to developing countries than the United States, measured as a percentage of GNP, whereas the United States contributed almost 50 percent more in terms of trade and 20–25 percent more in terms of foreign direct investment. The total contributions of each were about the same. EC aid, provided largely under the Lomé Convention signed in 1975 and renewed in 1980, went primarily to stabilize prices of commodities imported from developing countries and to finance EC manufactured exports to these countries. Japan aid also went largely to finance its exports to developing countries. To the extent Europe and Japan sought new initiatives, therefore, they called for large IMF quotas, expanded resources for IDA, the World Bank's soft loan affiliate, and in the case of the French and Canadians, even more radical schemes for large-scale debt relief. These were precisely the kinds of financial initiatives, in the absence of significant domestic or trade policy reform, the United States opposed. Whether a parallel set of incremental steps to expand trade and finance might have been possible, like the parallel progress sought between convergence and exchange market intervention, is hard to say. Hard

Table 8-1 U.S.–EC Comparative Contributions to Development: 1981*

	U.S.	EC†
Official development assistant to LDCs	.20	.51
Foreign direct investment in LDCs	.22	.18
Imports of manufactures from LDCs‡	1.20	.88
Total	1.62	1.57

Sources: OECD, GATT, and U.S. Trade Representative's Office.

*Totals reflect percentage of GNP.

†Excludes Greece and Ireland.

‡Excludes SITC 68, nonferrous materials.

interests stood in the way, but as in the case of the convergence–intervention initiative, so did a lack of good will.

The lack of good will came out most clearly in the case of the proposal for a joint meeting of trade and finance ministers. Objections were raised that trade-finance meetings, if they included other than summit countries, might interfere with existing institutions (such as the OECD) or, if they included only summit countries, would represent an undesirable bureaucratization of the summit process. The Germans and the Japanese let it be known quietly that such meetings would be very difficult to arrange between their trade (economics) and finance ministers, who were often bitter rivals domestically for jurisdiction over international economic policy. The initiative, in fact, unleashed a torrent of bureaucratic bickering, both domestic and international, culminating in French umbrage that the United States called the first meeting on French soil without the French being the formal hosts. The reactions were completely contradictory to the repeated calls by some of the allies, particularly the French, for a new Bretton Woods conference that presumably would have had to deal with both trade and finance.

Despite the objections, the United States decided in mid-April to go ahead with the first meeting. It took place in mid-May in connection with the OECD Ministerial Meeting in Paris. Beforehand trade and finance ministers met separately in late April. Trade ministers from the United States, Canada, the EC, and Japan gathered at the end of April for one of their so-called quadrilateral sessions. These sessions had begun in the early 1980s as a way to energize the GATT process. The quadrilateral countries saw themselves as an informal leadership bureau for the trading system and in 1983 began to meet periodically with selective developing countries as well, reflecting the progressive widening of the trading system to include key developing countries. It would have been a stroke of good fortune for the world economic system if these trade discussions had been linked up informally and in an ad hoc fashion with the much more elaborate apparatus that fostered discussions among finance ministers. The various financial groups (G-5, G-10, etc.), as well as the IMF Interim Committee and the IMF–World Bank Development Committee, had all met again at the end of April, as they do semiannually, to deliberate financial issues. The April G-5 meeting, as we noted earlier, focused on the exchange market intervention study, and the Interim Committee finalized approval of the IMF quota increase that had been agreed upon in January (see Chapter 7).

Thus, an opportunity existed to revitalize trade and trade–finance linkages. The communiqué from the OECD Ministerial Meeting on May 9–10 made not only the perfunctory appeal to halt protectionism but actually charged the industrial countries "to reverse protectionist trends," particularly with respect to measures introduced during the recent recession. It "invited the [OECD] Secretary-General to propose appropriate follow-up procedures" to implement this commitment.

The joint session of trade and finance ministers met immediately after the OECD Ministerial Meeting. Although the discussion was frank and helpful, it did not go well politically. The French refused to attend the meeting, and at the

OECD meeting the day before, had initiated an alternative proposal to give the OECD and the smaller countries a more active role in summit preparations. The French seemed to be engaged in a tactical move to reduce the significance of the Williamsburg Summit (Putnam and Bayne 1984, p. 182). From Mitterrand's call in early May for a new Bretton Woods conference to a summit meeting of socialist governments from seven countries held in Paris in late May, the French government used alternative fora to counter its isolation as the only major socialist-ruled country at the "capitalist celebration" coming up at Williamsburg (*NYT*, May 20, 1983). The effect of all this was to inhibit the initiatives at Williamsburg that might have resulted in a significant revitalization of the trade and finance systems.

At the final preparatory session for Williamsburg in La Celle St. Cloud, France, on May 11–13, the summit sherpas drew a link between trade and finance, but it was a retrogressive link for trade. The French delegation insisted on a link in the thematic paper between commitments on trade and commitments on an international monetary conference that France regarded as the key to both the exchange rate issue and the debt problem. The link itself was noncontroversial. That is what the United States and the others sought as well. But the specific link agreed to at La Celle St. Cloud set the trade round back a few notches and gave new life to the flickering idea of an international monetary conference.

The final thematic paper approved at La Celle St. Cloud invited the trade ministers to "define the conditions for improving the open multilateral trading system, including trade between developed and developing countries, and to consider the possibility of more frequent Ministerial meetings in the GATT to maintain urgency in the process." It then invited the finance ministers to "define the conditions for improving the international monetary system and to consider the part which might, in due course, be played in this process by a high-level international monetary conference." The identical wording of these commitments implied that the trade and monetary talks were at the same stage. In fact, preparations for the trade round, although still modest, were light-years ahead of the banalities that had been spouted by both the French and U.S. delegations about a possible Bretton Woods II monetary conference.

The Williamsburg Summit itself significantly upgraded the commitment on trade, reflecting the fact that the thematic paper was informal and nonbinding (in contrast to a prenegotiated communiqué). It committed the summit countries to "continue consultations on proposals for a new trade negotiating round in the GATT." This was the first mention of the new round in summit commitments. After Williamsburg, the issue would be when, not whether, a new round should begin. Nevertheless, the Williamsburg commitment actually weakened the stronger OECD commitment on trade, agreed to 3 weeks earlier. It added the qualifier "as recovery proceeds" to the OECD commitment to reverse protectionist trends. And it failed to endorse the role of the Secretary-General of OECD to follow up the OECD commitment.

In the area of finance, the Williamsburg communiqué adopted the same language as that of the thematic paper. The French, with little else to disguise their isolation at Williamsburg, insisted that the summit echo Mitterrand's call for a new Bretton Woods international monetary conference. The Williamsburg commitment inspired yet another study of international monetary reform, this one by the G-10 finance ministers, which would be issued in summer 1985 (see Chapter 9). Meanwhile, the summit communiqué said nothing about the more practical idea of future meetings of trade and finance ministers, although the thematic paper had called for continuing consultations in this area.

For the next 2 years the United States would struggle to secure a specific date for the new trade round. It would succeed in late 1986, but by that time unattended budget and trade deficits in the United States had triggered a highly visible U.S. diplomacy to intervene in exchange markets to bring the dollar down — without the fundamental domestic policy changes that the intervention study had said must accompany intervention if the latter is to have lasting effects. The United States would slip back again, in the fashion of the 1970s, to fiddling with the financial economy — exchange rates, interest rates, and monetary policy — rather than fixing the real economy — domestic budget and trade deficits and structural rigidities.

Nevertheless, none of this was evident in June 1983. Williamsburg concluded as a great success for President Reagan, and in many respects, as this analysis suggests, it was. (There were, of course, continued criticisms of U.S. policies at Williamsburg but without the passions expressed at Ottawa and Versailles). But Williamsburg also could have been more. The United States and the French, still the two main antagonists as they had been at Ottawa and Versailles, did not serve their own best interests by their reluctance to move ahead — the French on convergence and trade, the United States on intervention and finance. The United States took too much comfort from its stronger economic position, and France took too much satisfaction from its diplomatic panache. The United States was not aggressive enough, given its real economic situation; the French were too aggressive. The opportunity to restore the premises of efficient economic policies was missed. This opportunity would crest sometime during 1983 and 1984 and recede steadily after 1985.

The London and Bonn Summits

At the London and Bonn summits in 1984 and 1985 the administration kept its eye on the right targets but failed to make significant progress. At London in June 1984, the United States made a big push to get a starting date for the new trade round. The administration still perceived trade to be the leading edge of structural change in the world economy. But the desire not to rock the boat during the President's reelection campaign took some of the zeal out of the American campaign.

In addition, American trade policy continued to be suspect, both because of underlying U.S. macroeconomic policy and because of U.S. trade laws. In 1984 the administration was pushing several specialized trade initiatives on the Hill, including authority to negotiate a free trade agreement with Israel on a fast track basis (meaning an up or down vote on the agreement as a whole) and an import relief program for the carbon steel industry (Lande and VanGrasstek 1986, pp. 11–12). The carbon steel industry wanted a worldwide quota arrangement, such as the one that existed for textiles since 1971. In fall 1984, under election pressures, the administration ordered the negotiation of voluntary export restraints with leading debtor country steel exporters, to go along with similar arrangements that already existed with the EC and Japan (Destler 1986, pp. 127–131). It was a big blow for the new trade round, especially the idea of a North–South round to ease the debt crisis through trade liberalization. Steel, like textiles, was precisely one of the low- and intermediate-technology industries in which the NICs had comparative advantage. Now, even before the new round started in which the LDCs were supposed to be integrated in a more active way than ever before, the United States was taking the steel exports of these countries hostage. However hard the United States pushed for a new round to liberalize agriculture and services, it contradicted this posture by its trade priorities at home.

The administration never initiated legislation for new fast-track authority to participate in the new multilateral trade round (the old authority having expired in 1982); and although Congress eventually passed a reasonable trade bill in late 1984, thanks in no small part to the personal efforts of William Brock, the U.S. Trade Representative and former Senator, that bill provided negotiating authority only for bilateral, discriminatory agreements, while it further eased conditions for imposing import relief against dumping and subsidies and contained provisions to limit specific products, such as steel (Lande and VanGrasstek 1986, pp. 12–23; Destler 1986, pp. 78–83). The administration's policy did not reflect the position of a self-confident world leader with a view toward the overall well-being of the trading system but looked more and more like a tough bargaining strategy to implement narrow U.S. self-interests.

Of course, Europe and Japan, as well as key developing countries, such as Brazil and India, get no special award for their sullen attitudes toward trade in 1984. However ambivalent U.S. trade policy was, the United States was at least willing to risk an early trade round. The others were not. The French, in particular, stuck with their linkage between trade and money. Until U.S. interest rates and the dollar were brought down, the French vetoed any decision to launch the new trade round. The French and the Germans also stonewalled suggestions to work aggressively on agricultural trade. Brazil and India rejected liberalization of services. For a world in trouble with a hopeful recovery under way, it was a poor performance, to say the least.

The obstacles were not in trade policy, however; they were in the stalled domestic reforms, particularly in the United States, and in severe international imbalances in macroeconomic policies that emerged in 1983–1984. In early 1985 there were signs of hope. The administration seemed to want to address budget cuts (see discussion in Chapter 9), and the preparations for the Bonn Summit in May 1985 struck the right themes of domestic structural adjustment (i.e., removing microeconomic rigidities) and international trade.

In a speech at Princeton University in April 1985, Secretary of State George Shultz made the most comprehensive and persuasive statement to date of the administration's first-term international economic policy. He made it clear that a more active international economic diplomacy, which he had charged the administration to adopt in 1981 even before he joined the administration (see Chapter 7), was a supplement to, not a substitute for, aggressive domestic policy reform. He warned that "intervention in exchange markets addresses only the symptoms of the dollar's strength—and not at all successfully." Growth, he added, was a "result of the interaction of sound national policies." Thus, the United States must cut budget expenditures "*now*," Europe "should reduce . . . obstacles to change and innovation," Japan should "liberalize the Japanese capital markets . . . to channel Japanese savings more efficiently to both foreign and domestic uses," developing countries should concentrate on structural adjustments, and "all nations should support freer international trade" (Shultz 1985). It was the clearest and ironically the last domestic policy-oriented statement of U.S. international economic diplomacy in the Reagan administration.

The choice-oriented approach was reinforced in many ways by the study the OECD released in early 1985 culminating its decade-long work on structural adjustment (published in 1988; see references). This study deemphasized the differentiated, cyclical, fine-tuning approach to international economic policy coordination advocated by OECD experts in 1983 and stressed instead fundamental labor, capital, and product market reforms.

Thus, the themes of the original choice-oriented approach of the Reagan administration, which emphasized domestic reforms and market-based diplomacy, peaked at the Bonn Summit. As the *New York Times* reported, the "Bonn summit conference put its heaviest emphasis on making economic systems more adaptable and flexible." It "made clear that the governments represented . . . were particularly concerned about obstacles in the labor markets" (*NYT*, May 6, 1985). European unemployment had grown from 5.9 percent of the work force in 1979 to 10.9 percent of the work force in 1984 (OECD 1987b, p. 191). OECD studies suggested that this imbalance was far more than the passing consequence of cyclical economic conditions (OECD 1988). European labor and capital markets were too fragmented and overregulated. Eurosclerosis became a popular, if somewhat imprecise, term to describe these conditions (*WSJ*, July 9, 1985; *FT*, May 26, 1985). A more aggressive President of the European Economic Commission (the executive body of the European Community), Jacques Delor, the former Finance Minister in France, announced a new program to attack regulatory problems and to integrate the internal market in Europe. Between 1985 and 1992, the Community planned to address some 300 separate areas of economic policy, removing administrative and nontariff barriers within the Common Market, including financial markets and services, and creating for the first time a "genuine internal market without barriers" (*NYT*, May 6, 1985).

The other prominent issue at Bonn was trade. This time, the United States went all out to secure a precise date for launching the new round (see *Fortune*, March 18, 1985). It did not shy away from visible conflict on this issue. The headlines coming out of the Bonn Summit highlighted the differences: "Mitterrand balks at Reagan's plan for trade talks," reported the *New York Times*

(May 3, 1985); "U.S. fails to get summit backing for new global trade talks in 1986," lamented the *Wall Street Journal* (May 6, 1985); "France blocks agreement on start of trade talks," stated the *Washington Post* (May 5, 1985). The U.S. effort this time was solidly supported by others. Chancellor Kohl met repeatedly with Mitterrand to persuade him to accept a firm date, as did Reagan just before the negotiation of the final communiqué (*WSJ*, May 6, 1985). The initiative failed, perhaps inevitably, given the underlying macroeconomic conditions and the baffling Cabinet changes the U.S. administration made in early 1985, dispatching its well-known and highly respected trade minister to the Labor Department (see Chapter 9). But the initiative did not lack diplomatic zeal. Had this kind of effort been made in London in 1984, it might have succeeded.

The failure to get a date for the new round was a "political sour note" for the United States (*WSJ*, May 6, 1985). Despite that fact, however, the final communiqué did something rare. The other leaders isolated the French position by noting that "most of us think that [the start of a new round] should be in 1986." In diplomatese, as the new Secretary of the Treasury, James Baker, pointed out, "that is progress, believe me" (*WSJ*, May 6, 1985). The French did not win much sympathy for their obstinacy. The *Washington Post* editorialized about "those exasperating French" (May 6, 1985). In the end, the French did agree to convene, and the communiqué called for, a "preparatory meeting of senior officials [to] take place in the GATT before the end of the summer." This meeting occurred, and as it turned out, the new round began in September 1986, but without much forward momentum.

The Bonn communiqué seemed to express new resolve to address stalled domestic policy reforms—still uninitiated structural reforms in much of Europe and lopsided fiscal accounts in the United States (for text of Bonn Communiqué, see *NYT*, May 5, 1985). It did something again that had never been done before. It had each leader outline in some detail the domestic policy steps his or her country was committed to take. The steps, unlike the measures adopted at the Bonn Summit in 1978 or the stimulative, interest rate initiatives the United States undertook in 1986 and 1987, aimed at medium-term, not immediate, results and concentrated on enduring structural changes rather than on macroeconomic quick fixes. The second Bonn Summit in 1985 represented, in effect, a package deal of medium-term steps, comparable to the package deal of short-term macroeconomic stimulus adopted at the first Bonn Summit in 1978. From the point of view of a domestic policy-oriented approach, Bonn II represented progress; but Bonn II was quickly eclipsed by a sudden, new policy reorientation in Washington.

The Last Hurrah for Domestic Policy Reform

It was ironic that, just as Secretary of State George Shultz set forth the first comprehensive intellectual rationale for the new choice-oriented perspective of the first-term Reagan administration in April 1985, Washington was preparing to abandon the emphasis on domestic policy adjustments and shift the focus once again to exchange rate and international financial initiatives. The attempt to recapture the disciplined domestic and trade policy orientation of the original Bretton Woods system was discontinued in midstream, and U.S. economic diplomacy from 1985 to 1987 shifted in directions that looked more and more like a rerun of the failed policies of the late 1960s and 1970s — massive fiscal deficits, an easing monetary policy, a declining dollar, and more concern about international financial flows and new debt initiatives than domestic and trade policy reforms.

9

Retreat 1985–1989:
International Cooperation
Without Domestic Adjustment

As we discussed in Chapter 2, structuralist studies of international economic policy relations stress the role of international cooperation and institutions
to cope with domestic and international circumstances that are thought to be
beyond the control of domestic policy. In 1985, Reagan policies shifted in
structuralist directions. The value of the dollar, it was declared, was out of line
with the domestic fundamentals in industrial countries (i.e., did not reflect
substantial domestic reforms that had already taken place in these countries)
and could be brought down only by an active process of international cooperation to intervene in world exchange markets. Similarly, the world debt problem
was beyond the control of IMF programs and domestic reforms in developing
countries, which it was said had produced little more than austerity, and could
be solved only by fresh injections of new finance from international institutions and commercial bank syndicates. In fact, both the dollar and debt initiatives by the United States had less to do with following up already accomplished domestic policy reforms in either the industrial or developing countries
than accommodating the lack of political will to undertake or complete such
reforms. For the next 4 years, domestic fiscal, structural and trade policy
reforms slowed and, in some cases, moved backwards, while monetary stimulus pushed the world economy toward the limits of noninflationary growth and
opened up increasing policy conflicts among industrial, as well as developing,
countries.

If U.S. international economic diplomacy was insufficient to supplement
and consolidate domestic policy reforms in 1983–1984, it increasingly substituted for such reforms after 1985. The Reagan administration, following its
landslide victory in November 1984, had its best chance ever to complete its
domestic policy–oriented approach. This approach called for reducing government expenditures, maintaining a sound money policy, and seizing the initiative for freer trade without the handicap of massive budget deficits and associated dollar misalignments. Instead the administration failed to address

seriously any of the major expenditure programs (entitlement or defense), lost the initiative to Congress on both budget and trade policy, and encouraged a sustained easing of monetary policy in 1985–1986 to bring the dollar down.

The administration's refusal to consider tax increases was not the critical failing in this period. If, as the discussion in Chapter 8 suggests, a case can be made that the tax cuts contributed to substantial new investment in the early stages of the Reagan recovery, it would have made no sense to repeal these measures to reduce the budget deficit. In that case, new tax burdens would have slowed investment and growth, where budget deficits may have currently done so. The critical failing was the decision to ignore expenditure issues, because that meant, in the absence of tax increases, continuing large, indefensible budget deficits. Although the deficits were not the unmitigated albatross for the economy that some critics claimed, they certainly chipped away at private sector resources and capital formation over the longer run (Stein 1984, p. 289). Hence, they were not tolerable indefinitely, even if no one could predict exactly how long they could continue without evident economic harm.

Whether 1985 was the right moment economically to reduce the deficit may be debated, but it was clearly the best time politically to do so. (For an analysis that argued this point at the time, see Nau 1984–1985.) The recovery was still strong, and the politics were on the side of the administration. The President faced no further reelections and was thus free to make, if necessary, politically difficult budget decisions.

Inexplicably, the administration lost its nerve. It shuffled players in a way that suggested it had no stomach to complete the domestic economic reform program. The Secretary of the Treasury, who had spent his career in financial markets and knew much less about Washington politics, became the President's chief of staff. The President's chief of staff, who knew a lot about Washington politics and much less about financial markets, became the Secretary of the Treasury. The U.S. Trade Representative, who had labored long in 1983 and 1984 to launch a new trade round as part of a more basic solution to world debt problems and who almost singlehandedly nurtured a respectable trade bill through Congress in the fall of 1984 that strengthened the administration's negotiating hand, left to become Secretary of Labor. No one replaced him for 6 months, and protectionism mounted rapidly in the vacuum on Capitol Hill.

The personnel shifts reflected no apparent plan for the direction of U.S. international economic policy.[1] The new leadership in the Treasury Department did unleash a more active process of U.S. economic diplomacy than had existed in the first-term Reagan administration. Critics would acclaim this activity as a dramatic reversal of the neglect of foreign economic policy in the first term. But, in content, Reagan economic diplomacy in the second term was less consistent and integrated than it had been in the first term.

In the first term, the administration began on the right foot, giving priority to domestic reforms. It struggled but failed in 1983 and 1984 to complement

[1]Both Regan and Deaver confirm that the shifts were purely personal and procedural (Regan 1988, pp. 220–229; Deaver 1987, pp. 130, 134).

domestic reforms with a sufficiently integrated and aggressive international perspective linking domestic policy convergence, exchange market intervention, and long-term liberalization of international trade and financial markets. Now, in the second term, the administration fragmented these elements once again and reversed the emphasis. It relatively ignored domestic budget and trade policy reform, except in the narrow area of a revenue-neutral tax bill; it severed completely the link between trade and debt; it subordinated multilateral trade initiatives to highly combative unilateral and bilateral trade actions; it engaged in coordinated exchange market intervention to combat protectionism and lower the dollar but in the process encouraged cyclical monetary, rather than medium-term structural, adjustment of the U.S. and foreign economies; and it held out the prospect of greater finance to stimulate growth in the heavily indebted countries but simultaneously undercut cyclical policy reforms administered by the IMF and failed to launch convincing structural reforms through the World Bank, all of which dissuaded commercial banks from putting up the promised new money.

By the end of the Reagan administration, most of these initiatives had ground to a halt. The exercise to lower interest rates ended in early 1987, when it became necessary to stop the slide of the dollar and firm up U.S. interest rates. From this point on, the exercise to align exchange rates became more difficult, as markets doubted whether internal adjustments were adequate to maintain the existing level of the dollar. As a result, the dollar fluctuated — down after the stock market crash in October 1987, then back up about 15 percent in mid-1988, down again by the end of 1988 to postcrash levels 1 year before, and finally back up another 20 percent in mid-1989. The U.S. trade balance improved somewhat, partly because of delayed effects of exchange rate realignment, domestic restructuring in Japan, and a one-time reduction in the U.S. budget deficit due to tax reform. However, Congress passed a protectionist trade bill in August 1988, which led in early 1989 to serious confrontations with allies over the Super-301 provisions; the Uruguay Round of multilateral trade negotiations moved forward grudgingly in April 1989 after a failed mid-term session in December 1988; and the United States became the world's largest debtor nation. Debtor developing countries, Congress, and America's allies now increasingly called for outright debt relief to reverse the perverse flow of capital from developing countries to the United States and other industrial countries.

In the late 1980s, therefore, U.S. international economic policy initiatives and credibility sagged. The resulting mood fueled the gloomy prognosis that America's power and leadership had irreversibly declined — a mood captured by Paul Kennedy's best-selling book *The Rise and Fall of the Great Powers* (1987).

If America's power in 1989 was in decline, it was not inevitable or irreversible. The argument here is that American policy once again was significantly contributing to this decline. Second-term Reagan administration policies, instead of resolving outstanding first-term problems, such as the budget and trade deficits, pursued domestic and international initiatives that disguised and potentially exacerbated these problems. Tax reform in 1986, for example, did nothing in the long run to close the budget gap and achieved worthwhile cuts in personal income tax rates, which spur consumption, only by repealing industrial tax cuts that probably had something to do with the strong surge of investment and productivity in the early 1980s (see Chapter 8). Yet exchange rate realignment, the second-term administration's priority international initiative, could work only if the United States reduced domestic consumption and increased domestic investment. Domestic and international initiatives, therefore, seemed to be working at cross purposes.

To be sure, a greater willingness to intervene in exchange markets in late 1985 was appropriate, particularly given the extent of convergence of industrial country performance around low inflation that had been achieved by that time. But the public drama associated with this initiative was completely unnecessary and ultimately detracted from fiscal policy reform in the United States and disguised a potentially dangerous expansion of the money supply both at home and abroad. It bred the illusion that exchange rate realignment alone could alter trade account imbalances and that short-term macroeconomic, especially monetary, stimulus was a better way to achieve exchange rate changes than more fundamental fiscal, microeconomic, and trade policy reforms.

Realignment of exchange rates alone cannot alter fundamental trade and current account imbalances. That requires changes in the levels of domestic investment and savings, which in turn requires changes in fundamental domestic policies (see Chapter 1). The administration's initiatives were appropriate as a complement to such domestic changes in both the U.S. and foreign economies. But after 1985 these initiatives seemed to substitute for domestic policy changes and deal only with the symptoms of a lack of adjustment, namely, virulent protectionism in the United States and structurally retarded growth in Europe and Japan, rather than their underlying causes in the U.S. fiscal deficit and the allies' microeconomic and trade policy restrictions (see further discussion).

Thus, the second-term Reagan administration frittered away the economic capital it had stored up in the recovery of 1983–1984 to pursue partial and domestically unsupported foreign economic policy initiatives. It rode the crest of economic expansion as long as it could to reform the tax code and lower the dollar. But it left in place the investment–savings imbalance. And by the end of the decade, at the end rather than at the beginning of the business cycle, the chances were far greater that the imbalance would ultimately be resolved by a recession and less investment or by inflation and a lower cost of debt. The ultimate irony was that the United States and the world, over time, might get both recession and inflation. At least the chance existed that the Reagan administration would bequeath the world the same legacy of stagflation it had inherited.

Policy changes explain more of the developments from 1986 on than circumstances. Some critics of the Reagan administration argue that U.S. policy

changes in 1985 were inevitable, the necessary adjustments to irreversible structural trends that the first-term administration ignored (Oye, Lieber, and Rothchild, 1987). But what were these irreversible structural trends? Accelerating
inflation had yielded to policy actions in the early 1980s. So had declining
productivity, especially manufacturing productivity. Unemployment, trade
pressures, and debt had continued to mount, but these developments were
surely the consequence of policies, not circumstances, as even the critics would
assert. Moreover, exogenous changes such as the decline in oil prices in late
1985 did not occur until after U.S. policies had already shifted.

Nor is it easy to argue that domestic politics decisively constrained policy
choices at the beginning of the second term. Commentators in the period tried
to separate the President's popularity from the President's policies, arguing
that the elections, especially for Congress, showed no mandate for the administration's first-term program. But congressional elections are not based primarily on programmatic issues. After a narrower victory in 1981, the administration
put together a supporting coalition in a divided Congress, to implement far-
reaching initiatives. The second-term administration had a comparable chance
to do the same in 1985. That it chose not to do so was a voluntaristic choice. In
the end, policy decisions not circumstances or politics played the decisive role.

Domestic Reform Stalls in the United States

In spring 1985 the administration took one more, rather weak crack at reducing
budget expenditures and then abdicated fiscal policy to Congress and the
Gramm–Rudman–Hollings process. In September it attacked the trade deficit
through a more assertive and potentially protectionist bilateral policy but also,
for the first time in postwar history, left the drafting of trade legislation for a
new round of multilateral trade negotiations to the Congress. In 1986 it concentrated on revenue-neutral tax reform. And in 1985 and 1986 it relied primarily on monetary policy to bring the dollar down and help alleviate, at least
temporarily, the budget and trade deficits that it left festering on Capitol Hill.

Ducking Budget Bullets

The recovery and election campaign of 1984 had sealed the President's mind
against a tax increase. But in theory he was still committed to major expenditure cuts. The Democrats, who favored tax increases, were now stronger in the
House. But Republicans still controlled the Senate, and they were ready to cut
spending in 1985, including the vast area of the budget—some 40 percent—
made up by entitlements. The strategy, according to Budget Director Stockman, was to push a "spending-cut-only package" through the Senate, "bounce
it over to the House side," which would raise taxes, and then send a compromise bill with a "decent-sized deficit reduction package" down Pennsylvania
Avenue to the White House (1986, p. 388).

The Senate voted out a deficit reduction package in May 1985 by one vote, while President Reagan traveled in Europe to the Bonn Summit. The package shaved $56 billion from the 1986 budget and some $300 billion from the budgets over the next 3 years. The cuts included $24 billion in defense expenditures and $20 billion in domestic spending, including a freeze for 1 year on cost-of-living increases (COLAs) for Social Security recipients (*WP*, May 11, 1985). The way the package was put together, however, and the way the White House supported it suggested that it would not hold. In the House the package was blown to smithereens.

According to Stockman, the expenditure-cutting effort in the Senate was heroic. As he describes it, Senators Robert Dole, the majority leader, and Pete Domenici, the chairman of the Budget Committee,

> worked the strategy all spring. Day after day, we round-tabled in Dole's office, and this time it was the real thing. We marched through one program at a time, one Republican faction at a time, until we had gotten through the whole trillion dollar budget. Never before had the game of fiscal governance been played so seriously, so completely, or so broadly as it was in Bob Dole's office in the spring of 1985. Rarely before have two political leaders displayed such patience, determination and ability as did Bob Dole and Pete Domenici [1986, p. 388].

When all was said and done, however, the Senate budget cutters could muster only $20 billion cuts in domestic programs. "The awesome staying power of the Second Republic," as Stockman called the welfare state, was too much (1986, p. 388). Nevertheless, the Senate Republicans were willing to walk the plank and recommend cuts in Social Security. Politically, this was an important step, suggesting that middle-class America would finally step up and make its contribution to national solvency.

How was this "heroic" effort received by President Reagan? Reagan had promised in the campaign not to reduce Social Security benefits and in December 1984 had pledged specifically not to cut the COLAs. Now that the Senate had frozen COLAs, however, he tried to back away from this promise, but he did so with the same ambivalence and timidity that had characterized many of his actions on spending cuts. He explained: "I didn't have in my mind [during the campaign] that we were talking about potential or possible [Social Security] increases [for inflation]. But it was taken that way, and so okay I live with that." In the case of the Senate bill, he said, "I would suggest that I was faced with a mandate when 79 percent of the Senate, which means pretty much half-and-half, Democrat and Republican, demanded that we have some curbing of the COLAs" (*WP*, May 11, 1985). The most popular president in two decades, who no longer faced reelection requirements, was freezing Social Security COLAs because he was being forced to do so by the Congress. This was not the sort of leadership that might have rallied congressional supporters, defending thousands of domestic budget programs, to stand behind the President, accept collective sacrifices, and put the American budget process back on the road to sensible management.

It was little wonder, then, that when the package went to the House, it was ravaged not only by Democrats but also by Republicans. As Stockman records, "Jack Kemp joined Claude Pepper in leading the charge to save the COLAs of the old folks" (1986, p. 390). Don Regan, now in the White House and a firm believer that revenues alone would take care of the budget problem, agreed with Kemp (for Regan's views on the deficit and Social Security, see Regan 1988, pp. 174, 179, 199, and 280). In the end, "nobody was going to walk the plank on Social Security" (Stockman 1986, p. 390). COLAs came out, and the entitlements remained intact. The budget package fell apart, and Congress in desperation invoked the Gramm–Rudman–Hollings legislation. Without leadership in the White House or the Congress, the Gramm–Rudman legislation provided automatic, across-the-board budget cuts as the only means to trim the deficit and restore fiscal sanity. Even then, the Gramm–Rudman legislation would languish for a year, under challenge in the Supreme Court. It would be reconstructed in 1987 but with lower deficit reduction targets, putting the large cuts off into the political "nether world" after the 1988 elections.

It was an unfortunate abdication of presidential leadership. Ronald Reagan could have led a reasonable effort to pare back Social Security, Medicare, and pension programs without affecting the seriously needy among the recipients. He had supported substantial cuts in welfare programs for the poor. Now he backed off from cuts in entitlement programs that benefited mostly the middle class, including the elderly, who were better off than the poor. His purposes looked more cynical, imposing budget cuts disproportionately on welfare programs rather than calling on all Americans to pare back the spread of the public sector that had occurred during the 1970s. True, he had promised that he would not cut Social Security benefits. But he had also promised not to cut defense or raise taxes. So he had, in effect, abdicated any fiscal position other than one of drift.

Over the next 3 years, drift eroded all of President Reagan's first-term budget priorities. He did not avoid further political skirmishing over Social Security. Although rejected at the last minute, cuts in Social Security programs were again part of the deficit reduction talks after the stock market crash in October 1987. More important, defense spending declined in real terms after 1985 (about 12 percent in constant dollars), and tax receipts shot up significantly, from 18.4 percent of GNP in fiscal year 1986 to 19.4 percent in fiscal year 1987 and an estimated 19.1 percent in fiscal year 1989 (OMB 1989). The President got poor advice, to be sure, often from the same aides who counseled him to back off of spending cuts in 1981 (see Stockman 1986, p. 211). But he was not the victim of this advice; he got the advice he wanted. Stockman summarized aptly the President's choices:

> His stout opposition to a major tax increase obligated him to lead the fight to shrink the American welfare state's giant entitlement programs like Social Security, not just castigate "spending" in the abstract while ducking the real bullets. You couldn't have it both ways but ultimately that's how he usually came down [1986, p. 354].

Trade and Taxes

Failure to follow through on reducing the budget deficit shifted attention to the trade deficit. Congress began to see the trade deficit, not the budget deficit, as the central problem and began to call for direct measures to cut imports. The merchandise trade deficit burgeoned from $36 billion in 1982 to $122 billion in 1985 (*ERP* 1988, p. 364). Congress had been flirting for some time with reciprocity legislation to restrict U.S. markets in sectors such as telecommunications, where major trading partners denied access to U.S. products. Now, in summer 1985, three centrist Democrats on trade issues—Lloyd Bentsen in the Senate and Dan Rostenkowski and Richard Gephardt in the House—introduced legislation to mandate a surcharge on countries running heavy trade surpluses with the United States. The obvious offenders included Japan, Korea, Taiwan, and Brazil. Other protectionist legislation on textiles, shoes, copper, and other products was pending.

In this highly charged situation, the administration operated for 8 months with no trade representative or trade strategy. If the administration's trade policy in the first term had been ambivalent and weak, it was at least understandable in terms of the focus on domestic budget reform. But, at the beginning of the second term, as I. M. Destler notes, "its [trade policy] behavior . . . can only be described as bizarre" (1986, p. 104). In much less favorable political circumstances in 1973, the Nixon administration had used scarce political capital on the Hill to secure negotiating authority for the Tokyo Round on multilateral trade negotiations (see Chapter 6). In 1985, despite its international efforts to launch the next round of trade negotiations (see Bonn Summit discussion in Chapter 8), the Reagan administration did not ask Congress for new negotiating authority and even refused to cooperate with Congress when the latter began to draft trade legislation on its own. The administration turned instead to exchange rate initiatives to bring the dollar down and aggressive bilateral trade negotiations to open foreign markets on a discriminatory basis (e.g., free trade agreements) or close U.S. markets if necessary through retaliatory measures. In September the President appointed Clayton Yeutter as his new trade representative and announced a new "get tough" trade strategy with foreign competitors (*FT*, September 24, 1985).

Both expenditure and trade policy reforms thus went into neutral, while tax reform took center stage. The Economic Recovery Tax Act of 1981 (ERTA) had reduced personal and corporate income taxes and accelerated depreciation allowances for capital expenditures (the ACRS system). Tax legislation in 1982 and 1984 (known as TEFRA and DEFRA) had restored some corporate taxes but generally left the investment incentives of the 1981 legislation in place (Niskanen 1988, Chapter 3). Tax reform in 1986 simplified and reduced further tax rates on personal income but raised substantially effective tax rates on business and investment, essentially eliminating the incentives of the 1981 law (Niskanen 1988, pp. 101–106).

Thus, uncertainties and provisions associated with tax reform in 1986 contributed to a sharp decline in U.S. investment activity. Growth of real net nonresidential fixed investment in the United States fell by almost two thirds in 1985 and turned negative in 1986 (*ERP* 1988, p. 267). Tax reforms did even out the effective tax rate across different American industries where previous loopholes had allowed some major industries to escape taxes altogether. A more equitable industrial tax system promised a greater role for market forces in investment decisions in the future, and nonresidential fixed investment picked up again in late 1987 and remained strong through early 1989. But whether the tax reform would ultimately restore earlier high levels of investment and raise the level of national savings was unclear. In 1985 and 1986 it probably reduced both investment and savings because it eliminated the 1981 corporate tax cuts that had generated substantial new savings in industry in the form of retained earnings from faster depreciation allowances.

Monetary Policy and the Dollar Decline

With fiscal policy in neutral, the only lever left for adjusting macroeconomic policy was monetary policy. Money growth, which had spurted from late 1982 to mid-1983, had slowed once again from mid-1983 to late 1984. In early 1985, however, it began to accelerate again, and this time it appeared to go into permanent fast forward. M_1, the narrow monetary aggregate, grew in 1985 and 1986 at annual rates of 12.2 and 16.6 percent, respectively, rates that were completely off the chart of previous experience with M_1 growth (*ERP* 1987, p. 319). The Fed explained that the unprecedented growth was not inflationary because the velocity of money—the ratio of the growth of monetary aggregates to nominal GNP growth—had declined dramatically with the lowering of interest rates and deregulation of financial markets. Money in interest-bearing checking accounts, which had been kept before deregulation in certificates of deposit and other instruments included in broader monetary aggregates, such as M_2, was now showing up in M_1. In fact, larger money aggregates, such as M_2, did not show the same acceleration as M_1. M_2 growth was 8.1 and 9.3 percent, respectively, in 1985 and 1986, compared to 8.6 and 12 percent, respectively, in the preceding 2 years (*ERP* 1987, p. 319). The explanation seemed convincing, especially when inflation, partly at least on the strength of oil price declines, decelerated to a new low of 1.1 percent in 1986, the lowest level since 1961. (Niskanen, for example, agrees that money growth was not excessive in this period; 1988, pp. 181–184.)

Nevertheless, compared to historical experience, money growth in 1985 and 1986 was still high, even in the case of M_2, and was considerably higher in the case of M_1. What is more, beginning in early 1985, short-term interest rate differentials between the United States and key foreign countries began to decline, suggesting a relative easing of U.S. monetary policy compared to that of its major trading partners, especially Germany and Japan (IMFa 1988, p. 126). The dollar, which is sensitive to interest rate differentials, among other factors, peaked in February 1985 and declined steadily thereafter. As Figure 9-1 clearly shows, the dollar's decline did not begin with the exchange market intervention exercise initiated by U.S. officials at New York's Plaza Hotel in

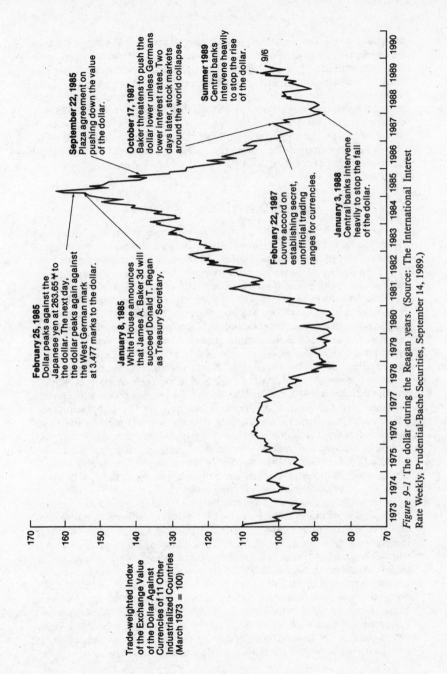

February 25, 1985
Dollar peaks against the Japanese yen at 263.65 ¥ to the dollar. The next day, the dollar peaks again against the West German mark at 3.477 marks to the dollar.

January 8, 1985
White House announces that James A. Baker 3d will succeed Donald T. Regan as Treasury Secretary.

September 22, 1985
Plaza agreement on pushing down the value of the dollar.

October 17, 1987
Baker threatens to push the dollar lower unless Germans lower interest rates. Two days later, stock markets around the world collapse.

Summer 1989
Central banks intervene heavily to stop the rise of the dollar.

February 22, 1987
Louvre accord on establishing secret, unofficial trading ranges for currencies.

January 3, 1988
Central banks intervene heavily to stop the fall of the dollar.

Trade-weighted Index of the Exchange Value of the Dollar Against Currencies of 11 Other Industrialized Countries (March 1973 = 100)

Figure 9–1 The dollar during the Reagan years. (Source: The International Interest Rate Weekly, Prudential-Bache Securities, September 14, 1989.)

September 1985. The dollar was moving in response to real economic policy developments. Whether it was responding primarily to monetary policy or to other real economic policy developments, such as the anticipated changes in the U.S. tax code (Treasury announced the first tax reform plan in November 1984), can be debated (see Niskanen 1988, p. 248). But with budget policy gridlocked, the only two economic variables that changed in winter 1985 were monetary policy and anticipated revenue-neutral tax changes. These changes in the United States and counterdevelopments abroad had as much, if not more, to do with bringing the dollar down after 1985 as the activist U.S. international economic diplomacy that began at the Plaza Hotel in September 1985.

International Diplomacy Accelerates

International cooperation to intervene in exchange markets is sometimes viewed in international economic diplomacy the same way arms control negotiations are viewed in nuclear diplomacy. Just doing it is considered to be helpful. When the United States halted intervention in exchange markets in 1981, it was the first interruption of such intervention in almost a decade. The noninterventionist policy did not sit well in diplomatic circles—finance ministries, central banks, and so on—that practiced intervention and identified it with the existence of international cooperation. But, as we noted in Chapter 1, exchange rate policy is only one element of international cooperation, and it is closely interrelated with other elements. As the interventionist study commissioned at the Versailles Summit in 1982 concluded, intervention to achieve exchange rate alignments depends, for sustained effect, on accompanying macroeconomic, structural (labor, capital, and product market), and trade policy adjustments. Without these adjustments, intervention can have no lasting consequences. Thus, international cooperation to intervene in exchange markets has to be judged in terms not simply of its existence but of its effects on related macroeconomic, structural, and trade policy changes. Does it help to achieve fundamental policy adjustments, or does it delay or distort these adjustments? In what direction, in short, does intervention influence monetary, fiscal, regulatory, trade, and other policies of the real economy?

It is easy to get lost in the blizzard of international financial initiatives that the U.S. administration launched from 1985 on to intervene in exchange markets, coordinate interest rates (in effect, monetary policy), and restructure international debt (the Baker Plan). The essential question to ask, however, is whether these initiatives helped to reduce, *on a sustained basis*, the American trade and current account deficit, the Japanese and German trade and current account surplus, and the debt-servicing burdens of the heavily indebted developing countries. Since current account deficits and surpluses are the direct consequences of savings and investment imbalances in individual countries, adjustments require an increase in savings or a decrease in investment in the deficit countries (e.g., the United States and the debtor developing countries) and the reverse in surplus countries (e.g., Japan and Germany).

The argument in this study is that U.S. policy initiatives from 1985 on achieved relatively little adjustment for all the international activity they generated and that what adjustment they did achieve was potentially counter-productive and hence unsustainable. In the United States, until mid-1987, domestic adjustment came largely at the expense of reduced investment rather than of reduced consumption. And although the U.S. budget deficit, and hence U.S. consumption, declined on a one-time basis in fiscal year 1987 and investment rebounded, the economy in 1989 came closer to full employment and the moment when the deficit and consumption would have to come down further or inflation would reignite. In Europe and Japan, adjustment until mid-1987 came chiefly through accelerated and potentially inflationary mone-tary growth. In late 1987, Japan initiated major supply-side reforms and Eu-rope stepped up its plans to deregulate the internal Common Market by 1992. Nevertheless, by mid-1989, continuing and even rising global trade surpluses in both Germany and Japan raised doubts about the adequacy of these internal reforms. Meanwhile, heavily indebted developing countries either severely re-stricted and subsidized external trade to meet debt servicing obligations (e.g., the Latin American debtors) or pegged their currencies to a declining dollar to sustain exports and avoid internal restructuring (e.g., the case with Korea and Taiwan until mid-1988). Except in a few cases, such as Mexico, internal reforms lagged badly behind.

The remaining sections in this chapter detail the record of the more activist U.S. international economic diplomacy after 1985, not just in terms of the process of international cooperation, which U.S. policy accelerated (and all commentators applauded), but in terms of the relatively meager substantive results it achieved (which too many commentators ignored).

International Monetary Conference

A hint of the different direction in which U.S. policy would move was sounded already before the Bonn Summit convened in May 1985. At OECD meetings in April 1985, the new Secretary of the Treasury, James A. Baker III, apparently with little preparation or discussion among other U.S. officials, proposed hold-ing a new high-level international conference in Washington to improve the world's monetary system (*WP*, April 13, 1985). The proposal was tied explicitly to the fateful (at least from a choice-oriented perspective) language of the Williamsburg Summit. If it was intended to soften up French opposition to the trade round by maintaining the semblance of linkage between trade and money that Mitterrand had insisted on at Williamsburg, it failed. The French did not sign off on a new trade round, although they welcomed Baker's monetary proposal as a "step forward" (*WP*, April 13, 1985). In fact, the monetary conference proposal probably worked against the administration's efforts to launch a new trade round. A French spokesman in Washington explained that "there is no major trade problem." "The real problem," he said, "is the erratic variation of the currencies" (*WP*, May 6, 1985). By raising the monetary con-ference idea, the United States focused attention on exchange rates, not trade.

The French view always had supporters, especially in Washington. The Institute for International Economics, headed by C. Fred Bergsten, a former high-level Treasury official from the Carter administration, had been advocating since the early 1980s the return to a more rigid system of exchange rates—a target zone system that allowed for wider variations than the old Bretton Woods system but called for frequent intervention to manipulate exchange rates within these ranges.[2] Now, this view was spreading on Capitol Hill. Congressional supply-siders had always believed that monetary policy was the great undoing of the supply-side experiment in 1981–1982 (see Chapter 7). They wanted less discretion for monetary policy both within the United States and with respect to international management of exchange rates. Jack Kemp, a leading exponent of this position in the House, and Bill Bradley in the Senate initiated a series of private international meetings to drum up support for a more managed exchange rate system. Kemp favored a gold-based system with considerably less flexibility than the target zone approach. But despite technical differences among them, all of these spokesmen put the problem of fluctuating exchange rates above the need for an early start of a new trade round.

Congress, turning more protectionist by the day, was becoming convinced that the U.S. trade problem was really a problem of the overvalued dollar. Later, after the dollar came down, Congress would redefine the trade problem as one of unfair trade and foreign barriers to American exports. In each case Congress was looking for an excuse not to have to deal with the real problem confronting American trade competitiveness, namely, the budget deficit. The administration was too, and from the summer of 1985 on, it geared its international economic diplomacy largely to assuage protectionist pressures in Congress while avoiding the central problem of the budget deficit.

None of this is to deny that the dollar was a serious problem in 1985. It had continued to rise throughout 1983 and 1984, even after the substantial early appreciation in 1981 and 1982. In trade-weighted real terms, it was 40 percent above the levels of 1980. But there were two ways to attack the dollar problem—head on, by reducing the budget deficit and increasing American savings, or elliptically, by intervening in exchange markets and raising taxes on U.S. investment, the net effect of which was to ease monetary policy and encourage consumption while discouraging investment. Inexplicably, the administration chose the latter route. Treasury officials, it must be remembered, were operating within constraints set by the President, with respect to both cutting expenditures, particularly entitlements, and raising taxes. Thus, the ultimate responsibility belongs to the President. Treasury officials were practicing the "art of the possible," given the administration's abdication in 1985 on both the budget and trade deficits (Funabashi 1988, p. 75).

With its proposal for an international monetary conference, therefore, the administration signaled a move away from the domestic and trade policy reforms it had emphasized in the first term toward exchange rate and financial

---

[2]Bergsten held the unique position of also having been a principal advocate of floating exchange rates in the 1970s. See his book (1975) and Chapter 5.

initiatives, which, although they were premised on further domestic and trade policy reforms, would not wait for such reforms. These initiatives would capture the imagination of the economic policy community for a while, inclined as this community was to equate exchange market intervention with international cooperation. But without domestic and trade policy reforms, the initiatives soon lost their credibility and accomplished, at best, only marginal improvements in correcting real economic imbalances.

Plaza — The Intervention Initiative

From the first through the third quarter of 1985 the dollar had already depreciated on a real, trade-weighted basis by 10 percent. In nominal terms, it had depreciated by 15 percent against the mark and 8 percent against the yen (*ERP* 1987, p. 365). The Plaza Accord in September 1985 did not initiate this turnaround of the dollar (see Figure 9-1 earlier in this chapter). As we noted earlier, that was done by a shift in U.S. monetary and perhaps tax (i.e., anticipated tax changes) policy in late 1984 and early 1985, accompanied by complementary monetary policy changes abroad (reflected in the German Bundesbank's rather large intervention in exchange markets in February and March of 1985; Funabashi 1988, p. 28). The Plaza Accord rode the dollar down; it did not drive it down. Appropriate fiscal and monetary policy changes could have accomplished the same thing without intervention. Intervention was important, primarily as a political device to pressure central banks to ease monetary policy, particularly given the reluctance of finance ministries to adjust fiscal policies. Intervention concealed certain domestic policy choices—in this case easier money to sustain the recovery in the face of fiscal gridlock. These choices became more explicit as the intervention initiative led to direct coordination of interest rate reductions (see next section).

Thus, the G-5 countries met at the Plaza Hotel in New York on September 22 to encourage an event that was already under way, not to cause it. The long and unusually wordy communiqué for finance ministers suggested that the ministers had agreed to realign exchange rates to bring them more in line with fundamentals, that is, basic macroeconomic and structural policies. The ministers believed, according to the communiqué, "that recent shifts in fundamental economic conditions among their countries, together with policy commitments for the future, have not been reflected fully in exchange markets." Policy adjustments, in short, had already taken place, yet exchange rates somehow remained out of line simply because of speculative or other mysterious reasons. The reasoning was confusing. What recent shifts in fundamental economic policies had occurred? The U.S. budget deficit rose in 1985 from $185 billion to $212 billion (and would rise again in fiscal year 1986 to $221 billion). If the Plaza agreement was an attempt to use intervention to help correct these fundamental fiscal policy imbalances, the communiqué was deliberately deceiving; if, on the other hand, the communiqué was correct and fundamentals were already in place, Plaza was merely an exercise to perpetuate existing policies (i.e., the large U.S. budget deficit). In fall and winter of 1985-1986, Plaza did little more than push U.S. domestic policies in directions they were already going—that is, larger budget deficits and easier money—and pressure other countries to follow suit, especially to ease money.

In 1985, growth of real total domestic demand in the United States declined substantially, from 8.7 percent to 3.8 percent, whereas it remained approximately the same in Japan (3.8 percent in 1984 and 4.0 percent in 1985) and actually declined in West Germany (2.0 percent in 1984 and 0.8 percent in 1985). Total domestic demand in the European countries as a whole remained about the same (IMFa 1989, p. 126). Slower U.S. growth was not offset by higher growth abroad and U.S. exports failed to respond to the lower dollar. The dollar came down largely as a result of a unilateral slowdown of growth in the United States.

Monetary policy in the United States eased in part to combat this slowdown. But its net effect in 1985 was not to lift growth but to shift it from investment to consumption. The strong investment-led character of the initial U.S. recovery petered out (see Figures 8-1 to 8-3). Real gross fixed investment in the United States slumped from 16.8 percent annual growth in 1984 to 5.5 percent in 1985. Meanwhile, real public and private consumption jumped from 9.2 percent in 1984 to 11.9 percent in 1985 (IMFa 1988, p. 113). Recovery was being sustained largely by stimulating short-term demand through the growing fiscal deficit and an easing of monetary policy.

In the industrial countries as a whole, real private and public consumption accelerated from 6.1 percent annual growth in 1984 to 8.0 percent in 1985, while real gross fixed investment declined from 8.7 percent annual growth in 1984 to 4.4 percent in 1985 (IMFa 1988, p. 113). Throughout the industrial world, therefore, growth in 1985 not only slowed but also shifted relatively from long-term investment to short-term consumption.

The Plaza Accord was hailed as a revival of international cooperation, but in fact it masked a sharp deterioration of domestic economic policies, on which sustained cooperation depended. Adjustment in the United States was taking place as a result of reduced investment rather than increased savings. The United States was no longer a smart borrower, as it had been in 1983–1984, importing capital to finance unprecedented levels of investment with which to repay future debts. It was now borrowing primarily to consume rather than to produce to repay future debts. The deterioration of domestic policy conditions showed up in the trade balance.

By the end of 1985, although the dollar was down now by 17 percent in real trade-weighted terms (in nominal terms, 20 percent against the mark and 24 percent against the yen), the U.S. merchandise trade deficit was still climbing in value from $112 billion in 1984 to $124 billion in 1985 (*ERP* 1987, pp. 365, 358). Part of this was due to the so-called J-curve effect, the fact that currency values change more quickly than trade volumes and thus in the short term a lower dollar raises the value of the same amount of imports while lowering the value of the same amount of exports. But, more important, part of it, as developments in 1986 would show, was due to a failure to adjust domestic policies to support the depreciating dollar. As the United States had argued repeatedly in the case of IMF-supported stabilization programs in developing countries, a currency depreciation required a shift of resources from domestic markets to exports. This generally meant macroeconomic tightening—restraining the growth of credit and reducing the size of fiscal deficits. Now, in its own case, the United States was doing exactly the opposite, expanding the money supply while maintaining a massive fiscal deficit. If the Plaza Accord is judged by real economic standards rather than political criteria of activist cooperation, the exercise was off to a very poor start.

Unwilling to reduce domestic demand directly by lowering the deficit or maintaining a disciplined money policy, the United States had two choices— accept a lowering of U.S. demand indirectly through slower growth, which happened in 1985, or turn to badgering America's allies to accelerate domestic demand abroad, even at the risk of abandoning medium-term structural reforms. This is the direction in which America turned in 1986, pressing for short-term monetary policy stimulus packages of the sort that characterized international cooperation in the 1970s.

Tokyo Summit—The Interest Rate and Economic Indicators Initiatives

By the end of 1985, Secretary of the Treasury Baker was impatient with the lack of results (Funabashi 1988, p. 43). The Plaza initiative had involved coordinated intervention in exchange markets but not direct coordination of monetary and hence interest rate policies of central banks.[3] As a result, in late October 1985, the Bank of Japan had actually raised its discount rate, a move consistent with support for the decline of the dollar but inconsistent with the stimulation of domestic demand in Japan. Baker now sought to achieve both objectives—to lower further the value of the dollar and to stimulate foreign demand. The only way he could do this was to press for easy money both in the United States and abroad. Interest rate reductions in the United States alone might precipitate a free fall of the dollar, while interest rate reductions abroad alone might prevent the dollar from falling further. Coordinated interest rate reductions, on the

---

[3]Exchange market intervention and monetary policy are separate policy variables and may be controlled in some countries, such as the United States and Germany, by separate institutions. Nevertheless, intervention may have direct consequences for monetary policy unless it is deliberately offset or sterilized by central banks. In the United States, for example, Treasury orders sales or purchases of dollars on foreign exchange markets to influence the exchange rate of the dollar. The Federal Reserve Board can offset these transactions and maintain existing monetary policy through open-market operations, which it controls independently of the Treasury. While the Plaza Accord in September 1985 did not deal explicitly with monetary policy, Funabashi reports that Treasury officials undertook more active intervention in the expectation that the Federal Reserve would not sterilize foreign exchange transactions through open-market operations but would in effect allow an easing of U.S. monetary policy (1988, pp. 35, 45–46). Other countries, such as Japan, combined intervention in fall 1985 with a tightening of monetary policy, a move that angered Secretary Baker and U.S. officials and led to the subsequent effort in January 1986 to coordinate the reduction of interest rates as well, ensuring that the allies too would ease monetary policy. See further discussion in text.

other hand, would keep the dollar coming down while also stimulating foreign demand. The consequence of these Treasury policies during the course of 1986 was to open up a wider rift between the Treasury Department and the Federal Reserve on U.S. monetary policy and to shift the focus from structural reforms to macroeconomic stimulus abroad. Both developments weakened prospects for sustained noninflationary growth over the longer run.

The G-5 met again in London in January 1986. They expressed satisfaction with the realignment of exchange rates but disagreed on whether to push re-alignment further through a coordinated reduction of interest rates (*FT*, January 20, 1986; *NYT*, January 25, 1986). Japan and West Germany felt the German mark and yen had appreciated enough and called for patience to allow exchange rate changes to have an effect. The United Kingdom agreed. Japan, France, and the United States pressed for simultaneous interest rate reductions, which would allow the dollar to fall further. What is more, the highly visible G-5 exercise stirred up ill feelings among other countries, particularly the G-10 countries, whose report the previous summer had been completely superseded if not repudiated by the Plaza Accord (Funabashi 1988, p. 37). Excluded G-10 countries now clamored to be a part of the Plaza group (*WP*, January 15, 1986).

Most important, the Baker initiative began to open up a growing rift be-tween the Treasury Department and the Federal Reserve (*NYT*, January 22, 1986). Although this rift was played down in the media—largely because Baker and Volcker were both politically popular and powerful in Washington—the split was unmistakable and led to an unprecedented situation in February 1986 when Volcker was outvoted by his board on a decision to reduce the American discount rate. (Volcker almost resigned over this incident; Funabashi 1988, p. 49.)

From the beginning of the Plaza exercise, Volcker worried about an uncon-trollable fall of the American dollar (Funabashi 1988, p. 15). Hence, he re-sisted further cuts in U.S. interest rates, unless key foreign countries also cut theirs. Otherwise, he argued, foreigners would at some point cease to invest in dollars and precipitate a sharp reduction in capital inflows needed to finance America's growing twin deficits in the budget and trade accounts. Even then, Volcker worried about continuing to ease monetary policy while sustaining the large fiscal stimulus. In explaining money growth to Congress in February 1986, he noted that there was some conflicting evidence that suggested that monetary policy may already be too liberal. "Domestic demand," Volcker pointed out, "continued to expand at a rate well beyond the rate of domestic output . . . [and] in that sense we continued to live beyond our means, at the expense of a widening trade deficit" (1986a, p. 4).

Volcker's concerns would have probably been even stronger had it not been for two intervening factors in late 1985 that seemed to justify continued mone-tary growth—the passage of the Gramm–Rudman–Hollings legislation and oil price declines. Gramm–Rudman–Hollings, which promised reduction of the budget deficit, later proved a disappointment as the federal deficit soared to an all-time high of $221 billion in fiscal year 1986. The unexpected drop in oil prices, on the other hand, proved to be real. Producer prices for energy products dropped by more than 30 percent in 1986 (*ERP* 1987, p. 316). This development helped bring inflation down in 1986 to 1.1 percent, its lowest level since the early 1960s, facilitating the lowering of interest rates and continued easing of monetary policy that occurred in 1986.

Secretary of the Treasury Baker had fewer concerns about the falling dollar and more desire to use the lower dollar or the threat thereof to stimulate domestic expansion abroad. The declining dollar represented powerful market leverage over the economies of Japan and Western Europe. Just as the United States had tolerated (some say used) a high dollar to disinflate the world economy in the early 1980s, Baker now sought to use a lower dollar to reflate the world economy in 1986. As the dollar declined and other countries' currencies appreciated, these countries' exports became less competitive. Either they could accept the loss of export-led growth, which had contributed significantly to overall growth in Europe and Japan from 1983 to 1985, or they could stimulate domestic demand to offset this loss. Baker obviously hoped they would do the latter. With the dollar down and no apparent change in the U.S. trade deficit, protectionist pressures continued to build in the United States, where officials needed a quick response from abroad. Thus, the U.S. emphasis shifted from the structural reform objectives emphasized at the Bonn Summit in 1985 to macroeconomic stimulus, urging foreign countries to imitate U.S. domestic policy and lower interest rates while expanding fiscal deficits.

In the early 1980s, when the United States urged other countries to imitate fundamental macroeconomic and structural policy reforms taking place in the United States, the U.S. approach was called ideological, exporting Reaganomics. But now that the administration urged more familiar policies from the 1970s of coordinated, short-term monetary and fiscal policy stimulus, its approach was acclaimed as a "rare example of timely [international] collaboration" (the *WP* editorial reaction to the coordinated interest rate reduction of early March 1986, March 8, 1986). It is hard to avoid the conclusion that coordinated exchange market intervention is associated with a preference for expansionary macroeconomic policies and relative neglect of microeconomic reform.

Some analysts worried about a hard landing for the dollar but few criticized the new Treasury Department strategy at the time for encouraging the very policies that could lead to such an outcome. The U.S. strategy sought to stimulate foreign demand for U.S. products and services; but even if foreign stimulus was sufficient to affect significantly the U.S. current account (and some doubted this—see Martin and Kathleen Feldstein, *WP*, July 8, 1986), this effect would follow interest rate reductions only with a lag. Meanwhile, the more foreign countries lowered their interest rates, the more the United States had to lower its rates to keep the dollar falling. Thus, the United States launched a self-reinforcing process of lower interest rates and a lower dollar that could only be arrested some day by a relative tightening of monetary policy in the United States to raise interest rates and finally stop the dollar's slide. That day, whenever it came, would mean slower growth and, if it coincided with the end of the business cycle, possible recession in the United States.

Volcker clearly shared Baker's desire to stimulate foreign demand but was skeptical about lowering interest rates in the United States, given the unwillingness in the administration or Congress to reduce the budget deficit. Nevertheless, because the Federal Reserve Board was divided throughout 1986 on the issue of interest rate reductions, Baker dominated the formulation of monetary policy. In late February, the Board, led by Preston Martin, the Vice-chairman, and other new Reagan appointees, voted against Volcker to lower the discount rate.

Volcker's enormous prestige abroad enabled him to postpone the rate cut until he had negotiated coordinated rate reductions in West Germany and Japan and thereby protected the dollar. On Thursday, March 6, the West German Bundesbank lowered the German discount rate; the Bank of Japan and the Federal Reserve followed on Friday.

This was the only time over the next 2 years that the United States was able to arrange simultaneous interest rate reductions among all three key industrial countries. The outcome reflected both effective Treasury and administration pressure on Volcker and Volcker's significant influence on central bankers abroad, particularly given his severe embarrassment in March 1986 if foreign banks had not cooperated. This combination of factors never occurred again. Nevertheless, as the media reported, "Treasury officials could scarcely contain their jubilation" over the new interest rate coordination exercise. "Coordination breeds coordination," crowed one senior Treasury official. "What we did on exchange rates on September 22 has now spread to the interest rate side." "It gives us confidence," he concluded, "that these things can move" (*WP*, March 6, 1986).

When Germany and Japan balked at further efforts to lower interest rates, the United States turned from its G-5 approach to other strategies — bilateral (G-2) and broader summit mechanisms (G-7) — to achieve continued movement. In April the European Monetary System (EMS) realigned currencies (*FT*, April 7, 1986). This moved the German mark to the bottom of its EMS range and made the Bundesbank extremely reluctant to lower domestic interest rates further, for fear of driving the German mark below its floor in the EMS (*NYT*, June 7, 1986). When German output spurted in the second quarter of 1986 to 3.3 percent real growth, Germany's fear of inflation also rose and its appetite for further interest rate reductions declined even more.

After the coordinated interest rate reduction of March, the Japanese yen continued to climb. By late April, therefore, Japan was ready for further interest rate reductions. This time, only the United States and Japan reduced rates. Thus, the momentum of international economic policy coordination was already slowing as the G-5 and summit countries met at the Tokyo Summit in May 1986.

Buoyed by the March success at coordinated interest rate reductions, U.S. policymakers readied a major initiative for Tokyo to upgrade the multilateral surveillance process. Treasury officials had ignored this process in 1985, discontinuing the special G-5 meetings with the IMF director that had been initiated at Versailles and that had highlighted medium-term policy convergence and structural policy reforms at the Bonn Summit in 1985. At Plaza and in March 1986, they had focused instead on intervention and monetary policy. Now Treasury officials saw an opportunity to use the process of multilateral surveillance to institutionalize and increase the urgency of the intervention and interest rate coordination exercise. They regarded the private, confidential peer group pressure of the original multilateral surveillance process as inadequate for this purpose. As Treasury sources explained, they wanted a more public process, a "means through which the seven summit nations . . . and perhaps a few others would meet perhaps four or five times a year and negotiate goals for their domestic economies." They would announce these goals, such that if any one country deviated sharply from them, the group would call an emergency meeting and "under the glare of headlines and television lights . . . the country that deviated would face political pressure to make domestic policy adjustments to bring its economy back into line." Among the goals set by this process would be "reference zones" for exchange rates, prescribing limits within which rates would rise and fall in relation to one another. Such a process, Treasury officials conceded, would not have the force of law, since no country would relinquish its sovereignty over economic policy. "But public pressure," they confidently predicted, "could have a similar effect" (all quotes from *NYT*, March 3, 1986).

The process that was announced at the Tokyo Summit in May 1986 was called multilateral surveillance or the objective indicator exercise. It had similarities in appearance to its predecessor from the Versailles Summit, although it was considerably more elaborate and formalized. Many of the indicators were the same ones that had appeared in the Monetary Annex at Versailles and had been discussed in the preparations for the Williamsburg Summit (see Chapter 8), and they were organized in terms of policy objectives, performance outcomes, and intermediate factors, just as the Williamsburg proposals had developed short-term policy and performance indicators to measure progress toward medium- and long-term policy goals.

Nevertheless, the Tokyo indicators exercise had a fundamentally different purpose than the Williamsburg convergence exercise. It was geared more to short-term policy adjustments, especially monetary policy adjustments, and automatic commitments to negotiate direct intergovernmental trade-offs among specific policy instruments—features more characteristic of international economic diplomacy in the 1970s than the patient, market-oriented approach advanced from 1982 to 1985 at the Versailles through Bonn summits. What is more, multilateral surveillance was to be conducted now in the glare of television lights and media attention. This approach was far more intrusive on national sovereignty than the Versailles approach. Although it stopped short of proposals in the 1960s for a world central bank (see Triffin proposals in Chapter 5) or in the 1970s for direct coordination of monetary policies (see Chapter 6), it was intended to override national parliaments, making policy changes

automatic if indicators moved outside a certain range. As one German participant in the Tokyo exercise pointed out, "What would be the reaction of the U.S.
Congress if the G-5 would decide that the U.S. deficit should be cut by $50
billion next year?" (Funabashi 1988, p. 134). The automaticity feature of the
indicators exercise was completely unrealistic, and it was stillborn at the Tokyo
Summit.

The major difference between Tokyo and the three prior summits was in the
content of the domestic policies being advocated. The earlier summits advocated policies to disinflate the world economy and promote structural change;
Tokyo advocated policies to reflate the world economy, largely through macroeconomic stimulus. To achieve this consensus, U.S. officials had to outflank
the conservative central bankers in the G-5 group, particularly the central
banks in Germany and Japan. Thus, at Tokyo, the multilateral surveillance
process was expanded to include Italy and Canada. Although U.S. officials
argued that this was necessary to associate the exercise with the summit process
and to supplement G-5 meetings, which were "chaotic," the reasoning made
little sense (Funabashi 1988, pp. 138–139). The surveillance exercise was already associated with the summits. The monetary statement annexed to the
Versailles Communiqué in 1982 committed the seven countries "to strengthen
our cooperation with the IMF in its work of surveillance; and to develop this
on a multilateral basis taking into account particularly the currencies constitut-
ing the SDR." Moreover, if the G-5 sessions were chaotic, adding Italy, Canada, and the media extravaganza of summit events to such sessions would only
make them more chaotic. The procedural reasons for expanding the G-5 group
were just camouflage. "The major objective," as one study concluded, "was to
induce growth and to bolster domestic demand in West Germany and in Japan"
(Funabashi 1988, p. 135). United States officials sought to change the substance of domestic policies. Although U.S. officials carefully avoided any identification of the exercise with the locomotive theory of the Bonn Summit in
1978, the Germans and Japanese suspected all along that this was the real
hidden motive of U.S. policies (Funabashi 1988, pp. 39, 94).

The stimulative content of policies at Tokyo made it easier to advocate
cooperation. As one supporter of the exercise put it, "nobody's being asked to
sacrifice for anybody else. . . . In adjusting to agreed-upon conditions, each
country is better off" (*NYT*, March 3, 1986). The same probably could not be
said of cooperation to disinflate domestic economies. The pain of such cooperation would certainly lead to charges that national sovereignty was being lost
and that domestic jobs were being given up to satisfy foreign pressures. It
may be, in fact, that the style of diplomacy practiced at Tokyo in 1986 and
Bonn in 1978 is suitable only under world economic conditions that invite
reflation. When world conditions call for disinflation, as they did in 1981,
choice-oriented diplomacy, which depends less on coordinated international
action and more on market-based, national action, may be the only politically
feasible alternative.

Despite the elaborate indicator proposal, the Tokyo Summit did not reach
agreement on policies for further stimulus (for Tokyo Communiqué, see *NYT*,

May 7, 1986). This time Japan as well as Germany refused to go along with further interest rate reductions. Almost before it was born, the new coordination process was being declared dead, even by its advocates (*WP*, June 29, 1986). Nevertheless, U.S. officials charged ahead unilaterally in summer 1986 to lower U.S. interest rates. After diplomatic leverage failed at Tokyo, market leverage of the lower dollar was all U.S. officials had left to put further pressure on Germany and Japan to expand.

In July and again in August the Fed unilaterally lowered the discount rate, now down to 5.5 percent, where the rate would remain until the ill-fated boost in early September 1987, before the October stock market crash. Prior to the July cut, the administration and Congress openly pressured the Fed to reduce rates. With congressional elections coming up in November, Robert Dole wrote a letter to Volcker urging lower rates (*WP*, June 24, 1986). In a deliberate double entendre, the White House announced that President Reagan wanted further cuts in interest rates but that he was not pressuring the Fed to do so (*WP*, June 20, 1986; *NYT*, June 20, 1986). That Volcker was reluctant to cut rates further but now found himself trapped by the Treasury Department strategy can only be surmised. After the last previous cut in April, Volcker had told Congress that rates had fallen enough (Volcker 1986b). And in summer 1986, he would worry aloud about the continued fall of the dollar (see next section). Nevertheless, lower oil prices and other economic factors gave him enough room to bend to Treasury objectives.

Germany and Japan did not follow either of the Fed's cuts. German experts now worried aloud about the shift of American policy to what they saw as an excessive reliance on monetary policy and an abandonment of structural reforms as the basis of domestic growth. Count Otto Lambsdorff, former West German Economic Minister and a highly respected commentator on international economic affairs, detailed his concerns in the *Wall Street Journal*:

> Every country must put its own house in order. . . . The U.S. budget deficit is and will continue to be a destabilizing factor on the foreign exchange markets. To a large extent *it shifts the burden of economic tuning onto monetary policy, which is too much for this one instrument to handle.* An inflationary policy is building up in the U.S.
>
> *The aim must be to strengthen West Germany's productive potential, which also implies strengthening investments and increasing the flexibility of the economy.* A policy of this kind requires a reduction in the share of public expenditure in the gross national product, tax cuts, privatization, deregulation and a cutting down on the bureaucracies. . . .
>
> *It would be much better if the Americans had clearly referred to this instead of simply calling for stimulation of domestic demand. I have always wondered why the U.S. did not present these suggestions for the German economy in the context of the OECD, at the G-5 meeting, or at the world economic summit.* The failure to do so has made it easy for the German government. As the American request [to lower German interest rates] was wrong in the first place it should be no surprise that the German response was not what was desired [August 18, 1986; emphasis added].

Lambsdorff was not alone at this point in noting the misdirection of American policy and the greater relevance in Europe of microeconomic reform rather than macroeconomic stimulus. In 1985–1986, Europe was taking another major step forward toward economic and monetary policy integration. This step involved further integration in the European Monetary System (EMS), emphasizing the conservative macroeconomic policies of the Bundesbank and German economic authorities, and a bold new initiative through the Single European Act of 1985 to dismantle microeconomic barriers and create a single internal market in Europe by 1992. Both of these policy developments were squarely in line with the choice-oriented priorities of the Reagan administration's first-term policies. Second-term Reagan officials had a golden opportunity to reinforce these developments. Instead they pursued policies that eventually brought them into public dispute with German authorities over macroeconomic policies (on the weekend before the stock market crash of October 1987) and that treated the single European market initiative exclusively as a threat to U.S. trade interests.

Thus, the campaign to lower the dollar was never, as officials asserted at Plaza, an attempt to adjust exchange rates to reformed economic fundamentals. It was instead an attempt to override unreformed fundamentals (namely, fiscal deficits in the United States and structural rigidities in Europe and Japan) with loose monetary policy, not only in the United States but also in other industrial countries. In the United States, real domestic demand, which needed to fall, actually rose, with the budget deficit reaching new highs in fiscal year 1986, and real fixed investment plunged further, from 5.5 percent in 1985 to 1.8 percent in 1986. Demand and investment, which needed to increase in the surplus countries, grew in 1986 in Germany but not in Japan (IMFa 1988, pp. 112–113). Exchange and interest rate coordination was producing few results. As the *Financial Times* pointed out in late summer 1986, greater cooperation "does not mean just coordination of interest rate cuts." "The pressing need," the newspaper went on, "is to correct the savings to investment imbalances which are the counterparts of the external surpluses and deficits" (August 23, 1986).

Louvre — Trying to Stop the Slide

From the beginning there was no agreement or even discussion among the G-5 about where exchange rates should be restabilized (*FT*, February 18, 1986). The Germans believed already in late 1985 that the German mark had appreciated enough (Funabashi 1988, p. 25). The Japanese found that *Endaka* (the high yen) was an extremely sensitive issue in the Japanese elections in summer 1986. Differences existed within countries as well as between them, as sharpening disputes between the Treasury and Fed revealed. As the time approached to stop the slide of the dollar, the question was not only where it should stop, but whether at that level enough adjustment had occurred in the fundamentals to hold the new exchange rates. The judgment of 1987 would be a negative one. Louvre stabilized the dollar for only a spell, and then it fell again.

By September 1986, U.S. officials needed some new demonstration from abroad that the Tokyo strategy was not just a solo U.S. exercise. With both the new G-7 and the old G-5 process stalled, the United States adopted a new strategy of "aggressive bilateralism." Germany was awaiting its elections in January 1987. And the United States had held off until Japan completed its elections in summer 1986 (earning favors which the Japanese would repay in the U.S. election in 1988). But now Japan had a popular prime minister going into an unprecedented fifth year of office. It was time for the United States to turn the screw on its Pacific ally.

Thus, Baker met with Kiichi Miyazawa, the new Japanese Finance Minister, in San Francisco in September. The United States offered stabilization of the dollar in return for further Japanese interest rate cuts. After intense negotiations, the two countries agreed on October 31st to stabilize the yen but subsequently disputed the range, as the United States suspected the Japanese of manipulating the exchange rate before the announcement of the agreement. Japan cut its discount rate to 3.0 percent, the lowest in postwar history, and Miyazawa, who favored fiscal expansion, submitted a supplementary budget package (*WP*, November 1, 1986; *NYT*, November 1, 1986; *WP*, January 15, 1987; for details, see Funabashi 1988, pp. 151–167).

Now it was Germany's turn—Baker traveled to Europe in December but to no avail (*WP*, December 10, 1986). In January the dollar began to fall again. Volcker protested (*NYT*, January 15, 1987). Since summer 1986, he no longer saw eye to eye with Baker. In August he had said in no uncertain terms that the "dollar has arrived at that point, long sought, at which both traders and government officials ought to regard any substantial movement in either direction as unwelcome" (*WP*, August 11, 1986). Treasury officials denied the rift, but it was too transparent (*WP*, January 18, 1987). Some commentators, fascinated by the struggle between two popular and powerful titans of the Washington power game, actually argued that the rift was orchestrated, with Baker talking the dollar down and Volcker ensuring that the slide did not become a free fall (*NYT*, June 3, 1987).

Time was running out for the U.S. Treasury. If it could not get agreement soon to stabilize the dollar, its differences with the Fed would burst into the open and its credibility abroad would plummet. Markets were beginning to anticipate a free fall, as private capital flows into the U.S. bond market began to dry up.

After German elections in January, the German government announced new tax measures and a drop in the discount rate to 3 percent. The Japanese and German contributions were not considered enough, however, even though the United States was backing away from its own meager fiscal commitments at the very moment it was pressing its allies for new commitments. While Secretary Baker traveled to Paris to sign the Louvre Accord, administration officials caucused with leaders on Capitol Hill to discuss the possibility of raising the fiscal year 1988 Gramm-Rudman-Hollings deficit target of $108 billion. Eventually, the target was raised to $144 billion.

Despite the U.S. budget slippage and mounting disagreements over real economic policy adjustments, however, the United States and its G-5 partners met in the Louvre Palace in Paris on February 22, 1987, to sign the now famous (or infamous) exchange rate agreement. The participants warned that "further substantial exchange rate shifts among their currencies could damage growth and adjustment prospects in their countries" and promised "to cooperate close-ly to foster stability of exchange rates around current levels" (*NYT*, February 23, 1987; for details, see Funabashi 1988, pp. 177–211). The United States undertook an ill-fated commitment to reduce its budget deficit from an esti-mated 3.9 percent of GNP in fiscal year 1987 to 2.3 percent of GNP in fiscal year 1988. As it turned out, it came nowhere close to meeting this commitment.

Given the absence of fundamental policy adjustments, the Louvre pact got an early test. In March and April the United States announced new restrictions against both Japanese imports and investments. The dollar began to slide below the ceiling established at Louvre (around 150 yen to the dollar). The trade measures acted as an effective tariff on American imports, amounting to an equivalent depreciation of the dollar. United States trade and domestic policies were now both working against stabilization of the dollar. The only way the dollar could be stabilized was to tighten monetary policy relative to the yen and potentially raise interest rates in the United States. When the dollar continued to drop in April, that is exactly what the United States and Japan agreed to do. Only 2 months after Louvre, they rebased the yen, lowering the ceiling to 142 yen to the dollar. To stabilize at this new level, the United States tightened credit while the Bank of Japan loosened credit (*WSJ*, May 1, 1987).[4]

In April Volcker declared that he "absolutely and fundamentally" believed that the dollar should not fall further. Running out of words to express his falling out with the talking-the-dollar-down strategy, Volcker added the school-boy's oath: "cross my heart and hope to die" (*WSJ*, May 1, 1987). As far as Volcker was concerned, the dollar's slide was over. It would remain that way, until Volcker left. In July he resigned after two full terms as the Board's chairman. Shortly after his departure, however, Louvre collapsed, the stock market crashed, and the dollar continued its descent.

Black Monday

After Louvre the effort to stabilize the dollar became expensive. In the absence of fundamental policy shifts, foreign investors lost confidence. Initially, the bond market collapsed, as private investors, particularly in Japan, began to withdraw from the dollar. Central banks stepped in with heavy intervention, ultimately financing about two thirds of the U.S. current account deficit in

---

[4]From May to June, growth of M_1 and M_2 in the United States decelerated on a 6-month sliding scale from 11.8 and 4.7 percent, respectively, to 4.5 and 2.9 percent, respectively. On a current basis, M_1 actually declined after mid-May by 1.1 percent (Federal Reserve Bank of St. Louis, 1987). For the year as a whole, growth of M_1 averaged 3.1 percent, whereas that of M_2 averaged 3.3 percent. By these measures at least, U.S. monetary policy was tighter in 1987 than it had been in the disinflation years of 1981–1982 (*ERP* 1988, p. 325).

1987 (reported by Bank for International Settlements, *FT*, June 14, 1988). The consequence was a significant loosening of monetary policy in both Japan and Germany as central banks sold their own currencies to support the dollar. Just as the dollar's slide had begun by a relative loosening of U.S. monetary policy vis-à-vis that of Germany and Japan, the dollar's stabilization depended on a relative tightening of U.S. monetary policy vis-à-vis that of Germany and Japan.

Interest rates crept up in the United States. In one of his first moves in office, the new Fed chairman, Alan Greenspan, raised the discount rate in September, both to follow market rates and to signal his Volcker-like determination to combat inflation and defend the dollar. Reportedly, Secretary Baker opposed this move, because it might reduce U.S. growth unless it was accompanied by further interest rate reductions in Germany and Japan (*NYT*, September 24, 1987). Although these reports too were denied (*WSJ*, September 25, 1987), the Federal Reserve and Treasury Department continued to worry about different things—the one about the dollar and inflation, the other about the slow growth of foreign demand and recession.

United States policy was now caught between the Scylla of inflation and Charybdis of recession. With fiscal policy gridlocked throughout 1985 and 1986, monetary policy had kept the recovery going. But monetary ease could not go on forever without dooming the dollar, and a free fall in the dollar was a prescription for accelerated inflation, rising interest rates, and eventually recession. For the first three quarters of 1987, inflation became a more prominent concern, as U.S. interest rates rose and commodity prices recovered, including oil, whose price rose from a low of $12 a barrel in fall 1986 to $18 a barrel in spring 1987. The consumer price index accelerated from the trough of 1.1 percent in 1986 to 4.4 percent in 1987. Although still low, inflation prompted both the Fed's move to increase the discount rate and a new proposal by the Secretary of the Treasury at the IMF–World Bank meetings in September to add a commodity index, including gold, to the indicators exercise designed at Tokyo (*WSJ*, October 1, 1987). The Secretary's "gold gambit" was part of the continuing high-wire act to stabilize the dollar (*WSJ*, October 12, 1987; *WP*, October 18, 1987). It was intended to calm inflationary fears abroad, especially in Germany, where, in the effort to support the dollar, the Bundesbank had consistently exceeded its monetary targets and faced growing domestic discontent with the policy of short-term monetary stimulus.

The Secretary's gold proposal failed to ease Germany's inflation psychosis and 1 month later it looked very ill advised as world economic concerns shifted rather suddenly in the wake of the stock market crash from inflation to recession. Worried about the massive overshoot of monetary targets, German authorities continued to raise key interest rates to match U.S. increases. On October 16 and 17, Secretary Baker issued stern warnings, calling the German moves inconsistent with the Louvre Accord and threatening to push the dollar down further if Germany did not cooperate (*NYT*, October 16, 1987; *WP*, October 17, 1987). Over the weekend of October 17 and 18, the markets witnessed what appeared to be a public trashing of the Louvre Accord by the U.S.

Secretary of the Treasury. Convinced the dollar was heading down again, the U.S. stock market reacted. On Black Monday, October 19, it plummeted 508 points, the largest absolute and percentage drop in market history. Suddenly, all the economic indicators swung from inflation to recession.

The immediate threat was financial collapse. The Fed stepped in on October 20 and pledged to make available whatever liquidity was necessary to keep the specialist traders on the floor of the stock exchange and the stock market afloat. As U.S. monetary policy loosened once again, unilaterally, the dollar began to fall. In effect, the falling dollar took the pressure off the bond and stock markets, which since February had borne the burden of stabilizing the dollar at levels that were now recognized as being unsustainable, given the lack of fundamental economic policy changes.

The stock market crash gave U.S. officials one more opportunity to cut the budget deficit. The deficit had dropped sharply from $221 billion in fiscal year 1986 to $150 billion in fiscal year 1987 (still 3.3 percent of GNP, however), due to one-time revenues from tax reform, asset sales, and budget gimmicks that moved revenues forward and expenditures outward. Before the stock market crash, therefore, both a tighter fiscal policy and the Fed's increase in the discount rate appeared, for the first time since 1985, to move U.S. macroeconomic policy in the appropriate direction to support the dollar's depreciation. But markets were not convinced. The deficit was expected to be the same or go up in fiscal year 1988 and beyond. (It ended up in fiscal year 1988 at $155 billion and in fiscal year 1989 at $152 billion— see *WSJ*, October 30, 1989.) Congress fixed the Gramm–Rudman–Hollings legislation in September to overcome Supreme Court objections, but in the process, stretched out the deficit targets to 1993 and raised the target for the election year of fiscal year 1988 from $108 to $144 billion (*WP*, September 24, 1987).

In October and November, therefore, the administration and Congress had one more chance to convince the markets that the United States was serious about policy adjustments. But once again the administration showed no stomach for tough budget issues, such as Social Security. The President, as before, said he would look at such cuts if Congress proposed them, but neither party was willing to move first on politically sensitive items (*WSJ*, November 23, 1987). Eventually, the compromise announced on November 20 cut the budget by only a little more than the automatic Gramm–Rudman–Hollings cuts of $23 billion (*WP*, November 21, 1987). Indeed, if gimmicks were removed, the real cuts were actually below the Gramm–Rudman targets (see Martin Feldstein's analysis, *WSJ*, December 1, 1987). The markets were unimpressed, but U.S. officials had to make the most of it.

Secretary Baker maintained the pressure on German authorities to expand. From the beginning of November on, German authorities eased interest rates. In early December they cut the discount rate once again, this time to 2.5 percent, the lowest in the history of the Bundesbank (*FT*, December 4, 1987); they also announced a package of subsidized credits to spur domestic investment. Other European central banks cut rates (*WSJ*, December 4, 1987). On December 22, the G-7 released a new statement, this time negotiated by telephone rather than in the glare of media lights. The statement departed significantly from the Louvre Accord. There was no longer a commitment to support exchange rates "around current levels." Rather, the statement said: "The ministers and governors agreed that either excessive fluctuations of exchange rates, a further decline of the dollar, or a rise in the dollar to an extent that becomes destabilizing to the adjustment process could be counterproductive by damaging growth prospects in the world economy" (*FT*, December 24, 1987). United States sources explained that the operative word was *could*, implying that no firm ranges had been set, as in the case of the Louvre Accord (*WP*, December 24, 1987).

The U.S. problem was simple. To set a firm floor on the dollar would imply a commitment at some point to raise U.S. interest rates. So soon after the October stock market crash and with an election year approaching, the Treasury Department did not want to raise interest rates, for the sake of the dollar or for any other reason. Without such a Treasury commitment, the dollar was vulnerable and was immediately tested. Heavy intervention was needed in January to arrest the dollar's slide. Even then the dollar dipped in early 1988 as low as 1.56 marks and 120 yen before rising by mid-year to levels well above the presumed ceilings of the December 22nd statement and falling again by year's end to around 1.72 marks and 120 yen. Exchange rate flexibility was apparently back in vogue, as long as the United States was unable to make convincing fiscal or monetary policy commitments.

Cooperation Flags

The stock market crash eased inflationary fears and the U.S. economy, which still had some unused capacity, rebounded in 1988. The Fed used the opportunity to raise interest rates once again. At G-5 and G-7 meetings in April and at the Economic Summit in Toronto, Canada, in June, officials trod softly, repeating the vague statement from December 22, 1987, about exchange rates and the dollar, and quietly adding commodity indexes, including a role for gold, to the indicators exercise as a demonstration of concern about renewed inflation (*WSJ*, April 14, 1988; *FT*, June 22, 1988). On the last day of the summit, the Bundesbank announced new interest rate increases. While the sequencing and conditions were different, the German move to tighten was not unlike the one in the early fall of 1987 that triggered U.S. anger. This time, however, U.S. officials kept quiet. Secretary of the Treasury Baker ignored the German moves, unlike his public outburst in October 1987.

For the moment the danger of recession, which U.S. officials feared in fall 1987, was remote. Real growth in the United States, topped off at 3.9 percent in 1988, well above even the most optimistic forecast in late 1987. Moreover, real growth in the Western industrial nations, taken together, was 4.1 percent in 1988 (IMFa 1989, p. 126). By mid-1988, the concern had shifted once again to controlling inflation. The Federal Reserve continued to push up interest rates throughout 1988, partly to cool off the economy as it neared full employment (unemployment dropped to 5.0 percent in early 1989, the lowest since 1974, and capacity utilization rates rose from 82 to around 84–85 percent), and partly to maintain the dollar, which began to sink once again in the second half of 1988. The question remained, however, as in 1987, how long the Fed could continue to tighten monetary policy without precipitating slower growth or, if it decided to loosen monetary policy, whether such loosening might precipitate a further fall in the dollar and possible inflation.

By mid-1989, the Fed's tightening produced signs that the U.S. economy was slowing, even though inflation accelerated during the first six months of 1989 at an annual rate of 5.9 percent, compared to 4.4 percent for both 1987 and 1988 (*WP*, July 20, 1989). Both slower growth and slightly higher inflation should have reinforced a lower dollar. But, paradoxically, the dollar, after falling by the end of 1988 to levels around those prevailing after the stock market crash in 1987, rose again in the first half of 1989, appreciating by some 15 percent against both the German mark and the Japanese yen. The dollar's rise threatened to halt and reverse the long-awaited adjustment on the U.S. current account. The new administration of President George Bush openly urged the Fed to reduce short-term interest rates, which had climbed three percentage points over the previous year, and help bring the dollar down (*WP*, January 26, 1989).

Meanwhile, foreign authorities in London, Bonn, and Tokyo worried more about the overheating of their economies. Growth in these countries in 1987–1988, as a *Financial Times* editorial noted, had been a consequence of "expansionary policies that followed the February 1987 Louvre Accord [and] attempted to put a floor under the dollar" (February 24, 1989). Central banks in these capitals, therefore, raised interest rates in late 1988 and early 1989, which helped to stem the flow of German marks and Japanese yen into U.S. dollars, but also threatened to spread an economic slowdown to other countries.

In the first half of 1989, domestic and international priorities seemed to coincide. The major industrial countries had a common interest in tightening monetary policy, just as they had had a common interest in loosening it in 1985. But just as conflict sharpened during the course of 1986, and eventually erupted in 1987 when the United States had to tighten money to stabilize the dollar and the allies did not want to continue easy monetary policies (which was the necessary parallel action on their part to stabilize the dollar), so too, at some point, their common interest in tighter money would end. In July 1989, the Fed, seeing the economy start to slow down, began to loosen money policy, at first cautiously. It knew that if it had to loosen money further to avoid a recession, major foreign partners would also have to loosen money policy or the industrialized countries would risk a precipitous fall of the dollar. Foreign partners, however, might be reluctant to loosen money at some point for fear of overheating their own economies.

As if anticipating these conflicts, the G-7 countries showed considerably less ardor in 1988 and 1989 for international economic policy coordination. Commentaries prior to the Paris Summit in July 1989, even among those who had applauded uncritically the Plaza-Louvre exercise, suggested that the "vaunted Group of Seven coordination process that blossomed under former

Treasury Secretary James A. Baker III had fallen on hard times" (*WP*, June 11, 1989). The leaders at Paris had so little confidence that existing economic fundamentals were adequate to support stable exchange rates, particularly previous commitments concerning the dollar, that they omitted any specific commitment to keep exchange rates stable or to maintain a certain (unspecified) range for the value of the dollar (*FT*, July 17, 1989).

The reasons for such flagging enthusiasm and confidence were not hard to find. If one looked beneath the surface of the drama and media-hype of G-5 and G-7 sessions, the actual internal adjustments that had been made in fiscal, trade, and structural policy reforms by the G-7 and, particularly, the G-3 (Japan, Germany, and the United States) countries to support external realignments had been minimal. At the Toronto Summit in 1988, the summit leaders finally acknowledged the important structural adjustment factors that the OECD had emphasized already in 1985 and added an annex to the Toronto Summit Communiqué that widened the scope of the indicator exercise to include such basic structural reforms as labor market deregulation, tax rate reform, education, and training (for text of communiqué, see *NYT*, June 22, 1988). But, because these reforms take more time to produce an effect on external balances (unlike monetary policies which act rather quickly on economic events), the opportunity for timely structural reforms had been lost already at earlier summits, particularly the summits from 1984 through 1987.

Stalled Adjustment

The data for 1987–1988 did begin to show, for the first time, some real economic adjustment between the United States and its major industrial partners. To the extent that this adjustment followed from the delayed effects of the realignment of exchange rates, the intervention and interest rate coordination initiatives could at last claim some successes. Although, as Figure 9-1 shows, the Plaza exercise did not initiate the decline of the dollar, it did sustain pressure in 1986, particularly on central banks, to keep the dollar falling. From mid-1987 on, U.S. exports began to surge at annual rates of 20–30 percent and the U.S. trade balance declined in 1988 by some 20 percent from $160 billion to around $125–130 billion. Nonresidential fixed investment, mostly in manufacturing, rose again, in 1988, to about 9 percent on an inflation-adjusted basis, compared to just 1 percent in 1987 (*WSJ*, June 2, 1988). Now that the dollar was down, American manufacturing was displaying some of the competitive strength it had developed through the strong investment and cost-cutting activities of the early 1980s.

Intervention had helped perhaps to bring the dollar down, but had it done so on the basis of sustainable macroeconomic, structural, and trade policies? Table 9-1 suggests that, by this standard, the intervention exercise was much less successful, especially given the expectations aroused by international economic policy coordination. The U.S. budget deficit declined from 5.3 percent of GNP in 1985 to 3.3 percent in 1988, but it was still well above the 2.3 percent for fiscal year 1988 that the United States had committed itself to at the

Table 9-1 Adjustment Among Principal Industrial Countries, 1985–1988

		Budget Deficits (% of GNP)	Current Account (% of GNP)	Monetary Agregates (annual % change)	
				Narrow Money (M_1)	Broad Money (M_2)
U.S.	1985	−5.3	−2.9	9.0	9.0
	1986	−4.8	−3.3	13.6	8.3
	1987	−3.4	−3.4	11.6	6.6
	1988	−3.3	−2.8	4.3	5.1
Japan	1985	−3.9	3.7	5.0	8.4
	1986	−3.4	4.3	6.9	8.7
	1987	−2.7	3.6	10.5	10.4
	1988	−2.1	2.8	8.4	11.2
Germany	1985	−1.3	2.6	4.3	4.9
	1986	−1.2	4.4	8.5	5.8
	1987	−1.4	4.0	9.0	7.1
	1988	−1.7	4.0	9.8	6.4

Source: IMFa (1989), pp. 139, 141, 157.

Louvre, and the U.S. economy was now at nearly full employment, when such a large budget deficit was clearly unjustified. Meanwhile, in trade surplus countries, where, according to the Plaza–Louvre strategy, budget deficits were supposed to go up, budget deficits as a percentage of GNP actually went down in Japan, from 3.9 percent in 1985 to 2.1 percent in 1988, or increased only slightly in Germany, from 1.3 in 1985 to 1.7 percent in 1988. Current account imbalances also proved stubborn. The U.S. current account deficit, as a percentage of GNP, was about the same in 1988 as it was in 1985, after going much higher in 1986 and 1987; and the combined current account surplus in Germany and Japan, which was supposed to go down, actually increased from 6.3 percent of GNP in 1985 to 6.8 percent in 1988.

After 4 years of a falling dollar, therefore, the adjustment was grudgingly small. What is more, the adjustment that did occur came about largely through higher one-time tax revenues in the United States reducing the budget deficit in fiscal year 1987, and, as Table 9-1 shows, through significantly easier monetary policies in the United States, Germany, and Japan, at least through 1987. These policies were not sustainable, as the U.S. budget deficit bottomed out at $150-$155 billion for fiscal years 1987 through 1989 and as money policies in all three countries tightened in the course of 1988 and 1989.

The world economy in 1988 and 1989 remained robust. World trade grew in 1988 by about 9 percent in volume, well above levels of 4.4 and 6.1 percent in 1986 and 1987 (IMFa 1989, p. 16). Investment surged in practically all the major industrial countries, holding out the prospect that overheating might be contained by new additions to supply capacity. But trade disputes multiplied, particularly when the threat of Europe 1992 became apparent in the United

States in 1988 and when the United States targeted Japan and other countries in spring 1989 to reduce trade barriers under the Super 301 provisions of the new trade bill or face retaliation from the United States within a specific period of time. The debt crisis persisted and perhaps grew worse, even with robust world growth (see discussion in next session). The weakness of the world economy, however, was not in its immediate performance; it was the fact that this performance depended mostly on the use by industrial countries of one policy instrument, namely, monetary policy, and on a narrowly-based and fragile process of international policy coordination that focused attention on exchange rates rather than underlying fiscal, structural, and trade policy reforms.

The Baker Plan Initiative

If, from 1985 on, real economic adjustments were slow to materialize in industrial countries, they barely materialized at all in the developing world. The fault lay equally with the lack of credibility of American leadership on the debt problem and the lack of political capability in developing countries to reform entrenched interventionist regimes in domestic economic and trade policy areas.

The ad hoc, short-term debt strategy of the United States in 1983–1984 had a weak link. One of the five points of that strategy called for structural reform in the developing countries (see Chapters 7 and 8). From 1983 to 1985 the United States had tried unsuccessfully to make trade liberalization (i.e., a North–South round) the cutting edge of that reform, opening up the markets of developing countries to more competitive forces while securing reciprocal access for developing countries to industrial country markets on a predictable GATT basis (involving multiyear commitments such as tariff bindings rather than single-year GSP benefits). Now, in 1985, it launched a new debt initiative ostensibly to complement the trade round and to encourage other structural reforms in developing countries. But the Baker Plan, as the new initiative was called in honor of the new, activist Secretary of the Treasury, ran the same risk as the intervention and interest rate initiatives undertaken with the industrial countries. Unless it was coupled with decisive actions on the U.S. budget and trade deficits, the U.S. debt initiative would have the effect of diverting attention from real economic reforms and putting the emphasis, once again, as in the 1970s, on purely financial solutions.

The failure to address domestic and trade policy reforms in the United States after 1985 crippled the Baker Plan and destroyed the credibility of U.S. diplomacy to encourage such reforms in developing countries. The Baker Plan also weakened the IMF by criticizing it for imposing austerity and by turning to the World Bank and structural reforms as an alternative rather than a complement to the macroeconomic policy reforms of the IMF. In the end, the Baker initiative shifted the emphasis to finance rather than real economic reform. And, ironically, although predictably, it failed even to get the finance, because private commercial banks, on which the initiative depended to provide most of the new money, were unwilling to put up new money as they watched IMF programs being scuttled one by one and World Bank programs being reshuffled in a baffling reorganization of personnel and missions under the new Bank President Barber Conable.

Secretary Baker announced his plan at the IMF–World Bank meetings in Seoul, South Korea, in October 1985 with the same fanfare he had orchestrated at the Plaza Hotel in September. The plan called for new financing for some 15 of the most heavily indebted developing countries totaling $29 billion over 3 years, $20 billion of which would come from private sources and $9 billion from the multilateral development institutions. In return, the recipient countries would undertake structural reforms under the auspices of the new structural adjustment loans (SALs) of the World Bank. The latter were program rather than project loans to facilitate basic reforms in specific sectors over a multiple-year period—agriculture, transportation, and so on. They had started in 1981 and were regarded as the appropriate means to fill in between the Bank's project loans, which were long term in focus but slow in paying out, and the IMF's stabilization loans, which were fast in paying out but short term in focus.

The Baker Plan was well aimed, but it was unintegrated with the other pieces of an effective longer-term debt strategy and eventually focused attention almost exclusively on new ways to finance debt. The plan had three principal shortcomings. At a critical moment in 1985, it diverted attention from domestic macroeconomic reform in developing countries, just when developing countries were becoming more open to market-oriented alternatives. By implying that IMF programs fostered austerity rather than growth, the Baker Plan emboldened indebted countries to break out of IMF programs. In 1987 Brazil became the first major debtor to declare a moratorium on its debt-servicing payments.

Second, the Baker Plan severed the link between debt and trade. Secretary Baker showed little interest in the trade round, while a new U.S. trade representative pursued the new trade round with a more aggressive, if not belligerent, American trade policy designed to appease a protectionist Congress. The World Bank, on which the Baker Plan relied for trade liberalization, was no substitute for the GATT as a means to liberalize trade in developing countries. The Bank could not ensure that, if developing countries liberalized on their own, they would, in return, get improved access to industrial country markets because the Bank has no programs in industrial countries. Eventually, without such reciprocal concessions from industrial countries, developing countries would encounter stronger and stronger domestic and political opposition to liberalization.

Third, the Baker Plan promised mostly somebody else's money and tried to jawbone private commercial banks to provide the money without the domestic or trade policy reforms in developing countries that might have motivated the banks to provide the money. To its credit, the Baker Plan did not try to override the necessary reforms with massive amounts of public aid, such as the government-encouraged recycling of petrodollars had done in the 1970s. But by failing to bring about the necessary reforms in developing countries, it opened the way for more generous, publicly financed debt relief proposals that became popular in Congress by 1988.

The shortcomings of the Baker Plan could not be attributed to external factors. In 1986, in particular, the world economic environment was broadly favorable for heavily indebted developing countries. Oil prices and interest rates declined steadily over the year. A few oil-exporting countries in the Baker group, such as Mexico, suffered reverses, but the adverse effects of reverse oil prices shocks were selective (IMFa 1987). In 1987, both oil and commodity prices would recover somewhat.

The absence of significant domestic macroeconomic reform in heavily indebted developing countries was evident throughout the 1980s. Monetary and fiscal policies deteriorated in most of the heavily indebted countries. Annual growth of broad monetary aggregates in the 15 Baker Plan countries quadrupled from 60 percent in 1981 to 246.3 percent in 1988. In the same period inflation in these countries almost quintupled, from 53.8 percent per year to 240 percent per year. Meanwhile, central government budget deficits in these countries averaged 4.2 percent of GDP in 1981 and 5.0 percent in 1988, after declining to 3.4 and 2.8 percent of GDP in 1984 and 1985 (IMFa 1989, pp. 136, 143–144).

There is no evidence in these numbers of sustained austerity. Nor is there evidence that the debt problem and debt-servicing obligations were the primary cause of poor economic performance in heavily indebted countries. As William R. Cline concludes with respect to the major Latin American debtors, "the principal cause of stagflation in 1987 and 1988 was from domestic policy distortions and high inflation in particular, not the debt problem." "Nor," he adds, "was inflation caused primarily by the debt burden" (Cline 1989). IMF programs had failed in some cases, to be sure, but they had failed for internal policy reasons not because of external circumstances or pressures that IMF critics and Baker Plan advocates emphasized. From 1982 to 1986 Brazil renegotiated some seven agreements with the IMF, each time because it had failed to meet the targets of the previous agreement. When Brazil refused to sign an agreement altogether in 1986, the IMF's authority was gone. The critics were literally beating a dead horse. The IMF had died because its authority was too weak, not because it was so strong that its programs had imposed austerity.

None of this is to say that the Baker Plan countries had not adjusted. They had adjusted as much as, if not more than, their industrial counterparts, such as the United States. Some, such as Chile, Colombia, Venezuela, and Mexico (after 1986), had adjusted with appropriate macroeconomic and trade policies to control inflation and promote competition. But others had adjusted primarily through restrictive trade policies (and exchange rate devaluations; see later discussion), not through IMF-recommended restraint in monetary and fiscal policies. After 1981, the heavily indebted countries sharply reduced imports and expanded exports. In volume terms, imports in these countries contracted from a positive growth of 3.0 percent in 1981 to a negative growth of 16.3 and 21.3 percent in 1982 and 1983, respectively. Imports then remained flat from

1984 to 1987, surging again to a positive growth of 7.7 percent in 1988. Meanwhile, exports soared in volume terms from a negative 5.0 percent growth in 1982 to a positive 5.6 and 8.7 percent growth in 1983 and 1984. Export growth slowed down in 1985 and was actually negative in 1986 but then climbed again to growth rates of 5.6 and 10.8 percent in 1987 and 1988, respectively (IMFa 1989, pp. 150–151).

The Baker Plan, rather than encouraging trade liberalization, actually saw the heavily indebted countries impose increasing import restrictions and export subsidies. These trade policies accelerated inflation and budget deficits in many Baker Plan countries and stimulated retaliatory actions in industrial markets. At the end of 1987, for example, the United States and Brazil were locked in a potential cycle of retaliatory restrictions over computer software. The trading system was going backward.

The heavily indebted countries also achieved significant adjustment by exchange rate realignments. In the three biggest debtor countries—Argentina, Brazil, and Mexico—real trade-weighted exchange rates declined from 1981 to 1986 by about 10 percent in Brazil, by 40 percent in Mexico, and by 35 percent in Argentina (Cline 1987, p. 69). As in the United States, however, exchange rate devaluations had limited effect in the absence of adjustment in domestic policy fundamentals.

Given these patterns of adjustment in the Baker Plan countries, it was no surprise that commercial banks did not provide significant amounts of new money. U.S. banks actually decreased their exposure in the 15 Baker Plan countries by approximately 10 percent from the end of 1984 to the end of 1986 (Cline 1987, p. 3). Altogether, from 1986 to 1988, private banks provided only about two thirds of the $20 billion in new loans which the Baker Plan envisioned (Cline 1989). Most of this went to the big debtor countries. The agreement between Mexico and the private banks in September 1986 provided $7.7 billion in new bank lending over 18 months. And the new package for Brazil in 1988 provided $5.2 billion in new bank loans. But in the process of approving the Mexican agreement, some 25 percent of the banks, all small ones, dropped out of the syndicate.

The Baker Plan was going nowhere. In late 1986 and 1987 the discussion turned increasingly to debt relief. If there was no way to encourage new money, old loans should be forgiven. Senator Bill Bradley proposed a plan over 3 years to forgive 3 percent of principal and three percentage points of interest annually for eligible debtors who undertook serious policy reforms. Other schemes proposed even more generous relief. Meanwhile, more and more loans were being sold at discount in the open market or were being swapped for full value in terms of local currency and equity. Some debtor countries did not wait for relief; they simply suspended payments.

The U.S. administration had no credibility to deal with these issues. Since the middle of 1986, the United States had become the world's largest debtor, owing a total of some $500 billion by the end of 1988. Although the U.S. debt was small in terms of GNP (about 10 percent) and the U.S. economy could hardly be compared to those of debtor developing countries, the symbolism was perverse. The country that had led the world in 1979 toward domestic policy reform was now leading the world in 1987 toward spiraling international debt.

In addition, the focus on finance as opposed to domestic policy reform began to revive an old nemesis from the 1970s. The developing countries, particularly the heavy debtors in Latin America, raised again the specter of North–South confrontation. In November 1987 the heads of state and government of eight Latin American countries met for the first time outside the U.S.-led Organization of American States (OAS) to press their case for debt relief. Although not formally establishing a debtor cartel, they condemned the rich countries, said the South could not be expected to pay for the disequilibrium of the North, and called for reconstructing the world economic system. As *The Washington Post* reported, "the occasion was marked by a revival of some of the North–South rhetoric that flourished during the 1970s, when industrialized and developing countries held inconclusive negotiations revolving around Third World demands for a 'new international economic order'" (November 28, 1987). One year later, the so-called Group of 8 Latin American debtors called for a summit meeting with the Group of 7 industrial nations (*WP*, December 22, 1988). And, in the same week as the Paris Summit in July 1989, French President Mitterrand hosted a gathering of the heads of state and government from developing countries that called for new attention to the North–South dialogue (*Economist*, July 15, 1989). North–South lines were being drawn once again.

Summary

After 1985, therefore, the administration turned in directions that looked more and more like the misguided policies of the 1970s and that began to reproduce some of the external features of that earlier era. The international initiatives themselves were not unhelpful or fundamentally inconsistent with the administration's efforts in the first term. But they became unhelpful as they increasingly substituted for macroeconomic, structural, and trade policy reforms in both industrial and developing countries and pushed the world back toward the easy monetary and financial solutions of the 1970s.

The Reagan Years Reassessed

The Reagan years witnessed an attempt to reintegrate American purposes, policies, and power and restore a sense of shared political community and efficient economic performance among Western countries. The Reagan administration acted instinctively to sharpen moral and political differences with totalitarianism and to reconstruct the domestic economic foundations of a stable, open, and growing international economy. But it would never fully understand conceptually what it was about or care enough to explain its policies consistently and patiently to its own public or to its allies.

The challenge, to be sure, was considerable. The intellectual literature, as we have noted in this study, tilted heavily toward structuralist interpretations of the world economy and would now criticize the Reagan program for being inconsistent with irreversible trends toward declining American power and the need for more managed national economies and international trade. Nevertheless, the opportunity was there. Public sentiment was ready to turn; it could have been led.

The Reagan administration, however, did not do enough to lead it. Unlike the Truman administration in 1947, it did not bring its message into the arena of nongovernmental debate to shape concepts and coalitions for sustained policy changes. Reagan's policies, in fact, accomplished more in reality than they did in altering perceptions. The 1980s witnessed the longest economic expansion since the early 1960s. They also witnessed considerable progress in alliance relations with the successful deployment of intermediate nuclear forces (INF) and more progress in U.S.–Soviet relations, especially arms reductions, than in any previous postwar period. Yet there was no persuasive explanation for this success. It was attributed to luck and change in the Soviet Union.

Between its instincts and its commitment to economic and military strength, the Reagan administration never developed an effective domestic or international economic diplomacy. It oscillated from a preoccupation in the first term with domestic problems and international markets to an activism in the second term that gave international diplomacy priority over domestic and international markets. In terms of international diplomacy it looked like the Nixon administration of 1971 in the first term and the Carter administration of 1979–1980 in its second term. An excessive unilateralism in 1981–1982 gave way to an excessive multilateralism in 1985–1986.

In one important respect, however, the Reagan years differed from the 1970s. U.S. domestic economic performance improved. Inflation dropped dramatically, markets became more flexible, and investment and manufacturing productivity grew more strongly. The difference suggests that U.S. policy in the early 1980s was potentially on course. If this domestic improvement could have been linked with a more private and patient international diplomacy (rather than the highly public, dramatic diplomacy of 1985–1986) and if the administration had followed through on its own domestic economic policy convictions and lowered the budget deficit in a timely fashion, then U.S. policy might have succeeded in leading the world community back toward the efficient economic policies of the 1940s and 1950s. It may still be possible in the early 1990s, as we examine in the Conclusion of this study. But the obstacles are now much more formidable than they were in the early 1980s, when the United States and the world were looking for new answers.

V

The Limits of
Bretton Woods

10

Managing East–West Trade:
Denial, Detente, and Deterrence

The choice-oriented perspective of this book emphasizes the importance of shared political community as the basis for extensive economic exchanges. Throughout the postwar period, U.S.-Soviet relations, and East–West relations more generally, lacked sufficient shared political purpose to support growing and interdependent East–West markets. Domestic purposes differed, with the East favoring totalitarian political and command economic systems over democratic and market-oriented ones, while diplomatic relations reflected the shifting balance of power to protect fundamentally different domestic societies. Today, events in the East suggest that this may all be changing. But if it is, the choice-oriented perspective suggests how truly dramatic the changes in the East will have to be if East–West markets are to develop the minimal elements of efficient economic exchanges—market prices, convertible currencies, and noncoercive political goals—that could expand East–West trade significantly beyond the 5 percent or so of Western trade that it accounts for today.

Denial and Detente

Since World War II, U.S. trade policy toward the Soviet Union has varied, depending on America's definition of shared political community with the rest of the world and its relative strategic power vis-à-vis the Soviet Union. During the Bretton Woods period (1947–1967), when the United States defined community largely in terms of containment of Soviet Communism and enjoyed strategic superiority over the Soviet Union, the United States pursued a policy of denial in East–West trade. Export controls blocked goods of economic as well as military value to the Soviet Union and other Eastern bloc countries (including China). In the late 1960s and 1970s, however, when Bretton Woods collapsed, the United States redefined political community in terms of detente with the Soviet Union and acknowledged a position of relative strategic parity with Moscow. Export controls were relaxed, and trade with the Soviet Union and other Eastern bloc countries was promoted, even to the point of subsidizing some trade and credits, particularly by allied governments in Europe and Japan (Nau 1976).

After the Afghanistan invasion by the Soviet Union in 1979 and the election of the Reagan administration, U.S. strategic policy returned to a hard line. United States East–West trade policy swung back again toward denial, particularly after the Polish crisis of December 1981. In the process it precipitated one of the most serious alliance disputes in the postwar period — the firestorm over the Soviet gas pipeline to Western Europe and the supply of Western equipment and credits as well as gas purchases by the West associated with that pipeline.

From 1985 on, however, U.S.-Soviet relations improved. Gorbachev's *perestroika* (economic reform) and *glasnost* (political reform) in the Soviet Union, and an arms reduction agreement on intermediate nuclear forces (INF), reduced tensions. United States policy swung back again — once more toward detente. Governmental and nongovernmental trade talks resumed. The Soviet Union decentralized aspects of foreign trade policy and promulgated new laws encouraging private cooperatives at home and joint ventures with Western firms. The United States subsidized grain sales to the Soviet Union and removed controls on energy equipment and technology sales and personal computers. The Reagan administration also undertook a far-reaching liberalization of trade control with China. Meanwhile domestic trade pressures in the United States generated growing support for a broad-scale reduction of export controls, particularly for shipments within the West.

As the 1980s ended, U.S. East–West trade policy seemed to be suspended between calls to liberalize further economic relations with Eastern Europe and the Soviet Union and proposals to restrict trade with China. The repression of student demonstrators in China in June 1989 brought about a reappraisal and tightening of export control policies. While the Bush administration in the United States struggled to limit the sanctions, Congress pressed for stronger action. Disputes with allies threatened, as Japan showed greater reluctance to condemn the Chinese action. What would happen if China continued harsh repression? Or how would the United States and the West react if repression were used at some point to contain political and economic instabilities in Eastern Europe or the Soviet Union? The pendulumlike oscillations of U.S. East–West trade policy reflected the absence of a conceptual basis for policy between the two extremes of denial and detente. The United States needed a more balanced policy that recognized the continuing political differences between democratic societies in the West and totalitarian societies in the East but also responded to the growing interest in the East in economic efficiency and political reform. The Reagan administration had attempted in 1981 to define such an intermediate policy, known as the prudent approach. But this policy did not survive the Polish crisis and the swing after 1981 back toward policies of denial.

Nevertheless, the conceptual basis of a prudent approach, defined more broadly as deterrence (rather than denial or detente), provides a useful vehicle in this chapter for critiquing the Reagan administration's East-West trade policy and for proposing an alternative for U.S. policy in the 1990s that may be more appropriate to changing conditions in the East. Deterrence defines a shared political community with the Soviet Union that is broader than containment yet does not confuse glasnost in the Soviet Union with pluralist democracy in the West. At the same time, deterrence also expects less from East-West economic relations than either denial or detente and thus provides a more modest but sustainable basis for growing East-West economic exchanges.

Deterrence

Policies of denial and detente both overstate the role that economic relations can play in strategic conflict. On the one hand, policies of denial emphasize fundamental and irreconcilable differences between the Soviet Union and the United States, including the way the two societies value and organize economic activities. Yet these policies assume that the Soviet Union, like the West, values economic exchanges so highly that denial of these exchanges will produce significant changes in Soviet foreign or domestic policy behavior. Denial of economic exchanges is seen as a significant instrument for waging and, ultimately, winning geostrategic conflicts (Walinsky 1982–1983). This is not likely to be the case, however, because the Soviet Union will more readily absorb economic costs to pursue other objectives in international relations, just as it more readily absorbs these costs to pursue other purposes in domestic politics (e.g., party control). Policies of detente, by contrast, assume that economic exchanges can eventually achieve sufficient volumes to help overcome fundamental differences and reduce military expenditures. Economic interdependence is seen as a decisive means to resolve geostrategic conflicts. Yet these conflicts derive, at least in part, from domestic values and institutions that inhibit economic exchange and engender military distrust. Without a convergence of political perspectives, therefore, the value and impact of economic exchanges will remain limited.

In reality, the role of economic relations in East-West conflict is much more modest. As Table 10-1 shows, economic ties with Eastern Europe and the Soviet Union are a relatively small proportion of total exports or imports for all the major Western countries (in 1984, less than 8 percent for any country). Although these ties may be important for individual sectors or firms (e.g., wheat for France and the United States, iron and steel plates for Italy and Germany, metal-working machine tools for France, Germany, and Italy), their potential consequences for geostrategic relations remain small. Thus, East-West trade is primarily useful neither in waging geostrategic conflict (denial) between East and West nor in resolving such conflict (detente) but rather in managing it (deterrence). In this sense, economic relations between East and West have a limited but nevertheless important role to play in deterrence.

A deterrence perspective entails elements of both denial and detente. It holds that fundamental political conflicts between East and West are deep-

Table 10-1 Major Western Countries' Trade with the East, 1966-1984*

	1966		1972		1974		1976		1978		1980		1982		1984	
	X	I	X	I	X	I	X	I	X	I	X	I	X	I	X	I
Belgium–Luxembourg	2.1	2.2	1.8	1.9	3.3	2.1	2.6	1.9	2.5	1.8	2.0	2.3	1.7	3.2	1.8	4.1
Canada	6.2	0.8	3.4	0.9	2.5	1.0	3.3	0.9	3.1	0.8	2.7	0.4	3.0	0.3	2.1	0.3
Federal Republic of Germany	6.0	5.9	7.0	5.9	8.4	6.4	8.4	6.6	7.7	6.5	6.5	6.2	5.7	6.7	5.4	7.0
France	4.6	3.5	4.1	3.2	4.3	2.9	5.8	3.5	4.1	3.4	4.0	3.9	3.0	3.7	3.2	3.7
Italy	5.4	6.8	4.8	6.3	6.1	5.0	5.9	6.1	4.7	5.6	3.5	5.3	3.3	6.0	3.4	7.2
Japan	6.2	7.5	5.2	5.8	7.4	5.8	6.9	4.3	6.6	4.8	2.8	1.5	3.2	1.4	1.8	1.3
Netherlands	2.1	2.3	2.1	2.1	2.6	2.4	2.2	2.6	2.2	2.6	1.9	3.0	1.5	5.2	1.3	4.7
United Kingdom	3.7	4.7	3.3	4.1	3.3	3.3	3.0	3.7	2.9	3.2	2.3	2.5	1.6	2.0	1.8	2.2
United States	0.7	0.7	1.8	0.6	2.3	1.0	3.2	0.9	3.1	1.1	1.7	0.6	1.7	0.4	1.9	0.7

Source: Compiled by the Bureau of Intelligence and Research, U.S. Department of State, based on Series A, *Statistics of Foreign Trade*, Organization of Economic Cooperation and Development.

*Exports (X) and Imports (I) to East as a Percentage of Exports and Imports to World.

From 1980, includes trade with U.S.S.R., Bulgaria, Czechoslovakia, the German Democratic Republic, Hungary, Poland, and Romania.

Before 1980, includes trade with the preceding countries plus Albania, People's Republic of China, other Asian countries (North Korea, North Vietnam, and Mongolia), and Cuba, except for 1978, which excludes other Asian countries.

Imports are c.i.f. (includes cost of insurance and freight) except for Canada and the United States, which are f.o.b. (free on board excluding cost of insurance and freight).

Exports are all f.o.b.

seated and unlikely to be affected significantly by economic relations for positive or punitive purposes, especially when economic ties are pursued in isolation from, or opposition to, larger political trends. On the other hand, deterrence views economic ties as useful channels for communicating political intentions and expanding limited areas of mutual economic interest.

Deterrence, therefore, calls for continuing defense to *manage* deep-seated differences (unlike extreme versions of detente, which expect differences and defense requirements to fade away) and developing dialogue to expand mutual interests (unlike denial, which eschews common interests). In defense areas, as it has been practiced in the postwar period, deterrence depends on a U.S. and Western technological superiority in strategic defense systems to offset a Soviet and Warsaw Pact advantage in conventional defense systems. In this context, deterrence requires the protection of U.S. and Western strategic technology leads through export controls. On the other hand, deterrence also depends on the capacity of superpowers to communicate quickly and credibly. Economic relations serve as one source of communications. Thus, deterrence also calls for market-determined levels of nonstrategic trade between East and West to serve as a barometer of foreign policy relations in critical situations of either escalating hostilities or significantly expanding cooperation.

From a deterrence perspective, however, economic relations do not lead to convergence of basic political values or even to significant leverage in bargaining between East and West. They remain useful primarily as a means to communicate growing trust or distrust between the superpowers. And they are most likely to build up trust when they take place on a foundation of balanced defense relations and mutual strategic confidence. Indeed, the erosion of balanced defense relations, as in the detente period of the 1970s, is precisely what leads, according to this perspective, to more radical and controversial uses of economic sanctions in East–West trade, often as substitutes for defense measures and in response to Soviet actions possibly precipitated, at least in part, by the prior lowering of Western defense vigilance. From a perspective of deterrence, therefore, the best way to ensure stable and useful economic relations is to maintain balanced defense relations. Table 10-2 summarizes schematically the basic differences between the deterrence perspective and those of detente and denial.

Prudent Approach: East–West Trade at the Ottawa Summit

The Reagan administration came into office in 1981 with strong substantive views on East–West trade. Within the first few weeks it identified East–West trade as a priority topic for the Ottawa Economic Summit scheduled for July 1981. An interagency group chaired by the State Department reviewed the subject but was deeply divided. The Carter administration had split between advocates of cooperation (e.g., in energy areas) and those who sought to use East–West trade for foreign policy leverage (Huntington 1978). The Reagan administration had no advocates of cooperation in 1981 but split between

Table 10-2 Perspectives on East–West Trade

	Detente	Denial	Deterrence
U.S.–Soviet conflict	Resolve	Win	Manage
Defense policy	Reduce defense expenditures — substitute economic and political means	Expand defense policies to include economic warfare	Maintain defense expenditures to preserve military balance — use economic and political means to communicate
Strategic trade controls	Minimum controls (what allies will accept)	Maximum controls (including unilateral controls)	Negotiate common controls aggressively within COCOM (add new technologies to controls while removing low-level technologies)
Nonstrategic trade	Promote (even subsidize)	Deny	Market-determined levels

advocates of total denial and somewhat less hardline views. To reconcile the differences, staff members of the National Security Council (NSC) drafted a paper in late March setting out the President's views. This was the origin of a paper entitled "A Prudent Approach." The President reviewed the paper in May 1981 at a meeting of the Cabinet Council on Commerce and Trade and formally approved it in July 1981 at meetings of the National Security Council. This more cautious rather than confrontational approach became the position that President Reagan advocated at the Ottawa Summit (*NYT*, July 19, 1981).

As developed in the prudent approach paper, U.S. policy differentiated among four areas of East–West trade: strategic trade, foreign policy uses to restrict or promote nonstrategic trade, economic security concerned with levels and conditions (e.g., subsidies) of dependence on nonstrategic trade (e.g., grain, energy), and economic influence dealing with the extension to nonstrategic trade of Western rules, disciplines, and institutions to minimize Soviet advantages as a monolithic supplier or purchaser of goods in the West. In each area the prudent approach called for different policies. In strategic trade it called for a review and tightening of strategic controls to protect the new investments being made by the Reagan administration in the defense field. In the second area, nonstrategic trade, it considered restrictions on trade for foreign policy purposes to be indispensable tools of deterrence in managing crisis situations but advocated contingency planning to ensure that these restrictions were applied, to the extent possible, multilaterally and within an agreed set of objectives and time table for eventual removal. To protect economic security, the third area, U.S. policy proposed a review of levels of Western dependence on Soviet exports and markets and development of collective measures to protect against vulnerability should the Soviet Union seek to exploit this dependence to influence the West. At this point, U.S. policy did not necessarily imply the restriction or even discouragement of specific levels of nonstrategic trade such as gas supplies, but called instead for safeguards to protect against the disruption of critical supplies by establishing back-up emergency supply arrangements among the allies and long-term allied cooperation agreements for the development of alternative supplies. It noted that unless the allies entered into discussions to deal with vulnerability, the levels of dependence themselves could become matters of controversy and undermine Western solidarity on East-West trade. Finally in the fourth area of economic influence, U.S. policy called for improved collective monitoring and review of East-West trade relations among the OECD countries to avoid competitive underpricing or subsidies in trading with the monolithic foreign trade organizations in the Soviet Union and Eastern Europe.

United States policy, as defined by the prudent approach, reflected many aspects of a deterrence, rather than denial or detente, perspective on East-West trade. It recognized the imperative to invest in and protect strategic technological capabilities as the underlying guarantor of deterrence (unlike the tendency under a detente perspective to substitute economic cooperation for defense). It also recognized the potential use of economic sanctions to influence the political and psychological management of deterrence, especially in crisis situations. At the same time, it did not extend policies of denial to general economic relations and actually held open the possibility of expanding economic ties, as long as these ties did not contribute to a leakage of strategic technologies or to excessive Western economic vulnerability in the face of Soviet threats to disrupt these ties.

Early Reagan administration decisions on East-West trade conformed quite closely to the guidelines of the prudent approach. The decision to lift the grain embargo, the priority at the Ottawa Summit for tightening strategic controls, the initiative to organize Western cooperation to develop alternative energy supplies rather than embargo equipment for the Soviet gas pipeline, and the secret meetings to plan contingency foreign policy sanctions in the case of a Soviet invasion of Poland—all were broadly consistent with this approach. That the approach did not hold after December 1981 was due both to the failure of U.S. and European officials to follow through on the various initiatives and to the inevitable boost that the military coup in Poland provided for those who advocated a return to the full denial approach to East-West trade.

Lifting the Grain Embargo

The decision to lift the grain embargo in April 1981 was dictated chiefly by domestic politics, but it was not inconsistent with the prudent approach (except in one minor way, considered later). The prudent approach allowed for, even encouraged, nonstrategic trade as long as that trade did not develop menacing levels of dependence and vulnerability on Soviet markets or resources. In 1979 the United States depended on Soviet markets for 16 percent of its total wheat exports and 21.2 percent of its total corn exports. This degree of dependence did not inhibit the United States from imposing the embargo (which blocked grain shipment beyond the level of 8 million tons per year guaranteed by the 1975 Long-Term Supply Agreement between the Soviet Union and the United States). The United States expected its grain exports to grow to other markets, such as Mexico and China, and did not feel a significant vulnerability from its dependence on the Soviet market (Paarlberg 1985, p. 172). Imposing the grain embargo, therefore, demonstrated American resolve in the face of Soviet aggression in Afghanistan at what was perceived to be a meaningful but acceptable level of cost. Moreover, this cost was expected to fall largely on the United States, not its Western allies, particularly not on Western Europe, which at that time exported very small amounts of grain to the Soviet Union (200,000 tons in the year before the embargo).

The embargo, as Robert Paarlberg persuasively shows, did not work because the United States was unable to halt grain exports to the Soviet Union by other suppliers, principally Argentina but also Australia, Canada, and the European Community. The United States contributed to this outcome somewhat by accelerating its own sales to non-Soviet markets. In effect, U.S. exports that might have gone to the Soviet Union went to third markets, and other grain exporters stepped up sales to the Soviet Union that might have gone to third markets. In this way the Soviet Union secured a "near-total replacement of sales embargoed by the United States." Paarlberg explains:

> As larger quantities of non-U.S. grain began to move to the Soviet Union, embargoed U.S. grain found compensating opportunities to move elsewhere. International grain shipments were substantially rerouted after the embargo announcement, but neither the Soviet Union—nor the United States—had been forced by the embargo into a dramatic downward adjustment in the total volume of its trade. On the contrary, U.S. grain exports as well as Soviet grain imports continued to grow during the embargo [1985, p. 197].

Whether the failure of the grain embargo was the result of perfidy on the part of the other suppliers, including the European Community, or of American haste to compete with other suppliers in non-Soviet markets may be debated. The fact that the embargo was imposed at all suggests that the United States took seriously its responsibilities for managing deterrence and was not primarily motivated by considerations of domestic politics or commercial advantage. Even if it expected to make up lost Soviet sales in other markets, it could not have been sure this would happen. It risked losing sales and was not asking its allies to take similar risks. The cynicism with which Europe greeted the lifting of the embargo, therefore, does not seem justified. The European Community had exploited the embargo to increase its own grain sales to the Soviet Union, albeit modestly when compared to some other grain exporters (Paarlberg 1985, p. 193).

Lifting the embargo was ill timed, to be sure. Secretary of State Alexander Haig opposed lifting it because he felt it sent the wrong signal to the Soviet Union, which was stepping up its pressure on Poland, and because he thought he could get something from the Soviet Union in return for it (Haig 1984, pp. 110–116). Thus, lifting the embargo muddled slightly the signals being sent to the Soviet Union and in that sense contravened the foreign policy guidelines of the prudent approach. But it also signaled correctly that the new approach was not the equivalent of denial. The United States would continue to trade with the Soviet Union in nonstrategic areas even as it tightened its controls on strategic trade. Ironically, the Europeans never chose to interpret the lifting of the grain embargo to make this point in the subsequent gas pipeline controversy.

The Pipeline Issue

The United States made a conscious decision going into the Ottawa Summit in July 1981 not to embargo gas pipeline equipment and technology to the Soviet Union. In 1978 the Carter administration had imposed license requirements on oil and gas equipment and technology to the Soviet Union for exploration and development purposes (but not for refining and transmission purposes); in January 1980, as part of the Afghanistan sanctions, Carter decided to deny future licenses for technology and to review licenses for equipment on a case-by-case basis. These decisions, however, did not affect equipment and technology for transmission (or refining) activities such as the Soviet gas pipeline. In summer 1981 President Reagan decided not to extend the Carter sanctions to cover transmission activities (i.e., the pipeline equipment), despite strong recommendations from Secretary of Defense Caspar Weinberger to do so (*NYT*, July 19, 1981; Haig 1984, pp. 240–241). This decision was taken in full recognition of the fact that the pipeline deal in Western Europe would be consummated by fall 1981 and that the Ottawa Summit represented the last opportunity to stop the project. The President chose instead to oppose the project in principle but to do nothing to impede it directly by export controls.

At this point the U.S. government, as a whole, viewed the pipeline more as a psychological than as a physical (e.g., generating hard currency earnings for the Soviets to purchase military technologies) threat. It was the wrong timing for a major East–West economic project when the West was trying to reverse the deterioration of the strategic balance during the 1970s by increasing defense spending and deploying new NATO nuclear weapons. The project might seduce public opinion into thinking that nuclear deployments were unnecessary, particularly in critical countries such as West Germany. Europe, on the other hand, viewed the project as essential to maintain economic contacts even as the West was beefing up its defenses.

The debate was a lively one but the United States believed its opposition to the pipeline would have an effect on the psychological environment and future large-scale energy projects with the Soviets even if the Europeans went ahead with the immediate project. The pipeline therefore might never have become a major allied controversy if the Soviet-inspired military coup in Poland had not occurred in December 1981 to alter radically the psychological environment once again. This conclusion is supported by the fact that, as late as 4 days before the Polish coup, the United States approved a shipment of Caterpillar pipelayers for the Soviet project, despite strenuous objections from the Defense

Department and some officials in the Commerce Department (Jentleson 1986, p. 205).

In 1981 U.S. policy gave priority to tightening strategic export controls, accelerating secret contingency talks with the allies to impose sanctions on the Soviet Union if the Polish situation got out of hand, and developing an alternative energy program with Western Europe to make future reliance on Soviet energy less attractive and less necessary. Had the United States pursued all aspects of this program vigorously, had the allies cooperated in this approach, and had the Polish situation eased rather than deteriorated, the United States might have succeeded in making the intermediate prudent or deterrence approach to East–West trade stick. In 1981, however, the allies did not want to discuss East–West trade issues, let alone cooperate with the United States in a prudent approach to these issues.

In the preparations for Ottawa, the allies rejected the idea of a summit discussion on East–West trade. They refused to engage the substantive issues and recommended at one point that the written presentation of U.S. views be reduced to a set of questions that the heads might or might not discuss at their discretion. The United States made clear that the President would address this issue, whether it was on the agenda or not. At the last minute, the topic was added to the end of the Ottawa agenda.

On the first day of the summit, the President informed the allies that he was sensitive to some of their concerns about his policies toward the Soviet Union. But he wanted to be honest and tell them clearly that he regarded the Soviet threat of expansion as real. It was essential therefore that the allies coordinate their East–West policies, whether in contingency planning for crisis, as had already begun in NATO (see later discussion), or in controlling the transfer of defense-related technology and equipment to the Soviet Union.

The President's message was clear and direct, but it went completely unanswered. No discussion of East–West trade issues occurred at the meeting until the summit countries convened on the last day to decide the final language of the summit communiqué. They had in fact finalized the communiqué and Prime Minister Trudeau was about to adjourn the meeting when President Reagan indicated that he had something to say about East–West trade. He spoke for about 10 minutes and then offered a couple of short paragraphs on East–West trade for the communiqué. Only one or two leaders responded, principally Chancellor Helmut Schmidt of Germany, who adopted, according to Reagan advisers in the meeting, a demeanor somewhat like a school master in downplaying U.S. concern about East–West trade issues. The entire conversation lasted less than half an hour. The subject was treated much more as an afterthought than as the serious discussion the United States had sought.

The Ottawa Communiqué ultimately called for "consultations and, where appropriate, coordination . . . to ensure that, in the field of East–West relations, our economic policies continue to be compatible with our political and security objectives." The summit countries also agreed "to consult to improve the present system of controls on trade in strategic goods and related technology with the U.S.S.R." This language became the basis for convening in January 1982 the first high-level meeting since 1957 of the COCOM organization, the informal Consultative Committee of NATO countries, minus Iceland plus Japan (and since 1989, Australia), that implemented multilateral strategic export controls (for text of Ottawa Communiqué, see *NYT*, July 22, 1981).

Altogether, the Ottawa decisions reflected American priorities. In private discussions President Reagan raised the pipeline issue, but he did not do so consistently or vigorously. At this point, the United States gave highest priority to strategic and foreign policy contingency controls, not concerns about economic security under which the pipeline was considered. The convening of the first high-level COCOM meeting in 25 years was considered a significant step forward, and the United States had initiated secret discussions with the key allies to coordinate economic sanctions in the specific event of an invasion of Poland by the Soviet Union (see later discussion). Moreover, after the Ottawa Summit, the United States dispatched a high-level mission to Europe to consider the cooperative development of allied energy projects as alternatives to Soviet energy supplies.

Nevertheless, the Ottawa Summit experience was unfortunate for allied relations on East-West trade issues and left a bad taste with some U.S. officials, especially the American President. The topic would smolder in allied relations from this point on. It might not have exploded, as it did, except for the Polish events of December 1981. But it was a sore point, both because the United States felt a certain frustration in getting the allies to take the subject seriously and because the allies could never be sure that the United States was, in fact, this time around, serious about the subject.

In the past the United States had blown hot and cold on the issue, sometimes simultaneously. During the detente period, Richard Nixon and Henry Kissinger had sought to promote trade and technology flows to the Soviet Union as part of the web of interdependence that would give the Soviet Union a heightened stake in the international system. At the same time, Senator Henry Jackson had engineered sharp legislative restrictions on trade and credit to the Soviet Union, until or unless the Soviet Union relaxed emigration rules on Soviet dissidents. In 1976 Secretary of State Kissinger asked the OECD to consider East-West trade issues on a more systematic basis (Kissinger 1976, pp. 78-79), but the Carter administration showed little interest in the subject until Soviet mistreatment of prominent dissidents and eventually the Soviet invasion of Afghanistan provoked harsh new trade embargoes against the Soviet Union (Brzezinski 1983, pp. 322 ff, 430 ff). It was little wonder, then, that when President Reagan addressed the subject at Ottawa, the allies were inclined to wait him out, to see if this was just another passing fad with the American leadership.

Swing Back Toward Denial: Poland and the Versailles Summit

There were real differences that accounted for conflicting views between the United States and its allies on East-West trade. The United States had primary responsibility in the Western alliance for both deterrence and defense. Hence, it was more likely than Europe or Japan to emphasize the control of strategic defense technologies to the Soviet Union and the use of foreign policy controls to manage deterrence vis-à-vis the Soviet Union. It had more strategic defense systems to protect, and in a crisis it was U.S. reactions to Soviet provocation that the Soviets and the rest of the world watched, not the reactions of the allies.

It was not that the allies cared less about defense or deterrence. After all, their security depended on the American nuclear deterrent and Western technological leadership to support that deterrent. At some level, the allies agreed on the need to protect strategic technologies and to react credibly to Soviet provocations. They simply disagreed in how far they wished to go in these areas. Germany and Japan, the two most advanced commercial technological powers among the allies, did not exercise independent responsibilities for their own security. Accordingly, their defense agencies were only marginally involved in strategic export control processes, and they were unlikely to think of East–West trade in terms of managing deterrence. Britain and France exercised independent security policies, but they tended to define their security interests less globally than the United States. They also defined a narrower range of technologies to protect or to restrict for foreign policy purposes than the United States did.

The differences in Western interests went beyond the strategic and foreign policy area. Economically, Western Europe and Japan were marginally more dependent than the United States on trade with the East, particularly in industrial products that were more likely to be controlled. Historically, Western Europe was closer to the East than the United States. The Soviet Union was still thought of as another European power (as Gorbachev recognizes today in calling for a "Common European Home"), and Europe had had centuries of experience dealing with the Soviet Union and Eastern Europe and felt that it was often better able to relate to these countries diplomatically than the more remote and less experienced United States (Hillenbrand 1988). Politically, Europe was also far more sensitive to East–West issues than the United States. From 1945 to 1989, the continent had been divided. Even after 40 years, as the world learned in 1989, this division was artificial. Germany, of course, was a special case. It had to contend not only with a divided region but a divided country and a perilous geographic location as well. It viewed economic and human contacts with the East, particularly with East Germany, as a matter of family ties and cultural identity, more so than as instruments of leverage to protect defense or manage deterrence. For all of these reasons, Europe and Japan were more likely to stress economic cooperation with the East and to oppose long lists of strategic controls or frequent use of foreign policy controls in East–West trade.

Basic differences of interest between the United States and its allies did not foreordain conflict, however. The lack of adequate institutional machinery to narrow these differences through negotiation also helped. By any reasonable standards, COCOM in 1981 was weak. It had lost some of its *raison d'être* in the detente period and showed no capacity from 1977 on to resolve tough issues on strategic controls, such as computers (Mastanduno 1988, Chapter 5; Nau 1988b). The Organization of Economic Cooperation and Development (OECD), which includes neutral countries, considered the subject of East–West trade taboo; and the North Atlantic Treaty Organization (NATO) was inhibited from addressing the subject because some members (e.g., France) rejected the extension of strategic cooperation into economic areas. The summits were reluctant to pick up the issue, as became evident again at the Versailles Summit in 1982. There were no formal structures whatsoever to discuss the imposition of trade controls for foreign policy reasons, other than bilateral consultations and periodic, secret meetings among the main NATO countries—the United States, France, Germany, and Great Britain.

As a consequence, the allies had few mechanisms to assist them when their continuing differences on East–West trade issues surfaced during the Polish crisis of December 1981. Part of the pipeline dispute that followed was precisely about the need for more regularized treatment of East–West issues in existing institutions such as NATO, OECD, and the International Energy Agency (IEA). At the time, the allies discussed East–West trade meaningfully only in COCOM. Accordingly, COCOM made progress over the next year and a half on strategic aspects of East–West trade, not without difficulty to be sure. But, on other aspects, the Polish crisis precipitated the worst alliance dispute over East–West trade in postwar history.

Polish Sanctions

The principal countries in NATO—the United States, Britain, France, and Germany—met secretly four times in the course of 1981 to discuss economic sanctions in the event of a Soviet invasion of Poland. The Soviet Union at this point had mobilized some 35 divisions around the borders of Poland and had put these divisions on alert status several times during the course of 1981, indicating a possible move against the Polish government if Warsaw did not contain the Solidarity movement. Although Soviet forces never invaded and some analysts drew the conclusion that deterrence ultimately worked, it was a massive and blatant use of force and intimidation by the Soviet Union. The United States pressed its allies to prepare for a firm response if the Soviet Union should invade.

Even under these circumstances, however, the allies were reluctant to agree to sanctions beforehand, arguing that a Soviet invasion was purely hypothetical and that it was unrealistic to specify economic actions before such an invasion actually occurred. The allies undoubtedly had good reasons not to cooperate. This was a period in which U.S. spokesmen were making loose comments about nuclear war, and the Reagan administration generally was viewed in Europe as belligerent. Nevertheless, the failure to cooperate weakened deterrence in the face of the most massive use of force in Europe since the Czechoslovakian invasion in 1968.

What is more, by not bargaining in 1981, the Europeans missed an opportunity to moderate the subsequent U.S. response. Had they been willing to commit to a collective response with the United States, they might have been able to persuade the United States to accept more moderate sanctions or at least sanctions that were more acceptable to the allies. Moreover, they might have been able to protect Europe's interests in the pipeline project. At this point, as we noted, the United States viewed the pipeline as a matter of economic security, not as a foreign policy sanction. The U.S. energy mission to Europe in October was not aimed at stopping the pipeline, the first contracts on which were signed in September. It aimed instead to minimize the size of the pipeline and associated European gas contracts and generally to redirect European thinking away from the detente-oriented impulses of the 1970s to projects of Western energy cooperation. Admittedly, the United States had little to offer to Europe to promote Western cooperation. Budget pressures in the United States made it difficult for the United States to contribute financially to alternative projects. And proposals to develop U.S. coal and Norwegian gas exports to Western Europe, the two main alternatives to Soviet gas, were not only expensive but long term (Jentleson 1986, pp. 184–190). Nevertheless, Europe had the chance in the fall of 1981 to commit the United States to a discussion that would have accepted a scaled-down Soviet gas project. By not cooperating, the Europeans left the pipeline project open to possible punitive sanctions and gave the hardliners in the United States, particularly in the Defense Department, a second crack at shutting the pipeline project down altogether.

When the Polish military staged a coup on December 13th and imposed martial law on Poland, Western economic sanctions were inevitable. The only questions were what would be included and whether or not the sanctions would be unified. The gas pipeline, like grain sales in January 1980, was a big-ticket item in East–West trade. Embargoing the pipeline, therefore, would catch the attention of the Soviets and the world and serve the purposes of deterrence.

On December 29, the United States extended licensing requirements to cover oil and gas equipment and technology for refining and transmission purposes. That brought equipment such as pipelayers and compressor stations used in the Soviet pipeline into the web of export controls. (Technically, Caterpillar did not need a license for the pipelayers that were approved for export 4 days before the Polish coup but sought approval anyway because of the controversial character of the pipeline project in 1981.) The United States then suspended all existing licenses for oil and gas equipment and technology, as well as for other high-technology products, and banned all new licenses. Other less significant measures were also taken, including postponement of the negotiations on a new long-term grain agreement with the Soviets.

The United States and its allies met in emergency session in NATO in January 1982 to decide on collective action. Although the allies joined the United States in condemning the Soviets for their active support of repression in Poland and agreed on the conditions in Poland that would permit a lifting of sanctions (end to martial law, release of political prisoners, and resumption of a dialogue with the church), they did not agree on a common set of sanctions. The allies felt the situation could have been worse involving a direct Soviet invasion. With its leadership responsibilities for deterrence, the United States felt the situation was bad enough. Thus, the U.S. sanctions went well beyond anything the Western Europeans or Japanese were willing to impose, especially in the oil and gas sector.

What is more, U.S. sanctions now interrupted European and Japanese sales of energy equipment to the Soviet Union for the Siberian pipeline and the Sakhalin project (a Japanese–Soviet oil exploration project on the Soviet island of Sakhalin in the Sea of Japan). Three major consortia in Europe had contracts for turbines and compressors to be used in the pipeline—Mannesmann (Germany)/Creusot–Loire (France), Nuovo Pignone (Italy), and John Brown (UK). Some of these consortia had received parts (mostly rotors) from American companies, particularly General Electric, that they now intended to put in compressors for shipment to the Soviet Union. The U.S. sanctions prohibited shipment of these parts in assembled compressors. At this point, the sanctions did not apply to parts produced entirely by American subsidies in Europe or European firms operating under U.S. licenses, which included most of the firms in the European consortia. The United States asked American subsidiaries and European licensees to cooperate, but European officials made clear that their pledge not to undermine the American sanctions did not include contracts to European-based firms or contracts for which American firms were merely subcontractors (Jentleson 1986, p. 192).

Europeans frequently argued in this period that the pipeline sanctions were mean-spirited because they attacked European exports to the Soviet Union while exempting American grain exports. (Haig records this argument, for example, made to him by Margaret Thatcher in late January 1982; 1984, pp. 255–256). The charge was indicative of the unconstructive character of U.S.–allied discussions on East–West trade throughout 1981. In fact, the charge was totally false. American firms, which had numerous direct contracts for the pipeline, were hit first and hardest. Caterpillar had contracts of $400 million at stake; Fiat Allis, $500 million; and General Electric, $170 million (Jentleson 1986, pp. 204–205). Altogether, American contracts at stake slightly exceeded those in Europe, each totaling between $1 and $1.5 billion (Commerce Department calculation at the time). The U.S. contracts were stopped immediately in December 1981 at a time when unemployment was just as severe a problem in the United States as it was in Europe. European contracts, by contrast, went forward, impeded slightly perhaps by parts that could no longer be shipped from the United States. The sanctions were hardly a favor to American exports. European charges to that effect suggested the ill will that is generated when sanctions are discussed only in the heat of a crisis.

The Buckley Group

When the United States went ahead with unilateral sanctions against the pipeline, it set a time bomb among its own industry. If the Europeans and the Japanese continued to trade with the Soviets, they would eventually take contracts away from American firms. Moreover, the Soviet pipeline would go ahead. Hence, American firms would soon become strong opponents of the sanctions, even though they were traditionally reluctant to challenge U.S. government actions on national security issues. At some point the sanctions would have to be lifted because of mounting domestic political pressure or they would have to be extended extraterritorially to block shipments from U.S. subsidiaries and licensees abroad.

In January, Secretary of State Al Haig urged the President to delay applying the sanctions to U.S. affiliates abroad until Haig had a chance to negotiate the issue with the Europeans (Haig 1984, p. 255). Later that month, the United States dispatched a team to Europe headed by Under Secretary of State James L. Buckley to negotiate a possible substitute arrangement on credits. The subsidization of credits to the Soviet Union was a particularly galling point to the Americans. It was not only the fact that the pipeline might lead to dangerous levels of dependence and vulnerability, but also the fact that the West was subsidizing the equipment and credits to achieve these levels of trade. The French were particular offenders in this regard. They had negotiated in 1978 a five-year agreement with the Soviets to provide an open line of credit at 7.8 percent interest, the minimum interest rate at that time under the 1978 OECD Agreement on Guidelines for Officially Supported Export Credit. They "remained reluctant to go along with either firm minimum interest rates or limits on total credits" (Jentleson 1986, p. 182). The Buckley group set out in spring 1982 to negotiate higher minimum interest rates for the Soviet Union under the OECD arrangement and to set volume limits on credits, including the amount of down payment on equipment purchases that could be financed.

The Buckley group met several times prior to the Versailles Summit on June 4–6, 1982. These negotiations were carried out separately from the preparations for the summit, which were conducted by the personal representatives, or sherpas, for the summit leaders. This separation either contributed to the confusion that developed over East–West trade issues at Versailles or was itself exploited by the allies, especially the French, to create such confusion. Either way, the American delegation helped the confusion along.

From the beginning, Secretary Haig and his Assistant Secretary of State for Economics and Business Affairs, Robert Hormats, who also served as President Reagan's personal representative for the Versailles Summit, operated on a different set of assumptions from officials in the White House and members of the NSC staff. As Secretary Haig recalls in his memoirs, he never believed that the December sanctions were meant to apply retroactively to export licenses that had already been granted for shipment of U.S. compressor parts to Europe. "I doubt that this was the President's intent either," he says (Haig 1984, p. 254). Hence, he assumed that if the United States could negotiate a reasonable agreement on credits, the President would lift existing sanctions, and he undoubtedly told the Europeans that this was the case. He also suggested to them, according to his memoirs, that the United States would be willing to intervene more actively in exchange markets if the Europeans, meaning especially the French, for whom intervention was the key issue (see Chapter 7), would adopt a more cooperative policy to limit future government-backed credits to the Soviet Union (Haig 1984, pp. 305–306).

This outlook was never the understanding of the National Security Adviser, Judge William Clark, or other members of the White House and NSC staff. As

Haig recalls, the decision to block existing contracts was taken administratively after the December 29 decisions. That is true, but the President fully concurred in this decision. For the White House staff, therefore, it was never a question of lifting sanctions on existing contracts for parts shipped from the United States, only a question of whether the President would extend the sanctions to cover equipment also shipped from U.S. subsidiaries and licensees abroad.

That would depend on the outcome of the Buckley mission. This message was clearly conveyed to the French and others in sherpa discussions before the Versailles Summit. Hormats, who led the sherpa team, was in a particularly difficult position. He was a trusted Haig associate, who would later resign when Haig resigned. On the other hand, he was also the President's personal representative to the summit, even though he did not always have the full trust and confidence of some members of the White House staff. His nomination to become the new Under Secretary of State for Economic Affairs, replacing Rashish, who resigned in January, was blocked at the time by conservative senators on the Hill.

Given these internal fissures within America's own representation to the summit, it is not inconceivable that the Europeans and Japanese heard several messages from the United States on the sanctions issue and chose to accept the ones they wanted to hear. The French, who were the hosts at Versailles, tried throughout the preparations to take East-West issues off the agenda. They argued that the Buckley group was treating the issues, and there was no need to bring the issues up among the summit leaders. Eventually, the French relented, but only at the last sherpa meeting, and then, as we shall see, they attempted to preempt and dispose of the East-West issue just before the summit began. The Japanese always considered their concerns about the Sakhalin project to be separate from the Soviet pipeline in Europe. Thus, they too may have operated under the illusion that existing contracts for Sakhalin would be exempt from U.S. sanctions.

The Versailles Summit

The planning for Versailles on East-West issues, therefore, was less than optimal. But there should have been no doubt about the U.S. position going into Versailles (especially the President's views—see later discussion). At NSC meetings before the Versailles Summit, the issues were thoroughly aired.

Haig felt the President should lift sanctions on existing contracts if the allies agreed to limit credits and reduce future gas purchases from the Soviet Union—specifically, no second leg of the pipeline, which the Europeans originally planned, and reduced purchases from the first pipeline, which was already happening in the market place as energy prices began to fall in 1982. Haig also wanted to exempt the Japanese contracts immediately. Frank Carlucci, Deputy Secretary of Defense, disagreed. Given the situation in Poland, he felt, it was no time to lift sanctions. Moreover, Carlucci doubted if an effective credit agreement could be achieved at Versailles. In the absence of agreement, the sanctions were the only means to delay and shrink the pipeline project and buy time to develop alternatives. Bill Casey, Director of the CIA, agreed.

Secretary of Commerce Baldrige and Secretary of Treasury Regan felt the outcome should depend on what Under Secretary Buckley could negotiate on credits. The last meeting of the Buckley group was scheduled a few days before the Versailles Summit. If Buckley got a good credit agreement, the Commerce and Treasury secretaries both favored lifting sanctions, although Regan felt this should be done separately after the Summit so as not to weaken the pressure at the summit for a good credit agreement. Under Secretary Buckley was not optimistic that he could get a good agreement, pointing out that others were opposed to it besides the French. He wanted to be able to lift the sanctions as leverage to achieve such an agreement, rather than to wait and lift the sanctions only after an agreement was reached. On the other hand, if the United States did not get a credit agreement, Under Secretary Buckley felt the United States would have to extend the sanctions to U.S. affiliates abroad.

In the President's view, the sanctions had been imposed because of Poland. He wanted the allies to support and strengthen the sanctions. If they agreed to a credit agreement, he would not extend existing sanctions to cover U.S. affiliates abroad. On the other hand, lifting existing sanctions depended on what happened in Poland. Walesa was still in jail. The President felt the allies should help persuade the Soviet Union and Warsaw to release him and meet other conditions agreed in NATO. If they did that, the United States would let the pipeline equipment go. The Soviets had to show deeds. For the President, lifting existing sanctions depended on what the Soviets did; not extending the sanctions to cover U.S. affiliates abroad depended on what the allies did — namely, the acceptance of a credible credit agreement.

There was no room for confusion in these discussions; nevertheless confusion prevailed at the summit. On the night the American delegation arrived, a rumor circulated that the French had agreed to raise the Soviet Union to a higher interest rate category in the OECD credit arrangement. That meant, so some members of the media were informed, that the East–West trade issue was settled and off the agenda. There was no substance to the rumor, and to avoid confusion, NSC and State Department aides jointly formulated talking points for the President to use in bilateral meetings the next day (the day before the summit officially opened) with his counterparts from Japan, Britain, Germany, Italy, and France. These talking points reiterated the President's position and went on to specify the exact terms of a credit agreement. The agreement, the talking points indicated, must include a commitment to restrict officially supported credits to the Soviet Union, where "restrict" meant to increase the cost or limit the volume of credit or both, not just to monitor credit flows or exchange information on credit flows. In addition, the agreement must provide a follow-up mechanism to flesh out the details of these restrictions by an early date (e.g., mid-July). The talking points then indicated that if the United States secured that kind of agreement, the President would definitely not extend the sanctions to cover U.S. affiliates abroad and would be as flexible as possible on lifting selectively existing sanctions, both for equipment already in Europe and for equipment going to Sakhalin.

When the talking points were shown to Secretary Haig, he recoiled at the rigid terms specified for a credit agreement and rebuked his own as well as NSC aides for coming up with such a document. When the talking points were shown to Judge Clark, he approved and added a note in the margin to the effect that the President might not wish to go so far as to say that he would do what he could to lift existing sanctions if he got the specified credit agreement. Clark knew that a credit agreement with teeth in it was a long shot, and he was reluctant to give up the President's leverage. Haig was ready to give up the sanctions for any face-saving credit agreement. The gulf between the President's two principal foreign policy advisers could not have been wider.

Haig reports that on the next day, the first full day of the summit, he secured an agreement among the foreign and finance ministers to restrain credits to the Soviet Union in return for a U.S. agreement to intervene in exchange markets and bolster the franc (Haig 1984, p. 309). If this was the case, it was a linkage that was never authorized at any poi nt in the summit preparations or at the summit itself and was apparently an agreement that, as Haig acknowledges, the U.S. Secretary of the Treasury did not participate in and was not even aware of. In fact, no one seemed to be aware of such an agreement, because when Mitterrand began the East-West trade discussions on the last morning of the summit, he seemed, according to Haig, "to ignore completely what the ministers had tortuously achieved the evening before" (1984, p. 309).

French participants at the Summit offer an alternative explanation (confided to the author later in 1982). According to them, the chief sherpas for the summit did negotiate a final text on East-West trade late in the night before the final plenary session. However, finance ministers were not involved in these discussions, and these ministers came up the next morning with a totally different text that did not say anything about exchange market intervention. Thus, the heads of state and government began the final plenary session with contradictory texts before them. According to this explanation, Mitterrand chose simply to ignore both agreements and to begin again from scratch.

In any event, the discussion of the East-West credit issue on the final day of the summit was ill prepared and confused, to say the least. (The author was notetaker in this session.) By this point it was apparent that no detailed credit agreement was in the offing. The French had also made clear that there was no agreement on the OECD credit arrangement, as earlier rumored. Thus, the debate in the final plenary session of the summit was largely about whether credits would be limited at all and if so how the word *limiting* would appear in the final communiqué language on East-West credits.

Three leaders—Trudeau, Schmidt, and Mitterrand—opposed the use of the word *limiting* applied directly to export credits. Reagan and Thatcher strongly supported it. Spadolini, the Italian head, and Suzuki, the Japanese head, were silent but did not oppose the language. The French were particularly worried that language to limit credits in the official communiqué would apply only to governments and hence government-supplied credits, which accounted for most of French East-West credits. Meanwhile, credits supplied by private banks in Germany, the United States, and other summit countries would not be included, because governments in these countries had no direct control over private lending. Mitterrand's concern was a valid one. But it was too technical to deal with at this late stage and among heads of state and government. Thus, Mitterrand sought to avoid any communiqué language that directly limited export credits. At this point in the discussion, the language still called for the "need for limiting export credits in light of commercial prudence." When Trudeau finally indicated that he could accept this language as long as it did not imply arbitrary cutbacks in credit, Mitterrand appeared to be out-flanked. At the last minute, however, Spadolini suggested rearranging the words to read the "need for commercial prudence in limiting export credits." Now the language called for commercial prudence, not directly for limiting export credits.

Mitterrand seized upon the new language and no doubt genuinely believed, as his comments at a press conference the next day suggested, that the language imposed no special obligations on France to limit export credits. In denying the next day that France had any obligation to limit export credits, Mitterrand referred to Secretary Regan's remark right after the summit that the United States had no obligation to intervene in exchange markets. As the preceding account of the final discussions at Versailles suggests, both men were technically correct. There had been no firm or clear agreements on either subject. Normally, the summit leaders might have taken refuge in ambiguity. That they aired their differences publicly, however, suggested the depth of feeling with which they held to their substantive positions. Their feelings were soon to erupt to the surface in one of the worst alliance brouhahas in the postwar period.

Extraterritorial Sanctions

The President traveled to Bonn, Rome, and London after the summit before returning to the United States on June 11. The sanctions issue followed him on his travels. Apparently, in Bonn, he made one more attempt to clarify his views to Chancellor Schmidt. According to at least one participant in the meeting, Chancellor Schmidt looked out the window when the subject of sanctions came up. The gesture, if it is true, was symbolic of the way the allies had dealt with the East–West issue for an entire year. Since the Ottawa Summit, they had done everything they could to ignore the issue. They had good reasons no doubt to feel that the American President was obsessed with the issue. But the President had equally good reason to feel that they were not listening. The issue would have to be escalated before it could be resolved.

The escalation came on June 18. The President, as he had said he would if he did not get a credit agreement with teeth, finally extended the pipeline sanctions to cover U.S. affiliates abroad. It was now illegal for American subsidiaries or licensing affiliates in Europe to ship pipeline equipment to the Soviet Union. The allies predictably rejected this invasion of their sovereignty. The British, French, Italian, and West German governments invoked national blocking legislation that protected their companies from American legislation and ordered them to proceed with their orders. When the European companies did so, the United States banned all American exports to these companies, subsequently scaled back to include only oil and gas exports. Secretary Haig resigned over this incident (although it was merely the straw that broke the camel's back); and the new Secretary of State, George Shultz, spent the summer months trying to find a way to stop the open economic warfare that now raged across the Atlantic.

Prudent Approach Revisited: From Versailles to Williamsburg

The Versailles Communiqué had actually borrowed language from the Reagan administration's initial prudent approach paper presented 1 year earlier at the Ottawa Summit. In the main paragraph on East-West trade, it read: "We agree to pursue a prudent and diversified economic approach to the U.S.S.R. and Eastern Europe, consistent with our political and security nterest." It went on to call for further work to control exports of strategic goods in COCOM (which could not be mentioned in the communiqué because its exi;tence is supposedly a secret), to promote information exchange in the OECD on all aspects of economic, commercial, and financial relations with the Soviet Union and Eastern Europe, and to handle cautiously financial relations w)th the East, including the "need for commercial prudence in limiting export credits." Just as its prudent approach was achieving some consensus, however, the Reagan administration swung toward a more restrictive policy of denial (for text of Versailles Communiqué, see *NYT*, June 7, 1982).

Differences within the alliance were just as much about the process of East-West trade discussions (i.e., where these issues could be discussed systematically among the allies) as about the substance of these issues. That point was brought home again at the end of June, when in the midst of the pipeline imbroglio, the allies agreed on a new OECD credit arrangement, which raised interest rates and put the Soviet Union into the highest category of rates. This agreement satisfied, in effect, a major objective of U.S. policy and was possible in part because the OECD credit arrangement, like COCOM, provided a forum where the allies could discuss these issues outside the glare and media attention of summits and crises. By contrast, although the Versailles Communiqué called for "periodic ex post review of East-West economic and financial relations", the Buckley group to discuss volume limits on credits was discontinued, and the OECD and International Energy Agency (IEA) were reluctant to consider systematically broader economic and energy issues in relations with the East. Any plan to patch up the pipeline dispute would now have to consider this lack of adequate negotiating mechanisms, or the substantive differences on East-West trade would continue openly to divide the alliance.

Flirting with Denial

From December 1981 to summer 1982, American East-West trade policy hardened. The United States identified some 22 categories of oil and gas equipment and technology that it now classified as strategic and for which it sought COCOM strategic export controls. Moreover, the United States adopted Defense Department arguments that buying gas from the Soviet Union increased Soviet hard currency earnings and helped the Soviets finance military technology acquisitions from the West. This argument, which went beyond the concerns of summer 1981 about the vulnerability of Western Europe to possible cutoffs of Soviet gas supplies, was in effect an appeal to discontinue all imports from the Soviet Union, since any Soviet export to the West earned hard currency and contributed to total Soviet resources. Together with expanded strategic controls on Western exports such as oil and gas, this approach amounted to a total ban on trade with the Soviet Union. The National Security Adviser, Judge Clark, outlined the new approach already in May 1982. He said, "We must force our principal adversary, the Soviet Union, to bear the brunt of its economic shortcomings" (*NYT*, May 22, 1982). The United States would flirt once again with the denial approach to East–West trade.

The denial approach is highly persuasive in the abstract but totally unworkable in practice (for the theoretical case, see Walinsky 1982–1983). It was also completely inconsistent with U.S. policy on grain sales. In the summer of 1982, in the heat of the battle over sanctions, the United States authorized a 1-year renewal of the Long-Term Grain Agreement with the Soviet Union, which expired in September 1982 and a multiyear extension, which had been postponed under the sanctions in December 1981. Although the argument was made that U.S. grain sales, unlike Soviet gas sales, absorbed rather than earned hard currency, this argument was completely specious. From a resource transfer point of view, grain sales allowed the Soviets to reallocate internal resources to exports and earn additional hard currency, no less than gas export sales. Nevertheless, the administration extended the Grain Agreement for 1 year in fall 1982, clearly with the November congressional elections in mind, and then signed a new long-term agreement in August 1983 that guaranteed future U.S. grain sales against embargo. It was an indecent contradiction of U.S. denial policy, to say the least. Zbigniew Brzezinski, NSC adviser to President Carter, whose administration had been severely criticized for weakness toward the Soviet Union, appropriately tweaked the Reagan administration when he observed on the signing of the new long-term agreement, "What is truly distasteful is Secretary [of Agriculture] Block crawling on his knees to Moscow" (*NYT*, August 26, 1983).

Despite these drawbacks, a policy of denial had powerful appeal in the early Reagan administration—particularly when compared with detente, the only other alternative that had been adequately conceptualized up to that point. Denial avoided any tendency for economic and political relations to substitute for military defense by calling for all policies, "diplomatic, political, economic and informational . . . [to be] built on a foundation of military strength" (*NYT*, May 22, 1982). The objective was, as one administration official put it, to develop a "long-term alliance strategy on East–West economic relations that has the coherence and depth of our military strategy" (Jentleson 1986, p. 194).

Nevertheless, in the end, denial and detente feed on one another because they both expect too much from economic relations between fundamentally different societies. Two decades of denial of all trade with the Soviet Union after World War II did not significantly handicap the growth of Soviet military capabilities. Nor did a decade of detente in the 1970s significantly alter Soviet domestic society or restrain Soviet foreign policy behavior. Only military and ultimately political means have sufficient consequence to stabilize and steer U.S.-Soviet relations in more peaceful directions. By pressing the case for denial, the Reagan administration only strengthened the subsequent case for detente. If economic ties could be that significant to punish the Soviet Union, they must also be that significant for improving cooperation.

Putting the Genie Back in the Bottle

George Shultz, who became the new Secretary of State in summer 1982, had well-known views on East-West trade policy. In *Business Week* in 1979 he had condemned "light-switch diplomacy" by which the United States sought to turn East-West trade on and off to signal approval or disapproval of Soviet policy (Shultz 1979). Thus, in summer 1982, with a host of other economic problems on the U.S.-allied agenda (see Chapter 7), Shultz approached the pipeline problem primarily to defuse it as quickly as possible without running into the same buzz saw at the White House and Defense Department that had terminated Al Haig's employment with the administration. The White House, on the other hand, was incensed by the way the allies had treated the President's views on East-West trade and was determined to exploit the pipeline dispute to pull the allies as far as possible toward a denial approach. The tug of war within the administration—particularly between the State Department and some elements of the Commerce Department (the trade promotion offices), on the one hand, and the NSC, Defense Department, and export control offices in the Commerce Department, on the other—was at least as intense as it was within the alliance. The tug of war even reached inside the White House, where public relations and policy officials clashed repeatedly over the next several months on the handling of the pipeline and East-West trade issues.

While the NSC and Commerce Department were busy in summer 1982 firing off denial orders to penalize European firms violating U.S. sanctions, George Shultz and the State Department got an early jump on maneuvering the dispute toward eventual resolution. At an informal meeting of NATO foreign ministers in La Sapinière, Canada, in early October, Shultz tabled a so-called nonpaper (a diplomatic paper that no one acknowledges officially) that sought to establish a set of criteria or principles for East-West trade, prescribe some common objectives for this trade, and create a framework for the study and negotiation of issues among the allies. This nonpaper became the basis of intensive discussion during the month of October between the ambassadors of the summit countries in Washington and Secretary Shultz. The NSC and other agencies were informed but not involved in these negotiations. Secretary Shultz shrewdly exploited his initial honeymoon period to preempt the East-West trade issue.

The State Department negotiations eventually produced a "Summary of Conclusions" that detailed allied agreement on the criteria, objectives, and work plan of a series of studies on East-West trade issues. On the basis of this agreement, the United States announced on November 13, 1982, that it had lifted the pipeline sanctions and the extraterritorial extensions of June 18th (*NYT*, November 14, 1982). The French disassociated themselves from the U.S. announcement because it linked the removal of the sanctions with the "Summary of Conclusions" and made it appear that the French had conceded something to get the sanctions lifted. At the same time, by announcing the agreement despite French repudiation, the United States made it appear as though it had gotten nothing in return for lifting the sanctions. Clearly, the agreement could not have been both things — so important that the French would not acknowledge it and so insignificant that the United States in effect got nothing for it. In fact, once diplomatic face was saved on all sides, the "Summary of Conclusions" represented a meaningful set of compromises among the allies. It created a framework that eventually put the pipeline genie and the larger East–West trade disputes back in the bottle.

The "Summary of Conclusions" established four basic criteria for East–West trade, which conformed generally to the categories and policy guidelines of the original U.S. prudent approach paper. Trade should

1. Not contribute directly to the strategic advantage of the Soviet Union.
2. Be conducted at prevailing market prices and terms.
3. Not result in situations where the West is overly dependent on Soviet resources.
4. Produce a balance of advantages for the parties involved.

On the basis of these criteria, the "Summary" set objectives for specific areas, such as limitations of energy imports from the Soviet Union, and then called for a series of individual and overall studies to achieve these objectives. The United States hoped that the study process would create an ongoing framework for discussion of East–West trade issues.

The November accord began a process that could have potentially carried the alliance back toward some middle ground between denial and detente. But it was clear that there was still a long way to go before this issue could be normalized. What is more, as a result of the Versailles Summit, the issue was still identified with heads of state and government. President Reagan would host the summit in Williamsburg in spring 1983. In November 1982 the French had not yet agreed to firm dates for the next summit, and Mitterrand had called the summit process itself into question (see Chapter 7). Thus, the East–West trade issue remained the tinder box that could set off another explosion within the alliance. It would have to be managed with extreme care in the preparations for Williamsburg. It is doubtful that East–West trade issues have ever been worked as intensively and aggressively, both within the alliance and within the U.S. government, as they were over the next 6 months.

East–West Trade Studies

In mid-December, in discussions between Secretary Shultz and French Foreign Minister Cheysson and President Mitterrand, French leaders made clear that they could agree only to conduct individual studies on separate aspects of East–

West trade. They could not accept the idea of an overall study, a new steering group or directorate to oversee such a study, or any new objectives or criteria that created binding commitments. France had always resisted binding commitments in NATO going beyond basic collective security pledges and had distanced itself from common institutional arrangements even in the security area (leaving the integrated NATO command structure in 1966). When U.S. Secretary of State Henry Kissinger tried in 1973 to negotiate an Atlantic Charter for both security and economic affairs, French Foreign Minister Michel Jobert was his principal adversary (Kissinger 1982, pp. 128 ff). In 1982 Jobert was Mitterrand's Trade Minister. He undoubtedly warned Mitterrand that the Americans were trying again to extend binding Atlantic commitments beyond the security sphere and to institutionalize these commitments in a new East–West economic directorate.

If U.S.–French differences were to be surmounted, therefore, the East–West studies would have to go forward without any preconditions. If the results of these studies came close to the criteria and objectives of the "Summary of Conclusions," the United States would be satisfied. And if there had been no precommitments, the French might then accept these results as emerging from the studies themselves and decide at that point to implement the recommendations of the studies.

Accordingly, the studies became the focal point for negotiations over the next several months, and preparations for the Williamsburg Summit became the overall, high-level diplomatic mechanism for forcing the individual studies and steering them toward some kind of overall conclusions. It was a lot of responsibility for the fragile summit process to bear.

Altogether, five studies were undertaken. In fall 1982, COCOM launched a new list review to determine new strategic products and technologies that should be controlled (done once every 3 years under normal circumstances but COCOM, because of its weaknesses, had been unable to conclude a list review since 1977). The United States had a lengthy list of new items to add to COCOM controls, including the 22 categories of oil and gas equipment and technology referred to as other high technology (OHT) products and a list of so-called emerging technologies (advanced technologies with no present but potential future military use) that the United States wanted COCOM to watch. Second, the allies convened the OECD export credit arrangement, that had been renewed for 1 year in late June 1982, to negotiate further provisions providing for automatic adjustment of interest rates to market levels, reduced levels of cover for credits and guarantees (i.e., higher down payments that could not be financed or guaranteed), and shortened maturities, continuing to place the Soviet Union in the highest interest rate category together with Western industrial countries. Third, the OECD initiated talks to monitor and assess the balance of advantages and disadvantages in overall East–West trade and financial relations, an assessment designed to protect Western buyers and sellers from monolithic trading practices of the Soviet Union. Fourth, the International Energy Agency (IEA) initiated an analysis that the United States hoped to use to limit future gas imports into Europe from any single, non-OECD supplier to 30 percent of total supplies. In addition, this study, the United

States hoped, would encourage accelerated development of alternative Western supplies, particularly the large Norwegian Troll gas field in the North Sea, and establish safeguard measures to protect against potential gas supply disruptions. And fifth, NATO launched a study of the security implications of East–West economic relations, which the United States hoped would become part of NATO's regular assessment of Soviet strategic capabilities and would help to upgrade the NATO Economic Committee, which seldom met at a political level.

Preparations for Williamsburg

Early U.S. planning for the Williamsburg Summit sought to draw these studies together and to steer them toward some conclusions at or before the summit. The objective was to satisfy President Reagan's deep concern with these issues without antagonizing the allies and producing a second misunderstanding like the one at Versailles. In January an interagency steering group was set up under State Department leadership to direct and coordinate U.S. participation in the allied studies. This group included members of the NSC staff who harbored deep suspicions that the State Department did not share the President's commitments on East–West trade issues and that the Department, like the allies, regarded the studies as little more than face-saving devices for having lifted the sanctions.

At the first substantive sherpa sessions in San Diego in mid-March, the United States reminded the allies that their two principal reactions to the East–West trade subject over the previous 2 years had been contradictory. On the one hand, the allies had consistently tried to shove the topic off the agenda; but on the other, they had complained that the issue, when it came up at the summits, had not been sufficiently prepared and that President Reagan's views on the subject had caught them by surprise. At Williamsburg, U.S. representatives said, the summit countries had a choice. They could either agree now in March, 2 months before the summit, to have the summit leaders discuss the issue, prepare it carefully, pursue maximum results in the East–West trade studies, and work for a constructive outcome at Williamsburg; or they could debate from March until May whether the topic should be on the agenda, assume that the trade studies were taking care of the issue, and run a large risk at Williamsburg of another serious public confrontation. The French continued to resist the topic for active discussion at Williamsburg. The British, on the other hand, saw the lurking dangers and strongly urged that the United States and France undertake immediately direct high-level contacts to determine exactly what could and could not be achieved in the various studies by the time of the Williamsburg Summit.

For the next month and a half, President Reagan and President Mitterrand carried on direct bilateral contacts, both through private correspondence and through their personal representatives for the summit meeting. These contacts set out the specific results that each country sought and could accept in the five separate East–West trade studies then under way. President Reagan assured President Mitterrand that if the studies produced the results that they agreed upon in their correspondence by the time of the Williamsburg Summit, the leaders could welcome these accomplishments at Williamsburg and otherwise keep the subject off the agenda. If the results were unsatisfactory, however, the President said that the summit communiqué, which the leaders would draft themselves at the summit, would have to express these differences candidly. The French President, in turn, made clear that the summit itself could not, under any circumstances, draw attention to the East–West trade topic. He expected the topic to be dealt with in other institutions. Both leaders therefore had a strong interest in resolving the issue before the summit. French support proved to be critical in reaching agreement in the other institutions such as the OECD.

COCOM, IEA, and OECD Meetings

COCOM met in high-level session again in April 1983. This meeting signifi-cantly expedited the list review, which would be completed over the next year and would tighten controls on electronic and communication equipment and include numerous new items the United States sought to control—computers, robotics, spacecraft, advanced printed circuit boards and manufacturing equipment for such boards, advanced aeroengine technologies, and so on (DOD 1985, pp. 79–87). It also agreed to develop a specific list of disembodied technologies and to study on an urgent basis a permanent inventory of emerging technologies. It charged the enforcement subcommittee to come up with spe-cific recommendations to harmonize national controls, started a program to improve cooperation with neutral and other non-COCOM countries, and authorized a strengthening of the COCOM facilities and staff. It also set up a group to look at the controversial oil and gas equipment and technology items that the United States sought to elevate from foreign policy to strategic con-trols. In the end, only two items from this list were placed under strategic controls. The U.S. program for wider denial of economic as well as military high-technology items did not succeed. But the strategic controls program took a big jump forward, and by most standards, the COCOM meeting met U.S. expectations.

The OECD and IEA Ministerial Meetings in early 1983 also produced substantial results, although not all that the United States wanted. The OECD credit arrangement was strengthened. Interest rates were raised to market levels and better monitoring of cover and maturities was introduced. But no auto-matic adjustment mechanism for interest rates or firm limits on cover or ma-turities was established.

The more novel development in the OECD was the completion of the study on balance of advantages and disadvantages and continuing studies of this sort in years to follow. The 1983 study concluded that trade prospects between East and West were poor, that state trading companies in the East could leverage private Western firms to twist the balance of advantage in favor of the East, and that Soviet efforts to expedite exports, including natural gas, or to increase borrowing in the West could have adverse consequences for the West (for the economic analysis behind these conclusions by a key adviser at the time to the

Secretary General of OECD, see Marris 1984). The only important conclusion the United States did not get that it wanted was an exchange of information on major industrial projects being negotiated between Western firms and Eastern partners.

The conclusions of the study on balance of advantages were reflected in general terms in the OECD communiqué and incorporated in their entirety in unpublished minutes of the OECD Ministerial Meeting. The acceptance of the minutes was particularly satisfying to the United States, considering the fact that the OECD includes many neutral countries that do not acknowledge any differences economically or politically in their relations with the East. On the last day of the OECD Ministerial, the neutrals rallied to pressure the French to call off the American proposal to adopt the minutes. The French stood firm, however. The minutes were adopted, and an ongoing OECD practice was initiated to prepare annually a report on the balance of advantages in East–West trade and financial relations. To appreciate the results, one need only recall that in 1976, after a major East–West trade initiative by Secretary of State Kissinger, the United States failed to get a single study out of the OECD on East–West trade, let alone an ongoing process. Prior to 1983, the OECD collected statistics on East–West trade and credit but nothing more.

The IEA Ministerial produced acceptable results on the controversial energy issue. Here the United States did not get agreement on a specific percentage limit — 30 percent — on imports from the Soviet Union by any one member. Instead, the communiqué acknowledged only the "potential risks associated with high levels of dependence on single supplier countries," and the members agreed to "ensure that no one producer is in a position to exercise monopoly control," calling for a "diversification of sources of energy imports." The IEA countries also gave priority to indigenous sources and pledged to develop "appropriate cost-effective measures suited to each country's situation to strengthen their ability to deal with supply disruption" (IEA 1983). Already, market conditions were forcing the European countries to scale back their plans for Soviet gas imports (Jentleson 1986, pp. 199–203).

Eventually, the Europeans would import substantially less than the original amounts contracted for, reducing purchases from $10 billion to about $5 billion per year. They would give up altogether plans for a second gas pipeline, and in 1986 they announced plans to go ahead with the joint development of North Sea gas supplies instead of further Soviet imports (*NYT*, June 3, 1986). Although these decisions reflected market conditions, they also reflected government priorities that influence market conditions. European governments, under U.S. prodding, decided to divert subsidies provided earlier for Soviet gas imports to the development of North Sea energy resources. In East–West trade, these government decisions often make the difference in market decisions.

The Williamsburg Summit

The Williamsburg Summit therefore had little work to do on East–West trade issues. By the time of the summit the results of four of the five East–West trade studies were in, and they were fully satisfactory from the standpoint of the U.S.

and other delegations. The fifth study in NATO was going well and would be approved at the NATO Ministerial Meeting in June. The East-West issue, therefore, was under control. The summit leaders discussed it briefly, and the communiqué took "note with approval of the work of the multilateral organizations which have in recent months analyzed and drawn conclusions regarding the key aspects of East-West economic relations." The communiqué called for "continuing work by these organizations, as appropriate" (for text of communiqué, see *NYT*, May 31, 1983).

The issue can be debated whether the trade studies and the Williamsburg Summit achieved meaningful and lasting progress within the alliance on East-West trade issues or simply dismissed these issues to avoid another high-level political conflict such as the one at Versailles. Bruce Jentleson, who examines the pipeline episode carefully, concludes that the "summit pronouncement . . . was empty . . . had no teeth, no specific policy content; therefore, while it projected an image of unity, it brought the allies no closer to true collaboration on trade controls" (1986, p. 199). Jentleson, however, considers few of the specific results that came out of the COCOM, OECD, and NATO meetings of 1983. These results were modest, to be sure, but they reflected substantial change by comparison with previous alliance cooperation on East-West trade matters. And although it would go too far to say that the pipeline episode reflected effective, albeit disruptive, American leadership within the alliance, it is probably closer to the truth to say that American leadership in 1981–1983 was instrumental in raising the level of seriousness with which East-West trade issues were treated among the allies and in creating or reenergizing mechanisms through which the allies could negotiate more systematically their continuing differences on these issues. Substantive differences continued, but procedural mechanisms were strengthened; and it would remain essential that these mechanisms be used extensively in the years ahead if the allies were to avoid further punishing disputes over East-West trade.

Detente or Deterrence: East-West Trade Policy After 1985

The improvements in allied treatment of East-West trade issues in 1983 derived, in part at least, from following a deterrence rather than a denial perspective. Sustaining these improvements will depend on retaining the deterrence perspective and not slipping once again, as U.S.-Soviet relations improve, to the other end of the spectrum, from deterrence to detente.

With substantial reforms being introduced in the Soviet Union after 1985, the United States is being tempted once again to act unilaterally to reduce controls. Trade legislation in 1988 called for a significant reduction of U.S. East-West trade controls. Effective controls, however, require cooperation. A study by the National Academy of Sciences in 1987 urged that the "United States . . . pursue the objective of developing a community of common controls in dual use technology among cooperating Western countries" (NAS 1987, p. 135). To achieve such a community, the United States and the allies have to give COCOM high priority within the alliance, find a way to deal in a more timely and systematic way with differences over the use of trade controls for foreign policy purposes (which, in the case of Poland, was the origin of the pipeline dispute), and avoid the excessive subsidization of trade and credits with the Soviet Union.

Common Market in Export Controls

A common market in export controls on strategic trade is a realistic objective, according to the National Academy of Sciences' study. To achieve it, the United States and its allies must narrow their differences on the list of strategic items to be controlled. Since 1982, COCOM has become a stronger negotiating forum for this purpose (Mastanduno 1988; Bertsch 1988). It has met in high-level session at least every 2 years, and in January 1988 the Deputy Secretary of State led the U.S. delegation to a COCOM session, the highest level of U.S. representation to a COCOM meeting in history. In 1984 COCOM completed its first list review since 1977, and in 1985 it agreed to conduct such list reviews continuously. In addition, COCOM launched major initiatives to harmonize export controls internally among the COCOM countries and to extend these controls, at least partially, to non-COCOM countries. In recent years the United States and other COCOM countries have concluded a series of bilateral agreements with European neutrals (e.g., Austria, Sweden, Switzerland, Finland) and newly industrializing developing countries (e.g., Singapore, India) to control COCOM and eventually perhaps indigenous technologies exported to or produced in these countries. Finally, a new group was formed in October 1985, known as the Security and Technology Experts Meeting (STEM), to upgrade military advice to COCOM members in the evaluation of export licensing, particularly for such countries as Belgium, Germany, and Japan, where defense ministries are not centrally involved in the national control process.

COCOM still has a long way to go, as illegal sales to the Soviet Union of submarine technology by Japanese and Norwegian companies revealed in 1986–1987. But it is moving in the right direction and the United States must continue to provide the necessary leadership. Critical to U.S. leadership is getting America's own export control house in order. In 1979 the Defense Department (DOD) was given a statutory role to review export licenses to Eastern countries. During the renewal of the export control legislation in 1983–1985, Congress debated whether to give DOD the same authority to review licenses to selective non-COCOM Western countries. DOD wanted this authority in order to exercise leverage over non-COCOM countries in bilateral negotiations to strengthen the export controls of these countries. DOD could delay or deny a U.S. license to a non-COCOM country, depending on whether or not that country tightened its controls. For this reason, of course, business groups, the Commerce Department, and its allies on the Hill opposed DOD's role. Eventually, when Congress could not agree, DOD acquired the authority by administrative order from the President.

As demonstrated by progress with non-COCOM countries, the Defense Department has used this new authority to good effect, although not without isolated abuses (Nau 1988b). It has computerized its review process and significantly reduced license processing time. Since 1987 Commerce has also streamlined and strengthened its review process (Freedenberg 1989). Both Defense and Commerce need to remain strong, and the White House (NSC) has to be willing to monitor Commerce and Defense Department disputes and resolve them quickly and decisively when they threaten to undermine U.S. leadership. The Reagan administration, despite all its professed interest and costly foreign diplomacy in this area (i.e., the pipeline dispute), did not exercise strong executive leadership over the interagency process on licensing issues. Future administrations must do so or U.S. leadership in COCOM will be compromised by bureaucratic deadlock or by defense and commercial interests dominating U.S. policy sequentially in the repeated swings back and forth between denial (defense overemphasis) and detente (commercial overemphasis).

Foreign Policy Controls

The allies have no mechanism to discuss foreign policy controls — with respect not only to the Soviet Union but also to Cuba, Nicaragua, Vietnam, Libya, Iran, South Africa, and so on. These controls, therefore, either are not discussed at all or are discussed only in the heat of a crisis, as happened in the case of Poland in 1981. The controls then spark controversy and spill over to disrupt cooperation on security controls or economic aspects of East–West trade.

The allies should find a better way. The Europeans and Japanese should recognize that they cannot have it both ways. They cannot resist more regularized discussions of foreign policy controls and still expect the United States to agree with them on not imposing these controls or on the precise controls to be imposed. There is no guarantee that more systematic discussions will yield agreement. The United States, as we have noted, bears the leadership responsibility for deterrence vis-à-vis the Soviet Union. It will always be more likely to use foreign policy controls and invoke stiffer measures than the allies. But the United States too has a stake in *common* controls. If the allies fail to cooperate, the deterrence signal sent to the Soviet Union is muddled or even worse, counterproductive, giving the Soviets an opening in the alliance to exploit. Thus, the allies should agree to discuss foreign policy controls on a regularized basis if the United States agrees to apply future controls to the maximum extent possible only on a common basis.

The deterrence approach raises the threshold for the use of foreign policy controls. Deterrence views economic sanctions not as decisive instruments to punish or reward the Soviet Union but rather as low-level instruments of conflict to manage differences with the Soviet Union. To be effective, such low-level instruments of conflict have to be backed up by the threat to escalate. Otherwise in a deterrence framework where their primary purpose is to signal intent, these instruments become meaningless. Thus, the United States and its allies should agree to apply economic sanctions only in situations where they are psychologically prepared to back them up with sterner political or military measures, either to increase the pressure on the Soviets to desist from an immediate provocation or to complicate Soviet calculations in contemplating any future provocation. In this way, economic sanctions would not exist alone but would always be supported by the threat to escalate if Soviet provocations persist or proliferate.

Thus, in the case of Poland in 1981, the United States and its allies would have imposed economic sanctions against the Soviet Union together with the threat to escalate those sanctions to political measures (e.g., recall of the U.S. ambassador in Poland or Moscow) if the Soviet Union did not moderate its provocative behavior within a reasonable period of time. Admittedly, attempting to achieve allied agreement on more serious, back-up sanctions as well as initial low-level sanctions would compound the difficulty of reaching agreement. But it might actually facilitate agreement because it would first ensure that the situation was serious enough to consider escalation (Poland unquestionably was) and second raise the chance that the allies would reach agreement on more credible low-level sanctions so as to reduce the probability of having to escalate to back-up measures. In any case, if the effort failed, the allies would be no worse off than they are at present without any deterrence criteria to guide the imposition of foreign policy sanctions.

Economic Security and Influence

What about the alternative temptation to exaggerate the benefits of economic relations? From a deterrence perspective, economic cooperation, like economic sanctions, offers limited benefits and largely again as a signaling device in connection with other, more significant political and military measures. For example, the policy of economic differentiation practiced by the United States toward Eastern Europe in the early 1970s was useful and probably effective because it went hand in hand with closer political relations in that period, not only with selective Eastern European countries but also with the Soviet Union. On the other hand, efforts to provide economic assistance to Poland in 1981, when the United States sought to support Solidarity, were probably much less effective because simultaneously U.S. political relations with the Polish government and the Soviet Union were souring.

In the early 1980s economic cooperation appeared more useful in relations with China. Here economic and political ties were clearly moving in the same direction. But, just because of that, there may have also been a tendency to slip too easily from a deterrence to a detente perspective and to expect too much from rapidly improving economic ties and, in some areas, military cooperation. Events in China in June 1989 showed that both the foreign and domestic policies of China are anchored too deeply in historical and traditional experiences to change *predictability* under the initial surface manifestations of economic liberalization. While recognizing the value of higher levels of economic cooperation, U.S. officials would also be wise to avoid inflation of both official and private expectations concerning such cooperation. This observation holds particularly true, it would seem, in the euphoric atmosphere generated in late 1989 by sweeping political and economic reforms in Eastern Europe and the prospect of massive Western economic assistance to support these reforms.

Two final observations apply to economic cooperation. The political benefits of nonstrategic trade are never so great as to warrant subsidizing this trade. In this respect the Reagan administration's decision in 1986 to subsidize grain sales to the Soviet Union is regrettable. It undermines efforts in the OECD to curb competitive subsidies on trade and credits to the East, especially within the OECD export credit arrangement. It coincides with a frenzy of European agreements in 1988 and 1989 to provide financial credits to support perestroika in the Soviet Union. In recent years the Soviet Union has doubled the potential debt it owes to Western banks, although it has not drawn down all of the credit lines available to it (*WSJ*, March 22, 1988; see also Robinson 1986). As Senator Bill Bradley has warned, these credits make little sense; the West "should applaud perestroika but not pay for it" (*WSJ*, November 14, 1988).

Subsidies also pose a further threat. Left to market forces, Soviet exports to the West or Western exports to the Soviet Union are never likely to become large enough to expose the West to blackmail should U.S.-Soviet relations deteriorate once again and the Soviets threaten to interrupt these exports. As we noted earlier, without the initial subsidies involved, Western companies turned eventually to North Sea natural gas to satisfy their energy requirements rather than larger Soviet gas imports. Subsidies, on the other hand, can lead to excessive Western dependence on Soviet resources or markets and amount to paying an economic and a political price (in terms of increased vulnerability) for something that is usually available for less cost and less risk in the West.

The vacillating U.S. policy toward East-West trade reflects the tendency to ignore the crucial role of political community in facilitating economic exchanges. When political community is weak, as it has been and remains between the United States and the Soviet Union (albeit with the prospect of growing stronger if Soviet reforms succeed), economic relations do not matter that much. When political community is strong, as it has become in the Western world throughout the course of the postwar period, economic relations do not have to be artificially stimulated to acquire significance. The lesson for East-West relations is to let the political dimensions of U.S.-Soviet relations lead the way toward greater economic interdependence, not to expect trade and credits to open the way toward superpower peace.

CONCLUSION

The Bretton Woods Policy Triad
in the World Economy of the 1990s

America's purpose and policy, more so than its power or politics, will shape the world economy of the 1990s. Although the relative decline of American power and the complex interdependence of international markets will define the context of future initiatives, national purpose and economic policy will define its content. This content is, in part, a reflection of institutions and interests, but as the evolution of postwar U.S. foreign economic policy in this study shows, it does not simply coincide with international structural alignments and power or with domestic political parties and interests. Ideas that cut across national institutions and political interests define the concepts of political community and influence the choices of economic policy that ultimately determine the shape of international institutions and markets.

From this perspective the essential lessons of the postwar international political economy are the following:

1. United States leadership was critical to the evolution of the open, efficient postwar economic system, not just to bear the costs of a liberal order, as the structuralist literature suggests, but to supply a purpose for that order. Without the purpose of defending Western freedom in central Europe, the postwar economic order in Europe and the Western world as a whole would probably not have progressed very far beyond the web of hundreds of restrictive trade and exchange rate relationships that existed in 1947.

2. The United States eventually funded the Bretton Woods system and the U.S. dollar, rather than a nonnational currency, became the principal reserve asset of the system because Europe and Japan accepted the broad containment objectives of U.S. foreign policy and, for that matter, still do. A foreign policy consensus is always essential to bring about the monetarization of an international economy.

3. The foreign policy consensus that underlies the international economic system derives both from the domestic nature of the societies participat-

ing in that system and from the diplomatic imperatives of defending these societies against external adversaries. Historically, it would seem, only open, democratic societies have taken strong initiatives to create liberal international economic systems, and they have been more likely to do so when they are also threatened by external circumstances. Great Britain liberalized trade in the nineteenth century largely as an expression of the competitive and colonial instincts of its internal society. The United States did so after World War II more as a defense against external threat (having retreated from the world in 1945, when no threat was perceived to exist). Liberal societies choose efficient international economic systems either because such systems organize international economic relations on the same competitive basis as domestic systems or because they maximize wealth in the pursuit of common defense goals (or because they do both). Generally speaking, the projection of domestic characteristics may be more important. Otherwise, we might expect illiberal societies (e.g., the Soviet Union) to create efficient economic systems when they mobilize to defend against external threats.

Thus, if the East–West conflict subsides over the next decade, the liberal or illiberal nature of the international economic system will depend increasingly on the character and compatibility of the domestic systems among the major industrial and developing countries. In this sense, the domestic economic and political features of such economically powerful countries or groups of countries as Japan, the European Communities (EC), and some rapidly industrializing developing economies, will become more important for the world economy. The spread of democratic principles and practices since 1945 to all industrial and many developing countries offers the best hope that the liberal international economic system may be preserved and strengthened in the 1990s.

4. The most important economic requirement for open and efficient international markets is noninflationary, market-oriented domestic policies. A liberal international economic system is not compatible with any and all types of domestic economic policy. This conclusion follows not only from economic logic (see Chapter 1) but also from postwar economic experience.

The Bretton Woods policy triad of price stability, flexible domestic markets, and freer trade clearly associates with the most prosperous periods of postwar performance. A case can be made that this triad is necessary and perhaps even sufficient to produce growth because it fosters, at the margins, progressively greater competition under stable price conditions, which ensures alternative calculations of resource use and hence the continuous movement of economically limited resources into activities that yield the highest returns. The critical ingredient in the triad is competition, not private versus public control. If government-owned entities compete in the market environment, government intervention is not incompatible with liberal international economic policies. On the other hand, experience shows that governments often have other objectives when they intervene in markets. Therefore, if government intervention accelerates sharply, as it did in the 1970s throughout the industrial and developing world, growth slows (see Chapter 6). Similarly, if governments seek national autonomy to pursue inflationary macroeconomic policies, price and exchange rate stability suffers and growth slows. The United States persuaded itself in the late 1960s that it could have both growth and complete national autonomy in macroeconomic policy. The results show that it cannot, although the United States continues to cling to this illusion even at the beginning of the 1990s.

5. Achieving a consensus, both domestically and internationally, to pursue the Bretton Woods formula of efficient international and flexible domestic markets may depend less on direct bargaining processes within domestic or international policymaking institutions than on convergence of policy ideas in the nonbargaining environment of the international marketplace and in what we have called the cocoon of nongovernmental organizations surrounding the immediate domestic and international policy arena. Markets provide an arena for indirect, anonymous bargaining among countries and may be able to help resolve some issues that international conferences alone cannot. International conferences and institutions today have become highly pluralistic and at times unwieldy. They may be more useful to review or to implement and monitor national policy decisions rather than to make such decisions in lieu of national institutions. Since the mid-1960s, the United States and other countries have resorted increasingly to markets, in addition to institutions, to exert leverage on one another.

Assertive or unilateral actions in the international marketplace are not necessarily unhelpful. If these actions serve to produce early results, such as a sharp drop in inflation in the world's leading economy (which Reagan policies achieved in the United States in 1981–1982), and push, otherwise stalemated international bargaining processes toward convergence around policy standards of efficient international markets, they strengthen the international community and world economy. On the other hand, if such actions push market adjustments too far too quickly, they may cause other countries to withdraw from international markets — as France and developing countries almost did in 1982 and 1983. Moreover, unilateral actions may push developments away from standards of stable and flexible markets, as U.S. policies did in the early 1970s and may be doing again in the late 1980s and early 1990s.

The lessons from the past provide a compass to guide U.S. foreign economic policy in the future. But are they relevant to contemporary circumstances? As in the past, many commentators argue that circumstances are different, that principles of freer trade and flexible economies are simply not compatible any longer with the changing character of comparative advantage, the pace of technological advance, and the diversity of the contemporary international system. They contend, in contrast to the perspective of this study, that policy should accommodate these circumstances, not seek to shape them (Drucker 1989).

The new circumstances of the early 1990s will obviously differ from those of the late 1940s. The Cold War appears to be ending, not beginning. America's economic dominance has clearly disappeared, and now its superpower military status may also decline as defense and security requirements diminish, especially in central Europe. The Pacific region has emerged as the most dynamic economic region, and Pacific nations like Japan and the NIEs (newly industrializing economies, such as Korea) have pursued domestic and international economic policies in the past that would be incompatible with the reinstatement of the Bretton Woods policy triad in the future. Capital markets have been liberalized and expanded; they now dwarf trade flows and appear to overwhelm exchange markets, precluding the stable currency alignments of the Bretton Woods era. Production and investment are being progressively globalized, eroding the control of national policy over trade and financial activities of multinational business and banking institutions. Unprecedented debt, both domestic and international, hangs over developing as well as industrial nations (including the United States) and threatens to depress demand and world growth, a situation unlike that after World War II or during the economic boom of the early 1970s when buoyant demand accelerated economic expansion. Many more special interest groups are now mobilized in all countries, including increasingly in Eastern Europe and the Soviet Union; these groups complicate, if not obstruct, coherent policy formulation as well as more fundamental structural adjustments.

As different as the 1990s will be, however, the coming decade has an overriding similarity to the world of the late 1940s. The world political community and international economic system are once again in fundamental flux. For the first time in 40 years, basic political purposes are changing, especially within the domestic regimes of Eastern Europe. Illiberal societies are evolving toward liberal ones. Threats that derived from fundamentally incompatible domestic political systems in central Europe are receding, and international economic and security relations, both within the East and West and between East and West, are shifting to accommodate these changes. The diplomatic task will be to safeguard and nurture further domestic political change in Europe and elsewhere without allowing this process to get out of control because of revived political fears or economic frustrations such as those that generated conflict in the 1940s and, at least theoretically, could do so again in the 1990s. Although the 1990s reflect a much more hopeful period than the late 1940s, stability is at stake as the Cold War unravels, no less now than it was when the Cold War began.

In the 1990s, therefore, the lessons of the earlier postwar period and the arguments of this book will be particularly relevant. The rest of this chapter looks at the choices America and other countries will have to make to ensure that the new world nurtures a growing convergence of domestic political systems around pluralist and domestic concepts and preserves as well as expands the scope of economic integration and prosperity in the world community.

Shaping National Purpose

The concept of national purpose in this study illuminates recent events in central and Eastern Europe. In contrast to structuralist concepts that predict growing conflict as dominant powers decline and economic competition accelerates, the concept of national purpose offers some explanation of the dramatic narrowing of political differences that the world is witnessing today, a narrowing which Soviet President Mikhail Gorbachev himself described at the Malta Summit in December 1989 as "a kind of movement bringing states, countries, and continents together" (*WP*, December 4, 1989). From the perspective of national purpose, threats do not emerge solely or primarily from fluctuations of wealth and power in the international system; they also derive from the conflict between domestic political systems that societies choose to define and organize their basic political principles, institutions, and processes. As Eastern European societies choose more open and pluralistic political systems, these societies pose less of a threat to Western societies; and as long as Soviet society moves in a similar direction, even if more slowly and less completely, Eastern European societies pose no additional threat to the Soviet Union. The pacing and management of these shifts toward more liberal societies in the East hold the key to stability and peace as the Cold War ends.

At the outset of this process, military power in central Europe is relatively evenly balanced, a consequence of forty years of costly, sometimes nervewracking, but ultimately highly successful efforts at deterrence, whereas economic power and political self-confidence are significantly tilted in favor of the Western countries. As domestic change proceeds, it will be imperative to preserve the military balance, even or especially as arms are reduced, and to avoid new economic and political threats that may arise both from domestic uncertainties in Eastern Europe and the Soviet Union and perhaps from well-meaning but potentially excessive efforts by the West to apply its economic and political advantages to assist the East. In a first phase as arms control negotiations succeed, the alliances (NATO and Warsaw Pact) can be transformed from military to political groupings (the original intention, for example, of NATO), while Eastern European countries and the Soviet Union concentrate on essential domestic economic and political reforms, without which they will be unable to compete confidently with more powerful and legitimate Western partners. Simultaneously, but then more vigorously in a second phase, the focus can shift to international economic and political arrangements. drawing the socialist countries more fully into the Western European (European Community) and international economic institutions (GATT, IMF, and World Bank) and engaging the two alliances in a more fluid process of cross-cutting (that is, transEuropean) political coalition-building within the framework perhaps of the Conference on Security and Cooperation in Europe (CSCE), the so-called Helsinki process. Just before Malta, Gorbachev seemed to embrace this political forum, which includes the United States and Canada, clarifying his earlier, more ambiguous references to a "European home" that might or might not include the United States. In a final phase, the international community can ratify new arrangements for a reassociated or reunified Germany, by then ensconced in a new and tested transEuropean environment that involves a better balance between East and West of political and economic capabilities and offers political and economic measures rather than military ones to safeguard security. Although actual developments in early 1990 may be moving too fast to allow for such gradual and sequenced phasing, the risks of a more precipitous and unstructured evolution should at least be recognized.

America's approach to these developments in the early 1990s will be critical. As this study demonstrates, how America defines its own national purpose determines its contribution to shared political community in the world economy and influences the economic policies by which it harnesses its physical power to political aims. Historically, America has defined itself in various ways. From its origins until the twentieth century, it saw itself as a unique society defined primarily by its separation from world affairs. For a brief period after World War I, it sought to make the world safe for this unique American system to spread, and after World War II it assumed that common democratic values had spread sufficiently to enable world politics to be organized around individualistic American principles of law and markets. In 1947 it discovered that differences remained and that defense of its values would require efficient economic policies to integrate the West and collective military measures to contain the East. After containment became militarized and exhausted in Vietnam, the United States redefined its national purpose to place renewed emphasis on cooperation within international regimes and institutions that encompassed a wider and wider range of domestic differences.

In the 1980s, America rediscovered the limits in world affairs of international cooperation without common domestic purposes but failed to find an alternative national purpose that avoided both the universal idealism of Wilsonianism, which the Reagan Doctrine easily lapsed into, and the great power politics of traditional military rivalries, which the Reagan defense buildup appeared to be. To establish a new link between domestic purpose and foreign diplomacy in the 1990s, America needs to come to terms with two extremes from its past: that its role does not matter and that its role is not over until every nation in the world is democratic.

After Vietnam many analysts counseled America to accept itself as just another ordinary power, with no more good or bad to offer the world than any other power (Rosecrance 1976). Although the Vietnam experience certainly recommended a new sense of humility for American policy, this counsel went too far. By historical standards, as we have seen, America's leadership in the postwar world made a special contribution to the improvement of the human condition in the world community. Peace has now been preserved for a period as long as any since 1870; wartime enemies have become close and empathetic, albeit competitive, allies; prosperity embraces a growing number of countries that, 40 years ago, were desperately poor; and democratic values and human rights are indeed more widespread than ever before. As we have noted, not all of this was due to American leadership, by any means; Europe, Japan, the developing world, and lately the Soviet Union and Eastern Europe made enormous contributions to this outcome. But it all happened on America's watch. America has to realize this accomplishment and take pride and strength from it. Not all our friends will share this confident self-assessment, but they will respect it, especially if it is made without indulging in the other extreme that has marked America's foreign policy behavior in the past, American messianism.

America does not have to live in a world in which all countries adopt America's domestic political and economic standards. In that sense, it is unnecessary to make the world over in America's image. The essence of what America stands for is freedom of choice. Other peoples have chosen throughout time to organize themselves around principles other than highly competitive economic and political systems, at least to the degree such competitive systems exist in the United States. Even within the Western world, much of Europe and Japan recoil at times at the atomistic nature of American society. All peoples do not seek individual freedom to the same extent we do and do not lack such freedom, as in the case of the Soviet Union, simply because their governments will now allow it (a popular view among those who believe the Russian people at large would choose democracy but for the oppression of the Communist Party).

On the other hand, America should reject the notion that there is a "wave of the future" in any other political direction, toward greater socialism or communism, let alone toward greater centralization or totalitarianism. As tempting as it may be to conclude, especially since the modern era, that history is unidirectional, that trends are inexorable and irreversible, this conclusion, whether on behalf of freedom or on behalf of socialist dogma, is the first step toward messianism and intolerance (Fukuyama 1989). Choices relating to political community at both the domestic and international level will go on continuously, and they may be made in either totalitarian or democratic directions, depending on the moral and political judgment of human beings. It is misleading to ignore the crucial role of these choices and to argue that modern technological developments, complex markets, or any other circumstances or structures constrain and channel political choices in one direction only.

From the preceding perspective, America can tolerate decisions of other societies, even if made undemocratically, to organize political life undemocratically (e.g., China, Nicaragua, South Africa). It can tolerate them, in terms of stopping short of direct intervention, as long as the totalitarian choice is not exported aggressively by overt or covert military means. But it can never approve these decisions. It must continuously and evenhandedly hold up the abuses of totalitarian governments on both the left and the right to the accountability of the international community and human conscience. The United States must use the United Nations and every other such institution in the contemporary international system to wage this debate to uphold a minimum democratic standard of human dignity and human rights. And from time to time it will have to supplement diplomatic debate in these institutions with bilateral and regional military assistance to "freedom fighters"—not for the purpose of directly overthrowing totalitarian governments but for the purpose of keeping alive in totalitarian societies the embers of free institutions and values.

One reason for avoiding the extremes of indifference and zealotry in American choices is, ironically, to ensure a continuing commitment on the part of the United States to play a significant role in international affairs. Defining America as an ordinary power is insufficient to win support for a world role from an American people that has a long tradition of noninvolvement in world affairs. Defining America, on the other hand, as the "beacon on the hill" of worldwide freedom is likely to lead either to unlimited defense commitments, which eventually violate the values of the America that is being defended (e.g., the experience in Vietnam), or to an erosion of defense commitments as policymakers look to inevitable or autonomous historical forces to resolve the issues of political community for them. (This is the case, for example, with the view that untoward events in the world, especially in the third world, would never occur if the United States simply stayed out of these situations.) Thus, to avoid the situations of advocating power without purpose or believing that purpose can succeed without power, America needs a self-definition that affirms freedom without imposing it.

Understanding this issue of purpose puts American power into perspective. Without doubt this power has declined since 1945. What this study questions is whether it has declined enough to weaken American leadership, especially if America achieves a consensus on appropriate purposes and policies. As the data in Chapter 2 suggested, there is no persuasive evidence that America's power has dropped below some magic threshold of dominance. But even if it has, power would still have to be assessed against the conditions it seeks to affect.

In that sense, America faces a more congenial world today than it did in 1945, at least within the industrial world and in some parts of the developing and socialist world as well. The industrial world, with developments in the 1980s in Portugal and Spain, is a seamless web of democracies. It is a more homogeneous political community than has existed among developed countries perhaps since the early part of the nineteenth century. Democratic openings in socialist countries took place in 1989 with breathtaking speed, including in late 1989 in East Germany. Although these developments have only begun and democratic trends in developing countries are also fragile, these events offer more hope for converging political perspectives in the world than has been the case at any other time in the postwar period. Thus, American power, even if it has declined, may not have to be as great today as it was after World War II to achieve the same purposes of expressing and defending democratic values. Over the past 40 years, America has traded some of its power for a more congenial political community, particularly within the industrialized world. As David Calleo notes, this trade-off represents not a defeat for American policy in the postwar period, but a "success in rebuilding a peaceful and prosperous world" (Calleo 1987, p. 10).

If democracy is more widespread in the contemporary world, America's expression of its own domestic purposes can be less strident and more self-confident. America does not have to speak for others; it can speak for itself, confident that others increasingly share its basic aims. Being more widespread, democracy also embraces greater variety, from essentially one-party systems, as in Japan, to practically separate provinces, as in Belgium. Thus, to express its national purpose in the contemporary world, America needs to focus on what is essential for individual dignity and freedom; that is what cuts across different democratic systems. This core of Western freedoms is what America then needs to defend in its relations with totalitarian powers.

East-West Relations

America's relations with the Soviet Union rest on two realities. Internally, the two societies have represented diametrically opposite expressions of political community or what constitutes the "good political life." In the Soviet polity until recently, what was left for private choice was relatively narrow (e.g., even religious choice was strongly discouraged), whereas what was public was decided by a relatively small number of party officials. In the American polity, the private sphere is large and public decisions are made by a broad electorate and highly fragmented Congress and executive branch. Although the situation in the Soviet Union today is changing, these basic domestic differences between the two societies will undoubtedly persist for a long time to come. Yet in international affairs, the two societies have much in common. They dominate the world community as nuclear superpowers and hold the key to peace and survival in the nuclear age. Neither of these two realities can be denied or freedom will be lost. Denying domestic political differences leads to moral surrender; denying international nuclear realities leads to political suicide. Under militarized containment America overemphasized domestic differences; under demilitarized detente it fixated on nuclear peace. In the future it will have to weave both realities together to attain a free and peaceful world.

The balancing of domestic differences and global commonalities in U.S.-Soviet relations means that human rights and political issues have to share the agenda with arms reductions and trade. As long as U.S. and Soviet societies differ substantially, tensions will exist. Arms are not the source, but the consequence, of these differences. Arms may, to be sure, exacerbate misunderstandings. But removing them will not end East-West disputes or do away with the problem of potential rearmament. Therefore, reducing arms must be done not just for its own sake but within the context of a strategy of deterrence.

In the euphoric atmosphere of early 1990, deterrence may seem irrelevant to some observers. Yet, as U.S. experience in 1945 and again in the early 1970s shows, indifference to deterrence may be just as harmful as preoccupation with it. From 1985 to 1988, American defense expenditures declined by 12 percent in real terms. Meanwhile, Soviet defense spending, as measured in constant 1982 rubles, grew by roughly 3 percent per year, almost double the rate of the Breshnev era (CIA and DIA 1989). Although some of this accelerated Soviet defense spending may have been tactical and transitional (Gorbachev's payoff to the military to support domestic reforms and international arms control agreements) and evidence emerged in late 1989 that Soviet conventional (not strategic) military expenditures were being cut back, the pattern suggests that differences and suspicions remain in U.S.–Soviet relations. It is much too soon to relax Western vigilance.

Because differences persist, the bedrock of U.S.–Soviet relations is deterrence — the management, not the resolution of conflict. For 40 years deterrence has rested on a Western strategy of the threat of offensive escalation. NATO doctrines of massive retaliation in the 1950s, flexible response in the 1960s, and escalation dominance in the 1970s and 1980s put the Soviet Union on notice that, should the Soviet Union try to exploit its conventional superiority in Central Europe or other areas of vital interest to the West, the West would retaliate by escalating the conflict to the next higher level of engagement, including ultimately strategic weapons where the United States held a superior advantage. The United States and NATO, in short, relied on a Western advantage in strategic technology to offset a Soviet and Warsaw Pact advantage in conventional forces.

As long as deterrence strategies of offensive escalation persist and the United States continues to lead the Soviet Union by a substantial margin in strategic technologies (as it does; see NAS 1987), the United States and the West have an interest in protecting their strategic technologies. Export controls and the informal Coordinating Committee for Multilateral Export Controls (COCOM) exist for this purpose. Weakening these controls unilaterally hurts deterrence and threatens free trade within the West, as U.S. sanctions against Toshiba in 1987 clearly demonstrated. On the other hand, strengthening and bringing COCOM "out of the closet," as this study recommends, makes it possible to negotiate controls down to the truly strategic items and to reduce licensing of strategic trade within the West. As a high-level U.S. official said in October 1987, "If this community of nations [COCOM] can form an effective barrier against diversion to the outside, there is no reason why we can't have more open trading [of high technology] within [the West]" (*WSJ*, October 16, 1987).

Deterrence may not always depend on Western strategies of offensive escalation. President Reagan pushed the Strategic Defense Initiative (SDI) and a vision of deterrence based on mutual protection against nuclear attack rather than mutual retaliatory capabilities. What the political benefits or drawbacks, technical feasibility, and economic cost of SDI might be are issues beyond the scope of this study. What is relevant here is that mutual defensive rather than offensive strategies would undoubtedly alter the logic of export controls. President Reagan signaled as much by talking about sharing defensive technologies with the Soviet Union. If only one side had these technologies, that side might be tempted to strike first, secure in the knowledge that it could successfully defend itself against a retaliatory attack. In this situation, therefore, deterrence, which seeks to manage conflicts rather than to win or resolve them, would depend on both sides having defensive capabilities, not preserving a Western strategic advantage to offset a Soviet conventional one, as in the case of offensive deterrence. SDI may still be a long way off or even prove to be a flight of fancy. But the SDI discussion reminds us that export controls are intended not to block strategic exports to the Soviet Union in principle, but to serve a specific deterrence strategy in practice.

Human rights and deterrence are the core of U.S. relations with the Soviet Union. Trade can round out a healthy political and military relationship, but it cannot significantly influence or, for certain, substitute for this relationship. Administrations in the past have expected far too much from trade in East–West relations. President Nixon saw it as a decisive carrot and President Carter, after Afghanistan, as a decisive stick to influence Soviet *foreign* policy. President Reagan saw it in 1982 as a decisive stick to exploit Soviet *domestic* vulnerabilities. In all these views trade is being asked to do too much between societies that do not even apply the same values to economic affairs.

Perestroika and *glasnost* in the Soviet Union and Eastern Europe have opened up, once again, significant hopes for expanding Western trade to the East and thereby influencing the reforms in socialist societies. It is inevitable and, in a limited sense, useful, as this study has pointed out, for nonstrategic trade to increase between the United States and the Soviet Union, especially as political relations improve. But it is politically irresponsible to subsidize this trade as U.S., EC, and other exporters were doing, for example, in grain in the late 1980s. It makes even less sense to subsidize credits and general balance-of-payments loans, which are directly fungible for military purposes. Yet Western banks were scrambling in the late 1980s to lend to the Soviet Union and between 1984 and 1988 more than doubled outstanding net real commercial debt held by the Soviet Union from $13.1 billion to $30.8 billion (CIA and DIA 1989).

The U.S. Senate passed a nonbinding resolution in June 1988 that warned against this frenzied lending and urged the U.S. government and its allies to take a firmer line on loans and subsidized credit to the Soviet Union (*FT*, June 20, 1988). The passage of the Senate resolution suggested the potential for political backlash in this controversial area, especially if political and economic reforms should go off track in the Soviet Union. In retrospect, it is clear that the U.S. and Western governments moved too fast to foster trade and credit relations with China. In the wake of events in Tiananmen Square in June 1989, the United States suspended military sales to China and opposed further lending to China by the multilateral development banks (e.g., World Bank). Congress pressed for still harsher steps. China, meanwhile, reacted defiantly. Predictably, trade and credit from the West, whether accelerated or denied, were simply not that important to the Chinese to get them to compromise significant domestic or foreign policy objectives.

For the same reasons the United States should moderate expectations for economic aid to the countries of Eastern Europe, especially large grants. Not only is it doubtful that such aid can be effective (Marshall Plan aid succeeded in Western Europe, for example, only because it was closely tied, as we have seen in this study, to converging domestic economic and trade policy choices among Western European countries), but the relationships of economics and politics in and among the socialist countries are extremely complex. The provision of large Western aid, particularly in the form of consumer products and subsidized credit, may actually diminish the motivation for economic reforms in Eastern Europe by encouraging current consumption rather than investment in more efficient production for future consumption. On his trip to Poland and Hungary in summer 1989, President Bush seemed to be aware of the benefits of a more prudent approach—one that emphasizes assistance to channel existing domestic savings into more profitable investments (e.g., to improve equity markets or to privatize some state-owned industries), to retrain and redeploy domestic labor, and to open Western markets more fully to Eastern European exports—rather than one that emphasizes new lending for imports of consumer products (Nau 1989a).

Security Relations with the Allies

In a period of transition from illiberal to liberal societies in Europe, the need for deterrence and alliances does not end; it merely changes form. As noted earlier, security arrangements shift, initially from high levels of military forces to lower levels, and eventually from military forces to political and economic arrangements such as those that prevail among countries that do not require military forces to manage their differences. The prospect of reducing military arms and lowering budget expenditures is a hopeful one, but it should not be pursued as a panacea for America's domestic economic, especially budget, problems. Nor should it be seen as a way to force greater burden-sharing on major allies, such as Germany and Japan—two countries whose internal political situations are likely to be tested as never before by the changes taking place in world affairs in the 1990s.

Military spending is no more the cause of the American budget deficit than spending for Social Security and entitlements, spending for domestic programs, or, for that matter, spending for so-called tax expenditures (i.e., tax reductions). Indeed as a share of GNP, U.S. defense spending has declined since 1970. What the U.S. government did, however, was to step up its domestic obligations dramatically. In 1970 the U.S. government devoted 39 percent of its budget, or 7 percent of GNP, to defense; in 1987 it devoted 27 percent, or 6 percent of GNP, to defense. Meanwhile, transfer payments, 75 percent of which are for so-called entitlement payments (i.e., transfers that are not based on need) rose from 29 to 38 percent of the budget, or from 5 to 9 percent of GNP (*ERP* 1987, p. 337). Lower defense expenditures after Vietnam did not guarantee lower budget expenditures overall. The Vietnam peace dividend was spent on domestic programs. Similarly, the new peace dividend in the 1990s may also be spent on domestic programs—environment, drugs, education, etc. America has to decide whether it wants to reduce its role in the alliances primarily to finance domestic goals, some of which, such as entitlements for the nonpoor, are not justified by human need in the United States, or whether it is willing to continue to carry sizeable defense obligations, at least through the coming transition period, because two of its most important allies, West

Germany and Japan, choose to rely on the United States substantially for their national security.

Despite some euphoric expectations to the contrary, a relatively high level of U.S. defense spending will probably continue to be necessary for some time to come. Even the dramatic cuts of $180 billion contemplated by the U.S. Department of Defense in late 1989 represented a cut of only 12 percent over five years because these cuts applied to a planned five-year military spending program of $1.5 trillion (Korb 1989). To achieve these cuts any faster would require either a more rapid pace of arms reductions in Europe, which might lead to hasty arms control agreements and hence be destabilizing, or a willingness on the part of West Germany and Japan to assume more responsibility and, hence, independence for their own security. For the United States to pressure the latter countries to do more to the point where it would make a difference for U.S. defense expenditures would undoubtedly risk setting into motion highly unstable developments, once again, in Japanese and German domestic societies as well as in their surrounding regions. In Japan's case, as George Packard notes, it might encourage "forces in Japan ready and willing to go much farther and faster down the road to an independent military force—with or without U.S. approval" (Packard 1987–1988, p. 357). In Germany's case, it might lead to either one of two developments, both of which would be harmful in the expected transition period—rapid demilitarization and possible neutralization of some sort with the East, which would be threatening to the West, or closer integration with Western military objectives, which would be threatening to the East.

America's commitment to its two alliances, with NATO in Europe and Japan in the Pacific, has safeguarded postwar peace and prosperity. At this point, there are no powerful forces in West Germany or Japan calling for America to leave these countries (as there were, for example, in the early 1950s in Europe and in the early 1960s in Japan). For America to leave on its own, therefore, would be a mistake of colossal historical proportions. This is especially the case in times of rapid change when American forces provide some assurance to both Western and Eastern partners that Germany and Japan will not alter their basic postwar defense and foreign policies quickly or disruptively. Certainly, both Germany and Japan can do more in some areas of defense and even more still in economic areas (see further discussion). But to devolve defense, especially strategic, responsibilities significantly to the allies is not a policy direction that will yield big savings for America's budget; it will only cause bigger insecurities and, in the end, when America is forced to return to Europe or the Far East to defend its interests, even greater military expenditures.

None of this is to say that the United States should not pursue aggressive negotiations to reduce arms. The Intermediate Nuclear Forces (INF) Treaty signed in Washington on December 8, 1987, though small in scope, eliminated an entire class of nuclear weapons. The Strategic Arms Reduction Talks (START), which seek to reduce nuclear weapons overall, seemed headed, in late 1989, after the Bush–Gorbachev Summit in Malta, toward an agreement to cut strategic weapons by a nominal 50 percent (one-third in real terms), perhaps as early as mid-1990. Gorbachev's announcement in December 1988 of unilateral conventional arms cuts in Eastern Europe and NATO's proposals in May 1989 for mutual cuts of superpower forces in Europe to equal levels of 275,000 troops (and related equipment) quickened optimism for early success in the talks on Conventional Forces in Europe (CFE), a smaller grouping of 23 nations that took over conventional arms talks from the earlier and unsuccessful grouping of 35 nations known as the Multilateral Balanced Force Reduction (MBFR) talks. While initial cuts in the CFE negotiations were not much below current levels of U.S. forces in Europe (around 320,000 personnel), further proposals in early 1990 envisioned mutual levels in central Europe around 195,000 troops.

Arms reductions make sense, however, not because they are inherently good or bad, but because they are consistent today with the changing threat in central Europe. The emergence of democratic governments in Eastern Europe removes the immediate threat of a massive conventional attack by Warsaw Pact forces, which is the threat that has driven all Western strategic responses since the early 1950s. These governments pose new political barriers to the extension of Soviet military power in central Europe, and once current Soviet forces are reduced in Eastern Europe, the West will have ample warning time to respond to Soviet attempts to reintroduce these forces and therefore will not have to station large, standing armies, as in the past, to defend against surprise attacks.

The decline of the conventional military threat in central Europe has reduced the salience of nuclear arms, by which the West has traditionally defended itself against superior conventional forces in the East. The INF Treaty left only short-range nuclear missiles in central Europe with ranges that made them useful chiefly on German soil. Understandably, the West Germans saw no need to modernize these weapons or even perhaps to keep them, since the weapons could be used only against other Germans who might shortly be a part of one country. Nuclear deterrence thus receded in importance in central Europe, although the U.S. nuclear commitment remained the linchpin of NATO and the raison d'être for the stationing of U.S. troops in Europe.

Some structure of deterrence (i.e., managing conflict), however, is still needed, whether in the form of much lower and tested balances of conventional arms (ultimately backed up, perhaps, by nuclear guarantees) or in the form of political and economic arrangements that eventually obviate the need for military forces. Threats have changed but not disappeared. The possibilities of economic and political setbacks in Eastern Europe or the Soviet Union (as occurred in China in 1989), the potential for ethnic violence, particularly in southeastern Europe and the Soviet Union, and the looming economic and, at least indirectly, military might of a reunited Germany all pose risks for the future of Europe and the United States. In addition, the Soviet Union remains a nuclear superpower. The CFE and START talks have only begun the task of converting the alliance structures, that for 40 years successfully deterred high levels of threat, to all European structures that can continue to provide confidence at much lower, though still dangerous levels of threat.

The political outlines of a reintegrated Europe or a "Europe whole and free," as American leaders have called for (Bush 1989), are also unclear. Much depends on how the two Germanies sort out their internal economic and political differences and whether more liberal governments in Eastern Europe and the Soviet Union succeed in achieving economic progress. The first outcome depends on how a reintegrated German society defines its national purpose and ultimately resolves the centuries' old struggle in German history between liberalism, now solidly rooted in West Germany over the past 40 years, and authoritarianism, which is all East Germany has ever known (whether from the right before the war or the left after the war). The second outcome is a function of economic policy choices that will be made in Eastern Europe and the Soviet Union. The United States and other Western countries will continue to influence the international context in which these choices are made. If they fail to find a new sense of shared political community to assist Eastern states toward greater political pluralism, or if they press economic competition among themselves to the point where it divides the Western world into regional economic and eventually conflicting political blocs, the opportunity to extend postwar freedom and prosperity to the peoples of the East will be lost and the prospects for growth in the developing world will be severely reduced. The challenge is not to discard the Western partnership of the past 40 years, but to adapt it to new purposes and new participants.

Choosing Efficient Economic Policies

The Western alliances underpinned the postwar international economic system. They supplied the reasons for which industrialized nations pursued open, efficient economic policies among themselves. Today these alliance arrangements are in the process of being transformed. Thus, other motivations will have to be found to sustain common purposes among Western nations if an open, efficient international economic order is to be maintained in the 1990s.

Two such motivations exist. First, competitive democratic and economic institutions are now deeper and stronger in the industrialized nations, including Germany and Japan, than they were in 1945. In this sense these nations are motivated to organize international economic relations along liberal and competitive lines because their own domestic systems are increasingly organized along these lines. Although Germany and Japan retain corporatist and consensual features in their domestic economies that distort international trade competition through nontariff barriers, the trends in both of these countries is toward greater internal competition (see further discussion). This does not mean that these countries will ever promote an individualistic, liberal international economic system with the same enthusiasm the United States did in the 1950s and 1960s. But they are also unlikely to use their considerable economic power to lead an effort to construct an alternative system, unless they are prepared at the same time to risk serious political conflicts with Western partners and to assume more independent responsibilities for their own military security.

A second motivation for the maintenance and extension of the liberal international economic system may lie in the changing attitudes of socialist and key developing countries. These countries are increasingly experimenting with more competitive internal systems, both economic and political. The command economies of the Soviet Union and Eastern Europe, as well as the old import substitution strategies of developing countries, find few advocates today anywhere in the world. The Western countries have an unprecedented opportunity to reinforce this shift toward market-oriented policies and to enhance global economic integration, not to defend Western pluralist systems but to assist pluralist development in the East and third world.

Viewed historically and comparatively, therefore, the broad trends are mostly in the direction of greater competition. These trends have emerged in socialist and developing countries even as the performance of the liberal trading order has declined and even as macroeconomic policies in the industrial countries have imposed harsh instabilities on world markets. They attest to the powerful pull of postwar prosperity under the influence of the Bretton Woods policy standards and raise the prospect that a return to these standards on the part of the United States, the principal architect of the liberal postwar order, could meet with unprecedented international support, even in the communist world to the extent that these countries succeed in developing greater competition within their own systems.

Thus, if it would be tragic for America to abandon its alliances, it would be even more tragic for the United States at this moment to pull back from its historic commitment to open, competitive international markets and abandon the socialist world, where citizens struggle for new freedoms, or the developing world, where 75 percent of the world's population still lives in desperate poverty. In terms of both threat and human need, the greater challenge lies in the third world. If the alliances are America's protection against the threat of instabilities in Eastern Europe and the Soviet Union as liberalization proceeds, the free trading system is America's protection against the threat of poverty and the scourge of revolutions, terrorism, and other instabilities that may come out of the third world. The choice for the United States in the third world in the 1990s, therefore, is not unlike the one it faced in central Europe in 1947. It can lead a collective attack against the human misery and political oppression of poverty in the third world, or it can abandon the third world to totalitarianism, a totalitarianism not necessarily imposed externally by a fading Communism but generated internally by resurgent despotism.

The challenge has many different characteristics in practice, as I shall detail. The Marshall Plan is not a specific model for U.S. policy toward the third world. Third world countries have neither sufficiently similar domestic economic systems nor sufficiently common security interests with the West to respond to a Marshall Plan in the same way the countries of Western Europe did in 1947–1948. But, in principle, the concept of a Marshall Plan—to open the Western, democratic system of freer nations and freer trade to the developing world as an alternative way for these countries to escape the closed cycle of poverty and oppression—has much to offer. What is more, the United States can do this today from a base of economically strong and politically homogeneous industrial countries that does not require the United States to carry the same disproportionate burden of costs that it had to bear in 1947. What the United States unquestionably must do is lead. And its leadership has to start at home with the realization that price stability, flexible markets, and freer trade are still the relevant and essential prerequisites for worldwide economic growth.

Restoring the Bretton Woods Policy Triad

The partial improvements in the U.S. and world economy in the early 1980s suggest that the policy triad remains relevant but that it must be more broadly understood and more consistently applied if the world economy is going to pull back from a disastrous repetition of the policy mistakes of the 1970s, this time with free trade being lost, along with price stability and flexible markets.

The Bretton Woods policy triad, it should be remembered, addressed medium-term objectives in macroeconomic and microeconomic (i.e., structural) policy, not the short-term, exchange rate and interest rate objectives often emphasized in the 1970s and highlighted again by U.S. initiatives after 1985. The essence of the triad is medium-term domestic price stability, progressively freer trade (no precise pace is required), and continuous structural adjustments (again no precise pace required) to ensure that markets become or remain flexible and therefore respond in a timely fashion to domestic macroeconomic signals and to international competition on the basis of comparative advantage.

The domestic roots of the policy triad are stable prices and flexible markets. Without these roots, freer trade will not only be more difficult to introduce and maintain, but also economically less efficient in terms of outcomes. Inflexible domestic economies prevent resources from being reallocated to meet external competition; prices rise unevenly to distort exchange rates if they are fixed or to cause them to fluctuate if they are flexible; misaligned exchange rates discourage market-oriented trade and encourage managed trade or nonprice arrangements to exchange goods. World markets become laced with bilateral trade and foreign exchange agreements, as they did, for example, in the 1930s and right after World War II in Western Europe.

The world, in short, has a choice: It can choose to protect and regulate domestic economies and, as a consequence, create unstable international markets that must then be managed by inefficient bilateral, nonprice-oriented trade arrangements; or it can encourage flexible domestic economies and, as a consequence, ensure relatively stable international prices and predictable, market-oriented trade relations. It is not possible to have both protected domestic markets and stable international markets, except at much higher levels of administered trade and exchange rate relationships and, therefore, much lower levels of overall efficiency and growth. In this sense, the postwar compromise of "embedded liberalism," in which, it is argued, liberal international policies were traded off against interventionist domestic policies, is no longer liberal if government intervention at home exceeds the point that allows internal resources to be reallocated flexibly to meet international competition.

The restoration of price stability is the crowning achievement thus far of U.S. and other industrial countries' policies since 1979. Sometimes confused with monetarism, this restoration grew out of the recognition by all of the principal countries, including the chronic high-inflation countries of France and Italy, that restraint of monetary policy and credit was the key to stable prices (by whatever monetary techniques). Monetary policy was restored to its rightful and coequal (rather than accommodating) role with fiscal policy, as it had been in the original postwar Bretton Woods consensus.

Two other domestic policy areas, however, escaped convergence in the 1980s, at least to the same degree. The two areas were fiscal policy and government policies in the microeconomy. The United States was the major offender in the first area, and Europe and Japan continued to be lagging performers in the second area. Continued divergences in these areas potentially threaten the world economy with renewed inflation and market rigidities in the 1990s.

Fiscal policy in the United States shifted toward expansion throughout the 1980s. Although the reduction in taxes in 1981 initially sparked new investment and helped to restore some flexibility in American labor and capital markets, the failure to reduce expenditures left America dependent on precarious foreign capital flows and sustained the recovery, after 1984, increasingly on the basis of an accommodating monetary policy and rising consumption. Meanwhile, fiscal policy in the other major industrial countries shifted toward contraction. From 1981 to 1985, while the deficit in the United States expanded by 2.9 percent of GNP, central government fiscal deficits in Germany, Japan, and the United Kingdom contracted by a total of 3.5 percent of GNP (IMFa 1988, p. 127). This divergence in fiscal policy among the major industrial countries represented an enormous shift and resulting imbalance in savings and investment flows between the United States, on the one hand, and Europe and Japan, on the other.

The fiscal policy contraction in Europe and Japan undoubtedly contributed to the allies' lagging performance in microeconomic policy areas, particularly the failure to reduce excessive government intervention in European labor, capital, and product (e.g., agriculture) markets and to stimulate greater domestic consumption in Japan. Relatively tight fiscal and monetary policies in Europe restricted growth and increased unemployment (Lawrence and Schultze 1987). This environment was not conducive to significant deregulation, particularly in agriculture, where reforms promised to displace additional labor. In addition, the public sector in Europe was much bigger to begin with than in the United States or Japan (see Chapter 6). There was greater reluctance to trim it back, to deregulate private industry, and to subject public entities to greater competitive forces. Privatization took hold in the United Kingdom and, to a lesser extent, France, but the transfer of public assets to private hands did not create many new jobs. Indeed, in the short run, it led to cutbacks to reduce redundant and inefficient employment.

Meanwhile, financial and trade policy deregulations proceeded slowly in Japan. Excess savings accumulated as much because Japanese consumers lacked opportunities to buy imports, new houses, and domestic leisure services as because the United States attracted these savings to exploit demand and investment opportunities in U.S. markets. Japan retained flexible labor and capital markets to respond to export opportunities but continued to restrict markets in Japan for demand and distribution of imports.

Meanwhile the United States deregulated more rapidly than either Europe or Japan, although not as much as Reagan administration officials hoped for or claimed, especially given the growing "reregulation" of the trading sector. Reagan policies moderated wage demands and subjected manufacturing industry to the withering competition of an appreciating dollar and cheaper imports. Although the United States failed to reduce the size of the public sector—indeed it grew because of defense and entitlement spending and, increasingly, interest payments on the national debt—it shifted the composition of spending to defense and other, potentially more productive outlays. Unemployment in the United States dropped dramatically from over 10 percent in 1982 to 5.0 percent in early 1989, and financial, especially venture capital, markets revived and fueled a burst of technological innovation and small business entrepreneurship.

The continuing domestic policy differences between the United States and its key industrial allies in macroeconomic and microeconomic areas bear critically on the ability of these countries to maintain open, efficient international markets in the West. As this study has consistently shown, stable and efficient international markets require broadly similar domestic policy practices and institutions if countries are to agree on what constitutes "fair" rules for economic competition among them. If America's propensity for irresponsible fiscal policies persists, Europe will surely withdraw to seek protection from a fluctuating dollar in the European Monetary System (EMS) and in the single European internal market of 1992. Similarly, Japan will accelerate economic and trade relations with regional partners and perhaps the Soviet Union. By the same token, if Europe's or Japan's greater interventionist or *dirigiste* tendencies persist or, even worse, increase, the United States will surely retaliate and reduce access to its market for European and Japanese products.

What the United States and its industrial partners need to realize is that none of these developments is preordained by unchangeable cultural or historical differences among the allies. Structuralist arguments like to make the claim that Japan is a fundamentally different society than the United States or even Western Europe and that Japan can never become an open, transparent political society or an open economy for imported goods (Johnson 1982; Krasner 1987; Prestowitz 1988; van Wolferen 1989). But this claim taken too far denies the possibility of change in structural characteristics of society and would have difficulty accounting for the evolution of more democratic institutions in Japan since the war or the incremental opening up of the Japanese market and society that has occurred over the last 20 years. In fact, as OECD studies of microeconomic policies in Western countries show, the United States, Germany, and Japan share important common internal economic features and contrast with other Western societies in that "governments in these three countries have been especially reticent to act as entrepreneurs; and, in a context of intense competition on the domestic market, firms have taken primary responsibility for maintaining traditional areas of strength and/or developing new ones" (1988, p. 36). Admittedly, corporate–government relations in these societies differ, but it would be hard to argue that these differences have grown larger rather than smaller over the past 25 years.

What is more, policy differences between the United States, Europe, and Japan can continue to narrow. Despite the dangers that linger from incomplete policy convergence, the progress that has been achieved is substantial. The United States today has a significantly more flexible domestic economy than in the 1970s, as reflected in unprecedented postwar rates of growth of manufacturing investment and productivity, although it threatens to re-regulate this economy by increasingly protectionist trade policies made necessary, in part, by continuing large budget deficits. The major industrial countries in Europe (France, Germany, and the United Kingdom) and the Pacific (Japan) have more disciplined macroeconomic policies and lower inflation than they have ever had in the postwar period (3.6 percent from 1983 to 1987, compared to 4.4 percent from 1947 to 1967 — see Figure 1-1), although they have made less progress in microeconomic reforms and behave no better and perhaps worse than the United States in trade policy. The key, as this study has argued, is to overcome the remaining divergences and reinstate the Bretton Woods policy triad more fully. America must rediscover the virtues of disciplined fiscal policies, Germany and Europe 1992 must succeed in liberalizing microeconomic barriers to spur internal growth without adopting macroeconomic stimulus or adding new barriers to trade with the outside world, and Japan must recognize that a society that saves too much is fundamentally anticonsumer and, to that extent at least, also anti-individual and perhaps antidemocratic.

None of these changes requires that these countries lose fundamental features of their cultural or national heritage. The changes are only incremental. They need not proceed to the point where, as some studies recommend, America reduces its current account deficit to zero (Williamson 1983, Bergsten 1988). If some countries, such as Japan, for cultural or other reasons, ultimately decide to save more than others, it is better that these savings flow to other countries, such as the United States (and create a current account deficit for the latter countries), than that they go unused in the surplus countries. The problem at the end of the 1980s was that no one believed that then existing levels of U.S.-Japanese or U.S.-German current account imbalances were sustainable or that the perverse flow of capital from developing to industrial countries was desirable. Hence, current account balances have to shift but not necessarily to any arbitrarily predetermined levels.

International Coordination of Domestic Economic Policies

In the preceding discussion we have been talking about the substance of international coordination of domestic economic policies but not the mechanisms for doing so. From the perspective of this study, the mechanisms or institutions of international economic policy coordination are much less important. Although structuralist perspectives focus almost exclusively on these mechanisms and almost always recommend more institutional cooperation, the choice-oriented perspective in this study shows that such cooperation may in some instances impede market-oriented adjustments and that the world economy may be better off with less, rather than more, institutional cooperation.

Figure C-1 arranges in a two-by-two matrix the approximate combinations of substantive domestic policy choices and international coordination mechanisms that the United States and other industrial nations adopted during the various postwar periods. Substantive policy choices and outcomes (which are related to one another according to the perspective in this study) appear along the horizontal axis, whereas international mechanisms fall along the vertical axis.

Except for the Bretton Woods era, and perhaps a brief period in the mid-1970s (box C), direct cooperation through international institutions has been associated with inflationary and interventionist domestic policies and outcomes (box D). Except for the first Reagan term (box A), so has indirect coordination through market mechanisms (box B). From this comparison, the existence or absence of institutional cooperation does not seem to matter much for the existence or absence of sound domestic economic policies, confirming the choice-oriented perspective's skepticism toward international institutional cooperation, especially as practiced in the 1970s.

The Bretton Woods and first-term Reagan administration experiences, on the other hand, suggest that both institutional and market mechanisms can contribute to sound domestic economic policies (boxes A and C). Both of these experiences are inadequately understood, however. The Bretton Woods experience is often cited to call for more institutional cooperation today, as in repeated calls for a Bretton Woods II conference. But institutions played their role most effectively in the late 1940s and 1950s, when international markets were weak and America dominated decision making in international institutions. The first-term Reagan administration experience, on the other hand, is often dismissed as unilateralism, but the impact of U.S. policies on world markets in this period, particularly in terms of bringing inflation down and sparking interest in market-oriented domestic reforms (e.g., tax reforms, privatization, and deregulation), was probably on the whole quite salutary. At the very least, this experience involved a different use of market mechanisms or leverage than U.S. policies employed in the Nixon period.

Institutional cooperation worked to bring about sound domestic policies during the Bretton Woods era for three reasons (see Chapters 3 and 4). (1) A domestic policy consensus, embodied in the National Employment Act of 1946, existed in the United States before international cooperation was

Domestic Policy Content/Outcomes

	Price Stability and Market Flexibility	Inflation and Microeconomic Intervention
Markets	Reagan (1981-1984) **A**	Immediate Postwar Period (1945-1947) Johnson (1967-1968) Nixon (1969-1974) **B**
Institutions	Ford (1974-1976) Bretton Woods (1947-1967) Earliest U.S. Postwar Economic Plans (1941-1942) **C**	Wartime Negotiations to Establish Bretton Woods (1943-1944) Carter (1977-1981) Reagan? (1985-1988) **D**

(left axis label: International Mechanisms)

Figure C-1 Models of international coordination of domestic economic policies.

launched, and this consensus called for domestic and trade policies that were fully consistent with open international markets and stable exchange rates. (2) The United States had the motivation, money, and diplomatic power to mobilize international institutions to promote this policy consensus in other countries; and these institutions were, in many cases, ad hoc and staffed by private sector personnel that did not seek to manage markets indefinitely but actually closed down the institutions once the initial tasks were completed. (3) Institutional cooperation could focus primarily on medium-term policy objectives—price stability, more flexible labor, capital and product markets, and progressive liberalization of trade—because international markets, especially capital markets, were not so large or deep that policymakers became immediately preoccupied with short-term feedbacks on exchange rates and therefore with fine-tuning macroeconomic policies to stabilize exchange rates. From time to time, as we saw in the operation of the EPU (see Chapter 4), the pace of liberalization of trade and foreign exchange restrictions was varied to buy time from external pressures and facilitate further domestic policy adjustments. This allowed exchange rates to be kept relatively fixed and to serve smaller countries as a stable long-term anchor for monetary policy while enabling larger countries to concentrate on more fundamental domestic and trade policy reforms.

For at least two of these three reasons, institutional cooperation was less appropriate in the early Reagan era. First, the United States no longer dominated international institutions. It was not able to achieve a negotiated consensus between countries that wanted to bring inflation down (e.g., the United Kingdom and the United States) and those that wanted to bring unemployment down even if that meant more inflation (e.g., France and Italy). Second, trade and financial markets were now so complex that any attempt to coordinate policies without a broader consensus would have undoubtedly focused very quickly on managing short-term exchange rates through fine-tuning macroeconomic policies.

By contrast to the Bretton Woods period, however, the United States had enormous power in the 1980s through the international marketplace. High interest rates and the high dollar worked to force other countries (e.g., France) to make earlier decisions whether to accept a depreciating currency indefinitely (which exacerbated inflation), erect trade barriers, or bring inflation down and spur growth by microeconomic reforms (e.g., supply-side tax reforms). Ultimately, U.S. market-based efforts came up short, not because they confronted overpowering external constraints, as structuralist arguments contend, but because they were not supported by a domestic policy consensus in the United States consistent with open and noninflationary international markets. Once the United States backed off efforts in 1985 to reduce the budget deficit and initiate free trade legislation, U.S. market leverage lost credibility and began to push other countries to fine-tune macroeconomic, especially monetary, policies rather than to reform more fundamental microeconomic and trade policies (see Chapter 9).

The United States can learn from the Bretton Woods and early Reagan experiences to improve the coordination of domestic economic policies in the 1990s. First, as both experiences suggest, the indispensable ingredient for success is a U.S. domestic economic policy consensus that is compatible with open and noninflationary international markets. If this consensus does not exist, as it has not since the late 1960s, neither institutional cooperation nor market leverage can be fully effective.

Second, market leverage is an important mechanism for achieving domestic policy convergence in today's complex and highly interdependent world markets. Governments other than the United States are increasingly recognizing this fact. The EC, for example, is employing the principle of mutual recognition in seeking to harmonize national environmental and technical standards in the Europe 1992 exercise. This principle says that products are acceptable in any EC market if they meet the national standards of the exporting country. This principle sets in motion a competition among national policies in the marketplace in which no country wishes to jeopardize the competitiveness of its own exporters and, therefore, seeks to harmonize its policies with those of the most efficient exporter. Minimum health and safety standards are still set by the EC (Calingaert 1988).

Structuralist perspectives are reluctant to recognize the advantages of market leverage or what they label unilateralism (Bergsten 1986). They emphasize the structural features of the world economy (e.g., the growing size of the external sector in national accounts) that compel the United States and other countries to moderate market actions and coordinate their policies directly through international institutions. They relatively ignore the content of government actions in either institutions or markets. They equate, for example, U.S. unilateral policies in the early 1970s with those in the early 1980s, mentioning only in a brief footnote the "big" differences in policy content and consequences between these two periods — one inflationary and interventionist, the other disinflationary and market-oriented (Bergsten 1986, p. 6). On the other hand, structuralist perspectives regard institutional cooperation as the optimum under all circumstances, applauding policies such as the Plaza and Louvre initiatives even when these policies result in significant instabilities (e.g., Black Monday).

The key lesson here is not that institutional cooperation is always good or that market leverage is always bad but that there is a consistent pattern in the historical record when these mechanisms are used and what they are used for. In recent decades at least, institutional initiatives seem to be associated with efforts to reflate the world economy, whereas market initiatives seem to be associated with efforts to disinflate the world economy. There is an obvious political reason for this pattern. Political leaders like to lower interest rates and increase fiscal expenditures, and they like to get credit for these actions. Visible international cooperation helps them gain credit and even helps them, in some circumstances, gain leverage over more conservative central bankers, who as a rule are more reluctant to lower interest rates, at least as much as politicians would like. Circumventing central bankers was clearly the effect, if not the intent, of the "politicization" of the G-5 process by the second-term Reagan administration. As Yoichi Funabashi concludes in reference to the Plaza–Louvre exercise, the "politicians basically got from the central bank governors what they wanted — lower interest rates" (1988, p. 231). On the other hand, politicians seldom if ever like to raise interest rates and lower fiscal expenditures. Hence, they are less likely to dramatize policy coordination when world economic conditions call for monetary and fiscal discipline. Indeed, as in the first-term Reagan administration, they consciously downplay international coordination and look for ways to suggest that markets, not policies, are responsible for rising interest rates and recession.

The world community, therefore, should be more suspicious of international economic policy coordination when inflation is low (as in 1977 and 1985) and less critical of unilateral initiatives when inflation is high or rising (as in 1981). Coordination, especially if it emphasizes primarily exchange market intervention, is particularly suspect when fiscal expenditures and deficits are also high. For just as central banks can sometimes be trapped into financing a government's domestic debt by purchasing government securities at low and stable interest rates even as inflation accelerates (the U.S. Fed, for example, before 1951 and again after the mid-1960s), they can also be trapped into financing cumulative international debt by a process of international policy coordination that emphasizes low interest rates. This temptation will be particularly great in the 1990s, when massive debt hangs over much of the developing world and, most important now, affects the world's central banker itself, name-

ly, the United States, which is the world's largest debtor. Because growth accelerated after the stock market crash in 1987, the United States was able in 1988,
without political costs, to push up interest rates to ward off inflation. But as
the U.S. economy nears full employment and growth inevitably slows, and
particularly if the United States cannot reduce its budget deficit, the Fed will be
under increasing pressure from domestic and international debtors to keep
interest rates down. German and Japanese officials have been particularly
sensitive to this possibility. For this reason they have been reluctant to stimulate
fiscal policy and thereby add to total world domestic and international debt;
and, in the case of Germany, Bundesbank officials have maintained a tighter
monetary policy than Germany's EMS partners (e.g., France) would like. The
United States seriously damages prospects for lasting economic coordination
among the major industrial allies when it misuses G-5, G-7, and summit processes to override these allied objections.

But a third conclusion is also clear from the Bretton Woods and first-term
Reagan experiences. Some type of international coordination process is necessary that goes beyond market mechanisms. Otherwise, there is no means to
cultivate the domestic consensus among principal countries on the Bretton
Woods policy triad, which, as our first conclusion suggests, may be the most
important prerequisite for open and noninflationary international markets.
The Reagan policies of the early 1980s failed to promote sufficient international cooperation to generate this policy consensus.

The test of the market-based approach in the early 1980s may not have been
entirely a fair one. The situation in 1981 represented a backlog of massive
maladjustments coming out of the late 1960s and 1970s. Had the market-based
approach been applied already in the mid-1960s, it might have worked much
better. The United States might have used its market leverage to promote further structural changes in Europe and Japan before the requirement for these
changes was compounded by the inflationary policies and oil crises of the
1970s. It might have also adapted the Bretton Woods institutions earlier to the
structural tasks that regional institutions (e.g., OEEC, EPU, and other institutions) implemented so successfully in the 1950s. Instead the United States
focused on liquidity reforms in the IMF in the 1960s, which proved to be
unnecessary, and encouraged the World Bank to turn exclusively to poverty
projects in the 1970s, which added little to macroeconomic or trade policy
reforms (see Chapters 5 and 6).

But the market-based approach to domestic policy coordination was moving too slowly in the mid-1980s, and it needed to draw on a broader collective
leadership—exploiting the exemplary policies of the United States in microeconomic reforms, of Europe and especially Germany in macroeconomic discipline, and of Japan in investment and financial areas.

A "Soft" Institutional Approach to Exchange Rate Stability

An improved process of international economic policy coordination has to
avoid the negotiated or "hard" institutional approach to cooperation that has
historically resulted in excessive exchange market intervention, fine-tuning of

macroeconomic policies, and inflation. This approach, which targets exchange rates, is unrealistic in today's world. Not only is it utopian to expect governments to agree, even among the G-3, to coordinate macroeconomic policies directly and primarily to stabilize exchange rates, but this process would take place without effective international political controls or accountability over finance leaders and central bankers. Central banks are highly sensitive political institutions inside countries. Issues of who controls these banks lie at the center of national politics. Who would exercise such control over a G-3 or G-5 world central bank? If monetary policy were institutionalized internationally, political institutions and security policy ultimately would have to be integrated as well. As this study has shown, common political purposes are never far removed from the existence of integrated monetary systems. The European Monetary System has recognized this fact and slowed the rush to establish a common central bank, making this the last step in a process that will proceed by open-ended rather than automatic decision making.

Moreover, although the United States still functions to a large extent unilaterally as the world's central banker, it is no longer free to ignore feedback on its policies from its own actions in the international marketplace, at least not for as long as it could in the 1950s and 1960s, when, for example, it could be indifferent to discrimination against its exports or the lack of microeconomic reforms in Europe and trade reforms in Japan.

A more complex world, therefore, requires a more flexible exchange rate system. But this system need not be a fluctuating, let alone volatile, one. That depends on the system's success in bringing about convergence among principal countries on noninflationary and flexible domestic economic policies. To promote more effective convergence, the present system needs three modifications leading to what I call a "soft" institutional approach to exchange rate stability and growth.

First, the multilateral surveillance process inaugurated at the Versailles Summit in 1982 and elaborated at the Tokyo Summit in 1986 needs to be pursued with more professional, and less political, enthusiasm and visibility. Rather than being invisible, as the process largely was before 1985, or a prop for dramatic meetings of finance ministers, as it became after 1985, the discussion of what constitutes convergence and what is necessary to achieve it should be taken up with more professional seriousness not only in government circles but throughout the private and quasi-governmental cocoon of international research institutes, political parties, and public policy organizations concerned with world economic affairs. Some private research and discussion of this sort take place now, but too much of this effort remains oriented toward current account, exchange rate, and demand management objectives, rather than toward flexibility of labor, capital, and product markets; tax reform; and trade liberalization (Bryant et al. 1988; Institute for International Economics 1982, 1987). In addition, the substantive record of multilateral surveillance meetings and other G-5 discussions among finance ministers and summit leaders should be made available, after an appropriate delay to protect confidentiality (say, 3–5 years), as raw material for analysts to study and use in professional discussions. In this way the discussions of government officials will be held more

accountable to long-run world economic and not just short-run national political standards, and the discussion and recommendations of economists and
other policy analysts will be made more realistic.

Second, while multilateral surveillance is being better understood profes-
sionally and being made more accountable politically, the system needs some
defense against the tendency for dollar policy (i.e., the exchange rate of the
dollar) and short-term exchange rate coordination to get all the public attention. In this respect, the best defense would be a U.S. policy that remains wary
of active exchange market intervention (Morris 1989). This policy would call
for neither the complete abstinence from intervention that was practiced in
1981–1982 nor the sometimes frantic intervention that has taken place since
1985. But it would consistently resist the tendency, both in the United States
and in many foreign countries, to believe that dollar policy is all that matters
and that exchange rates can be affected on a sustainable basis without changes
in underlying fiscal and monetary policies.

The focus on dollar policy in the United States is a direct reflection of the
loss of domestic consensus on other economic policies. Because there is no
agreement on what constitutes sound domestic policies, each school of thought
is willing to let the dollar rise and fall to compensate for the consequences of its
own domestic priorities. Thus, supply-siders and monetarists were willing to let
the dollar rise in the early 1980s because it accommodated tax cuts and a
tighter monetary policy. Fiscal conservatives, on the other hand, who saw the
rise as a perverse consequence of budget deficits, were willing to see the dollar
fall even below the levels of late 1988 if that was an inevitable consequence of a
decision to reduce the budget deficit (Feldstein 1988).

Yet surely if the dollar was too high in 1985, a good case could be made that
it was low enough at the end of 1988, when its trade-weighted value was
roughly the same as at the end of 1980, before the sustained increase of the
early 1980s (see Figure 9-1). This conclusion seemed warranted for two other
reasons. First, as we saw in Chapter 9, relatively little internal adjustment in
either the United States or its principal industrial allies accompanied the substantial lowering of the dollar after 1985, at least until the end of 1987. In 1989,
therefore, it seemed wiser to encourage further domestic adjustments before
deciding whether the dollar needed to fall further. Second, as was shown in
Chapter 8, there had been a relative improvement in the competitive position of
many American industries. That improvement should have been reflected in a
higher real exchange rate for the dollar in 1988 than in 1980. Conceivably the
latter factor played a role in the perverse rise of the dollar again in the first half
of 1989.

On the other hand, the perspective in this study does not recommend
propping up the dollar at any arbitrary level, including the dollar's value at the
end of 1988. If the dollar falls further as a result of lowering the budget deficit
further, as some analysts believe it will, that fall should be accepted. Any
attempt to target the dollar's value, especially if the United States does not
reduce its budget deficit further, only transfers dollar instabilities to other
areas, such as bond and stock markets (e.g., Black Monday).

Abroad, the belief that dollar policy is all that matters is deep-seated in the psyche of the allies. It is a residue of the "abject" dependence of the allies on the United States after the war, affecting, in particular, the attitude of France, and the continued dependence of some of the allies (Germany and Japan, in particular) on the United States in critical national policy areas such as security. It is a largely incorrect belief, as we observed in Chapters 7 and 8. European countries had many more options in the early 1980s than they acknowledged. They did not have to respond to U.S. policy only through monetary policy, that is, by raising their own interest rates and contracting growth. They could have loosened fiscal policy; or, failing that, concerned as they were with their own fiscal deficits from the late 1970s and U.S. fiscal deficits in the early 1980s, they could have initiated much earlier and more vigorous structural adjustment in labor, capital, and product markets that lagged in Europe and Japan in the 1970s. They had choices; they simply made certain choices while the United States made others.

As long as the dollar is the world's principal reserve currency, Europe and Japan will complain about American economic policy whether the dollar is up, down, or sideways. The United States should listen to these complaints but do more of its own complaining about European and Japanese policies. In this sense, the more aggressive policies of the second-term Reagan administration were a welcomed relief. Too often in the first term the allies did most of the attacking and the United States did most of the defending. This type of discussion only perpetuates the myth of dependence. But the United States should attack on the right issues and keep its own domestic responsibilities foremost in mind. In the mid-1980s, it lost sight of its own responsibilities and attacked the Europeans and Japanese on their exemplary macroeconomic policies rather than on their deficient microeconomic policies.

Thus, a third modification is needed in the present system to prevent the repeated tendency to slip back into coordination of short-term exchange rate and monetary policy objectives, that is, a mechanism to keep the allies and United States focused on medium-term macroeconomic and structural policy issues. In the Bretton Woods system this mechanism was provided first by the Marshall Plan programs to stimulate productivity growth, liberalize labor markets, and consolidate a consensus on moderate fiscal and monetary policies. Later, the mechanism was provided by trade liberalization, initially through the dismantlement of quotas under the OEEC, and then through the lowering of tariffs under the GATT. Today this function has to be played by several mechanisms. Trade liberalization must continue to play a substantial part (see next section), but it cannot carry the load alone, among either industrial countries or, especially, developing countries, particularly when fiscal and structural policy adjustments in the industrial countries lag and put enormous pressure on trade relations and when basic market infrastructures in developing countries are so weak.

Two other mechanisms are needed. First, the industrial countries need a coordination process that focuses on fiscal policy that is as intense and developed as the one that focuses on monetary policy. Finance ministers meet often and intensively to discuss monetary policies. In most countries, these ministers also control fiscal policies, but not in the biggest and most important industrial country, the United States. The Secretary of the Treasury controls revenue policy, but generally not, if the OMB Director is a capable actor, expenditure policy. In any event, it is the OMB Director who knows the intricate details and political minefields of expenditure policy, yet the OMB Director participates in no regular international meetings whatsoever with foreign counterparts. Moreover, the Chairman of the Council of Economic Advisers or the Under Secretary of State for Economic Affairs, not Treasury officials, has the lead role in most of the significant OECD meetings that deal with underlying microeconomic or structural adjustment policies. The Secretary of State, in fact, heads the U.S. delegation to OECD ministerial meetings. Thus, the Secretary of the Treasury, who dominates international finance meetings, shares responsibilities with other U.S. officials in international meetings that deal with fiscal and structural policy issues. Institutional interests inevitably lead the Secretary of the Treasury to give more emphasis to short-term monetary problems—interest rates, dollar value, and so on—than to medium-term fiscal or structural adjustment problems.

Second, therefore, the coordination process needs a mechanism to upgrade the consideration of structural adjustment policies. One way to develop these mechanisms for both fiscal and structural policies would be to create subgroups within the industrial country summit process. Currently, the financial sherpas form a subgroup for monetary policy. These individuals are the so-called G-5 deputies, who negotiate most of the technical issues for finance minister meetings and for economic summits among the heads of state and government. Conceivably, future sherpa groups might be created for fiscal, structural adjustment, and trade (see later discussion) policy discussions, the members of which would then simultaneously serve as deputies for trade and other economic policy officials when these officials meet separately from the summit process.

The OMB Director, Chairman of the CEA, and Secretary of the Treasury, known as the Troika (T-3), already coordinate internally within the U.S. government to do economic forecasting for budget purposes. The deputies for this group—the Under Secretary of Treasury for International Affairs (also the G-5 deputy), the chief economist of OMB, and a member of the CEA—might conceivably become more involved in the fiscal policy discussions of the summit preparatory process. Accommodating the OMB Director himself in this process would be more difficult to do. Because finance ministers in most other countries also control expenditure policy, it would be awkward to create a separate group from the G-5 just to accommodate the OMB Director. But the OMB head could function as a deputy to the Secretary of the Treasury in all G-5 discussions dealing with expenditure policy (with the chief economist of OMB playing the same role in G-5 deputies' meetings). A second deputy group for structural adjustment policies might be constructed out of current discussions in the OECD. The CEA Chairman traditionally participates in and chairs meetings of the executive bureau of the OECD's Economic Policy Committee.

For a number of years this group wrote the principal macroeconomic paper for summit preparations. This practice could be resurrected and shifted somewhat to focus more exclusively on structural reforms in economic policy, sustaining the outstanding work of the OECD in recent years on microeconomic policy (OECD 1988).

To give more attention to medium-term policies, the summit process might also, from time to time, when agreement can be reached, create temporary, ad hoc institutions to implement basic policy reforms in the tradition of the Marshall Plan and regional economic institutions in Europe that did these tasks after World War II and then disappeared after completing them. In the present situation of complex markets, these institutions could be quite specific, unlike the broad scope of Marshall Plan institutions, and they might actually be created as temporary units of existing organizations, such as the OECD, to reduce potential institutional rivalries. (The Paris Summit in July 1989 created such a follow-up organization to coordinate aid proposals for Eastern Europe.)

For example, the Williamsburg, London, or Bonn summits in 1983–1985, when structural issues were still high on the agenda, might have called upon the OECD to create a special, temporary unit to facilitate privatization activities or to develop and apply programs to promote greater labor market flexibility. In keeping with the model that worked so well under the Marshall Plan, these units would be headed by high-level private sector individuals from the OECD countries, with the OECD secretariat providing appropriate staff support. Such private sector leadership would not only ensure fresh, creative approaches, but also increase the likelihood that these units would disappear as soon as their task was completed.

The soft institutional approach advocated in this study, therefore, does not reject a role for international institutions, but sees these institutions functioning at either end of a broad spectrum. At one end, general, high-level conferences such as the annual economic summits among the industrial countries, and perhaps, eventually, Cancun-type economic summits between developed and developing countries (see further discussion), would work to establish and maintain a consensus on medium-term policy objectives such as those represented in the Bretton Woods policy triad. And, at the other end, specialized and perhaps more regionally organized institutions would take on more limited and time-bound tasks assigned by economic summit meetings to promote basic policy reforms. The approach in this study sees much less of a role for institutions in the middle of this spectrum, where general meetings, such as summits, would seek to deal directly with specific problems on a short-term basis, as in the model of the Bonn Summit of 1978.

Trade Liberalization—A Crossroads

If countries are ultimately unwilling to promote greater fiscal and microeconomic reforms to maintain flexible domestic economies, through whatever mechanisms of international economic policy coordination, they have no choice but to reduce and manage trade and exchange rate relationships to accommodate more inflexible domestic economies. A strong argument is being made in the late 1980s to do just that (Chaote and Linger 1988, Cohen and Zysman 1987, Prestowitz 1988). Trade liberalization, the last leg of the Bretton Woods policy triad, is about to be kicked out, just as the other two legs — price stability and some modest new flexibility in domestic markets — have been painfully, if only partially, restored. Trade liberalization is under attack as never before, including a theoretical attack against the long-standing theory of comparative advantage (Krugman 1986, Helpman and Krugman 1985).

The assault on freer trade is lamentable, given the strong historical coincidence between freer trade and economic growth (see Chapter 1). It is doubtful, as we saw in Chapter 4, that the Marshall Plan would have succeeded in doing much more in postwar Europe than restoring prewar industrial production if it had not moved on, in late 1949, to adopt the goal of integrating Western European economic resources through freer trade. The European Common Market then became, in many ways, the impetus for broader global trade liberalization, as the United States sought a way to accommodate a customs and potential economic union in Europe by lowering worldwide barriers and thereby avoiding permanent discrimination against U.S. trade. Similarly, Japan and the NICs could not have pioneered their export-led models of growth without free and mostly one-sided access to large "international" markets, chiefly in the United States. Keeping U.S. markets open, even after the United States abandoned stable domestic economic policies in the late 1960s, was probably the most important reason the global economic system held together in the 1970s and 1980s.

On the other hand, the assault on freer trade is also understandable. It is due, as we have seen in this study, to three principal developments. The first is the increasing macroeconomic and exchange rate instability the world economy has experienced since the early 1970s. The loss of international consensus on relatively disciplined macroeconomic and microeconomic policy objectives introduced enormous volatility and stress into the world economy and steadily distorted and weakened the trading system. Second, the success of postwar trade negotiations in lowering tariffs and other trade restrictions at the border made more visible the nontariff trade barriers inherent in varying domestic policy institutions and practices within individual countries. And, third, new participants arrived on the international trading scene, beginning with the Japanese and including the NICs, or as we noted earlier, now called NIEs (newly industrializing economies, to suggest that some of these entities, such as Taiwan, are not necessarily countries), whose domestic policy institutions and practices emphasized government intervention and thus exacerbated the problems with nontariff trade barriers. The domestic institutions of these countries were less pluralistic, and their trade policies emphasized aggressive exporting combined with equally resourceful import protection. All these developments revealed the limits of a free trading system if there were no limits on government macroeconomic or microeconomic policy objectives and intervention.

Throughout the postwar period, trade liberalization has been encouraged precisely because it increased competitive forces and kept the pressure on governments to limit intervention that hindered the flexible reallocation of resources to more efficient uses. The postwar free trade ideology did not rule out government intervention per se, but it did hold government intervention to a standard of internal, or, at least, external competition (the latter, for example, being important for the smaller states in Europe; see Katzenstein 1985). The new ideology of managed trade now turns the free trade logic on its head. It not only justifies extensive government intervention, but also suggests that greater competition, and, hence, freer trade, are not necessarily more efficient. The theoretical attack on free trade is particularly significant because, as this study has emphasized, new ideas lead and, if they succeed in convincing a broader group of national and international constituents, generate policy changes that in turn generate new circumstances.

Free trade theory is based on the existence of scarce resources and the concept of opportunity costs. Because resources are scarce, using resources for one purpose is always seen as taking resources away from another purpose that might yield higher returns. This notion of opportunity costs encourages more open international markets because wider markets take advantage of varying resource endowments (e.g., labor and land) and ensure the most efficient calculation of opportunity costs. Comparative advantage says that, even if the absolute costs of producing two products in one country is lower than these costs in another country, the opportunity costs of producing one product in one country, that is, the cost of that product in terms of the second product, will always be lower than these costs in the other country. Hence, both countries can gain by specializing their production and trading. The theory thus encourages governments to lower barriers to trade and increase domestic competition.

New theories start from the premise that technology dramatically reduces the constraint of scarce resources and that economies of scale, which derive from higher levels of output, are more important to increase efficiency than opportunity costs and the geographic distribution of resource endowments. Thus, anything that accelerates technological development or that enables companies to get into a market first and build up economies of scale leads to greater efficiency. Governments, accordingly, play a major role in this process. They can accelerate technological change through subsidies to research and development and the like, and they can promote market size for their own companies by protecting import markets and promoting exports. New theories thus encourage governments to raise barriers to trade, decrease domestic competition, and subsidize exports.

The new theories have always existed as exceptions to free trade theory (e.g., the infant industry argument, which said it was efficient to protect a new industry from imports until that industry achieved sufficient scale to compete in export markets). But now these theories are said to apply more widely. They have become the rule, it is argued, because intraindustry trade (i.e., trade in similar products based on economies of scale) is nearly as large in today's international markets as interindustry trade (i.e., trade in different products based on comparative advantage).

The debate between free and managed trade will go on, and it will have similar consequences for trade policy and, ultimately, trade outcomes in the 1990s that Keynesian arguments had for macroeconomic and microeconomic policies and outcomes in the late 1940s, the mid-1960s, and the 1980s. These debates, which go on in the arena, or cocoon, of nongovernmental organizations and therefore seem less relevant to immediate policymaking institutions, such as Congress, are ultimately far more decisive for outcomes because they motivate and legitimate policy changes in more immediate institutions. New managed trade theories are behind the growing unilateralism and bilateralism of American trade policy in the U.S. Congress.

What is noteworthy, from the perspective of this book, is that the new managed trade theories reflect structuralist arguments. Like structuralist arguments for more institutionalized coordination of domestic economic policies, they advocate more directly negotiated, bargained solutions for trade policy. They want to rely less on market competition and more on political deals to decide outcomes. They seek agreement on results, not rules (Prestowitz 1988). Although put forward as being more pragmatic and realistic approaches to new circumstances, managed trade theories, like proposals for highly institutionalized coordination of domestic economic policies, are totally unrealistic in today's more complex and pluralistic world.

As even some of the creators of the new trade theories acknowledge, the new trade theories are extremely difficult to apply in practical circumstances (Krugman 1987). This is true for both economic and political reasons. To intervene efficiently, governments have to have far more economic information than it is realistic to expect; and even if this information and the reliable economic models were available, the theories assume that pluralistic political institutions will, in fact, act rationally according to these theories and not according to special interest group demands.

Free trade theories, of course, also face certain economic and political limitations. But they require less information to be implemented by governments, and they coincide with practical economic results in the postwar period that are hard to ignore, except by arguing that freer trade policies have had nothing to do with unprecedented postwar economic growth. To implement freer trade policies, governments can limit their concerns to antitrust and broader infrastructure (e.g., education) problems. And although freer trade policies do require a degree of faith in the efficacy of competition, the postwar record would seem to justify such faith. Economies of scale emerge only in open international markets. If the latter are closed to exploit oligopolistic rents created by economies of scale, economies of scale themselves diminish.

The most important reason for maintaining and expanding open international markets, however, is to meet the needs of the developing world, especially the heavily indebted developing countries. These countries can only meet the requirements of servicing large debt obligations through larger exports. To date, as we noted in Chapter 9, these countries have met such requirements primarily through reduced imports and, hence, reduced growth. Significant debt relief will not change this situation very much because, at least in the short run, which may be expected to last a minimum of 5–10 years, debt relief will reduce new foreign loans, investment, and aid to these countries. If the postwar period demonstrates anything, it suggests that developing countries that have tried development strategies that cut them off from international markets have not fared very well.

What is certain is that developing countries, even the more advanced ones, will lose in a world in which trade outcomes are more directly negotiated and managed. Such a world substitutes political clout for market opportunity. It freezes existing market shares, as we have seen in the case of quantitative restrictions that have reemerged in the world trade system in such products as apparel and clothing, textiles, steel, automobiles, machine tools, and electronic products. The only market opportunities left are for new products, that is, those that will be developed through new technologies. The new trade theories encourage developing countries to move immediately into these markets, where they have few, if any, chances to succeed against advanced country competition. Indeed, to succeed, they must limit access to their markets, try to steal foreign intellectual property, and divert scarce domestic capital to promote exports in industries that advantage primarily a technological and propertied elite at home. Even if successful for individual countries, this alternative, followed by all industrial and developing countries, is a prescription for less imports (demand) and more exports (supply) in world markets, leading to lower growth and, ultimately, greater inequality both between and within countries.

As we have argued in this study, proportions of trade that countries conduct with one another are a good indicator of the political community that exists among them. As Table C-1 shows, trade patterns in the world since 1947 have become increasingly global and more balanced among industrial countries. This coincides with what one might expect, given the increasingly homogeneous political community that has emerged among these countries. The proportion of U.S. trade with Japan has grown to complement that with Canada and the Common Market, which has also grown slightly (read across U.S. row). The proportion of Japanese trade with both the European Community and Canada has grown, whereas the high initial proportion with the United States has declined (read across Japan row). The proportion of Canada's trade with Japan has grown substantially, whereas that with the United States and Europe leveled out after 1957 once the process of global trade liberalization got under way in an intensive fashion in the late 1950s (read across Canada row). Only the proportions of European Community trade with various regions of the world do not show an increasing globalization or balance of trade across regions (read across EC row).

Trade between industrial and developing countries shows less global patterns and therefore heightens the urgency expressed in this study to integrate the developing countries in a more significant way into the international trading system. The share of trade done by each of the major industrial countries or groups with developing countries dropped significantly from 1977 to 1986 after rising from 1967 to 1977 (read down column 6). This pattern reflects the vulnerability of developing countries to macroeconomic and financial instabili-

Table C-1 Share of Country Trade with Various Regions, 1947–1986*

From/To:		U.S.	Cnd	Jpn	EC	IndC	DevC
U.S.	1947	—	15.6	0.5	19.0	40.5	n/a
	1957	—	20.0	5.4	19.3	48.1	n/a
	1967	—	23.2	10.2	17.3	61.9	30.2
	1977	—	20.5	11.7	18.5	55.0	41.6
	1986	—	20.4	18.4	22.0	64.1	31.0
EC	1947	n/a	n/a	n/a	—	n/a	n/a
	1957	n/a	n/a	n/a	—	n/a	n/a
	1967	9.2	1.0	1.0	—	70.5	18.0
	1977	6.7	1.1	1.6	—	71.4	24.1
	1986	8.3	1.0	2.9	—	78.7	17.9
Jpn	1947	64.2	0.2	—	2.8	67.2	n/a
	1957	31.3	3.3	—	8.0	43.3	n/a
	1967	28.3	4.1	—	5.4	43.2	41.0
	1977	21.4	3.0	—	8.6	41.3	51.7
	1986	32.9	3.1	—	13.4	56.3	37.9
Cnd	1947	56.4	—	0.0	22.4	80.3	n/a
	1957	70.4	—	2.0	11.7	85.2	n/a
	1967	68.4	—	3.9	5.8	88.1	7.6
	1977	67.7	—	4.8	9.3	84.3	10.4
	1986	71.1	—	5.6	8.6	87.0	7.8

Sources: 1947 data from *UN Yearbook of International Trade Statistics 1950.*
1957 data from *UN Yearbook of International Trade Statistics 1958.*
1967 data from *IMF Direction of Trade Statistics Yearbook 1966-70.*
1977 data from *IMF Direction of Trade Statistics Yearbook 1981.*
1986 data from *IMF Direction of Trade Statistics Yearbook 1987.*

*Exports and Imports as Percentages of Total Exports and Imports

ties in world markets as long as these countries remain insufficiently integrated with the real economies of industrial countries. Inflation and the petrodollar recycle drove the expansion of North–South trade in the 1970s, but disinflation and oil price declines reduced capital and trade flows in the 1980s. Industrial countries and those developing countries most closely integrated with industrial country markets weathered these instabilities better than developing countries pursuing autocratic trade and development policies (Balassa 1983).

The failure to sustain and expand global free trade, therefore, will leave developing countries increasingly to their own devices or to more regionally limited trade relations with former colonial industrial powers. Bilateral and regional free trade agreements may be already accelerating this trend. The historic U.S.–Canadian free trade agreement that went into effect at the end of 1988 has been followed by more urgent appeals to conclude a similar agreement between the United States and Mexico, which, in any case, have been improving their trade and investment relations in recent years. Further free trade agreements have been proposed between the United States and Japan, Taiwan, Korea, ASEAN (Association of Southeast Asian Nations), Australia, and the

Pacific Rim countries as a whole (Schott 1989). In 1987, as part of its restructuring under the pressures of yen appreciation, Japan began to accelerate investments and imports in Southeast Asian countries. Japanese investments in individual countries in the region rose from 30 to 250 percent in 1987 over the previous year, and similar increases were estimated in 1988 (*WP*, January 1, 1989). The European Community (EC), as Table C-1 shows, has the least global patterns of trade, and was pressing ahead toward further regionalization in the late 1980s with its exercise to complete the single internal European market by 1992.

Meanwhile, global trade talks were stalled. The mid-term review session of the Uruguay Round failed to reach agreement in December 1988 and eventually moved forward in April 1989, but with little political momentum. Although considerable technical work and even preliminary agreements have been accomplished in a number of controversial areas, including the liberalization of trade in services, the talks remain troubled over U.S.–EC differences on agriculture and U.S.–developing country differences on services, intellectual property rights, and sensitive products such as textiles and steel. To be sure, it is not unusual for multilateral trade negotiations to move slowly until the deadline for concluding such negotiations approaches. Yet advocates of global free trade may well worry that, by December 1990, when the Uruguay Round is scheduled to end, global trade talks may have been superseded both by world economic events and by bilateral and regional trade developments.

The Reagan administration clearly gave the global trade talks a lower priority than its budget objectives, and it used unilateral and bilateral trade actions to appease a Congress that had been made increasingly protectionist by the administration's budget policies and with which the administration did not wish to negotiate budget compromises. In principle, the administration's use of market leverage in trade policy was no more inappropriate than its use of market power in macroeconomic policy in 1981–1982. In practice, however, it became inappropriate because the administration failed to relate bilateral actions to multilateral objectives. In Congress, where support for GATT in 1989 was all but nonexistent, the Super and Special 301 provisions of the 1988 trade bill seemed to place more emphasis on mandatory closing of U.S. markets, if bilateral negotiations failed, than on discretionary retaliation depending on whether multilateral negotiations succeeded. The Bush administration tried valiantly in 1989 to implement these provisions cautiously, but it did not appear to be succeeding in convincing either the Congress or its trading partners that it could both open markets and satisfy domestic political interests.

The United States can serve both national and international trade interests simultaneously if it gives the global trade talks higher priority and coordinates its unilateral and bilateral trade actions more closely with multilateral objectives. Giving global talks higher priority means negotiating more aggressively such issues as textiles and steel that raise the stakes in the multilateral talks for other countries, especially developing countries. This means, in turn, negotiating these issues more aggressively with Congress. Thus far in the Uruguay Round, the United States has shown no willingness to take on the tough domestic political decisions in products of interest to other countries, yet it continues to press the EC to do so in agriculture and developing countries to do so in services and intellectual property.[1]

Coordinating unilateral and bilateral trade strategies more closely with multilateral objectives means linking bilateral deals directly with multilateral reforms. Under this strategy, for example, the United States would take bilateral action under Section 301 of U.S. trade law against, say, Japanese rice policies unless or until Japan took specific positions in the Uruguay Round to back U.S. efforts to liberalize agriculture in Europe. Or the United States would retaliate against South Korean products in U.S. markets unless or until South Korea supported specific provisions tightening protection for intellectual property rights in the GATT. Meanwhile, the United States would insist that these countries modify national laws in line with the specific provisions agreed to between the two countries, even before the Uruguay talks achieved final multilateral solutions. Through this strategy the United States would achieve concrete results in bilateral trade relationships, temper somewhat growing anti-Americanism, which purely bilateral pressures exacerbate, especially in developing countries (see Kihwan and Chung 1989), and generate new momentum and confidence in the multilateral trade negotiations.

Debt and Development Crisis

As we have argued in this study, political purpose inspires markets and the creation of wealth. Today a larger conception of why the United States or the Western world as a whole should deal decisively with the debt and development problem in the third world is still missing. The debt problem is perceived largely as a matter of tinkering with the trickling spigots of finance to restore capital flows to indebted countries, and development is viewed by one school of thought as a matter primarily of markets and by another as one primarily of charity.

None of these approaches is likely to make a difference. Stretching debt out, which the Baker Plan emphasized, or relieving debt outright, which Secretary of the Treasury Nicholas Brady introduced in his plan in early 1989, can facilitate policy reform, but these plans can also delay it. Policy reform, such as trade liberalization, for example, can facilitate debt servicing by making exports more efficient, but it can also accelerate debt by expanding imports. Reforms and finance must work together to create efficiency and wealth. The Baker Plan, although well aimed, supplied too little of both market reform and financial assistance. Most important, it was piecemeal and lacked a larger political rationale. The same would seem to hold for the Brady Plan.

What would make a difference is a more convincing concept of shared political community between the industrial and developing worlds. Some inspi-

---

[1]In July 1989, the Bush administration extended U.S. bilateral quotas on steel for 2½ years during which time, it said, it would seek a global agreement on free trade in steel in the Uruguay Round. The decision was less protectionist than the steel industry wanted (an extension for 5 years), but it essentially just put off the tough decisions into the future.

ration for how this might be done could be drawn, at least in the case of Latin America, from the inclusion in 1986 of Spain and Portugal in the European Economic Community (EEC). These countries were included in the EEC primarily for political reasons, to strengthen their incipient democratic systems. Similarly, many of the Latin American countries, which also trace a good part of their heritage back to Spain and Portugal, might be included in the world trading community to strengthen their incipient democracies. How this might be done specifically, given the legacy of Western, especially American, political and military intervention in these countries is beyond the scope of this study. Politics would have to lead the way, however, and it could not be a politics of the old-style North–South dialogue from the 1970s which stressed intervention and the "new international economic order"; it would have to be a new-style politics of common domestic values that stressed more economic competition and political pluralism.

Defining political community with Asian and African developing countries that have non-Western traditions (e.g., Islamic, Confucian) may be even more difficult. In Asia the problem is alleviated somewhat by economic growth. But as we noted in Chapter 9, economic growth in countries like Korea, Taiwan, and even Japan, until recently at least, reflected much less emphasis on consumer welfare and much more emphasis on corporate and national goals than is the case in the West.

Although the United States has played a larger role in the economic development of the more advanced developing countries in both Latin America and Asia by absorbing over two thirds of the manufactured exports of these countries, Europe may have to take a more prominent role in the political approach to these countries. European countries with stronger communitarian traditions place more emphasis on economic and social justice and may be able to mobilize more enthusiasm and empathy to integrate developing countries more fully into the global economic system. If so, European leaders must take a more forthright position to reform European policies, such as the Common Agricultural Policy, and to ensure that Europe 1992 does not become inward looking (Fortress Europe), not because the United States demands such reforms and gains from them, but because it benefits developing countries and gives Europe a global leadership role in an area that is compatible with its social and economic values.

If political and economic reforms in Eastern Europe and the Soviet Union continue and result in early successes, Western European countries will face a powerful temptation to divert economic assistance to the development of greater Europe, devoting less resources and attention to developing countries and the needs of the global economic system. Although currently doing only a small share of their trade with the East (less than 8 percent for any one country; see Table 10-1), Western European countries, especially West Germany, are poised to pour massive amounts of capital and technology into the East. This possibility, along with existing uncertainties about Fortress Europe (the possibility that Europe 1992 might discriminate against U.S. or Japanese interest), could strengthen tendencies toward regionalism in world trade, leading to trading blocs in Europe (common European home), the Western Hemisphere

(U.S.–Canada–Mexico–Latin America), the Far East (Japan, China, and Southeast Asia), or the Pacific region (Asia Pacific Economic Cooperation – a grouping of 12 countries including Canada and the United States which held an inaugural meeting in November 1989 in Canberra, Australia). Regionalism, as we noted earlier, is likely to hurt the interests of developing countries the most. For the sake of developing countries, therefore, Europe has a special responsibility to retain a global role.

Europe's global role in development policy is also important as a way to moderate and bridge changing U.S.–Japanese roles in world finance. Because of its financial power, Japan has a leadership role to play in managing the world debt problem. This role brings it into conflict with U.S. leadership and the U.S. desire to preserve its veto powers in the world financial institutions. Twice in 1988 the United States rebuffed Japanese initiatives to go beyond the Baker Plan in dealing with world debt. Europe, through a more aggressive involvement both in the Uruguay Round and in the international financial institutions, could help ensure that U.S.–Japanese differences do not damage broader interests in the global community.

The politics of a new global initiative to overcome debt and development problems is more complicated than the economics. Although this study cannot examine all the details, it can emphasize two points. (1) The postwar experience in 1944–1945 and again in 1947–1948 suggests that without a new and more inspiring political motivation, the economics of a new debt and development initiative is not likely to match the size of the task. (2) This new political motivation will have to emerge from a combination of U.S., European, and Japanese leadership contributions. As the only economic and military superpower in the world, the United States still bears the primary responsibility, but Europe and Japan have much greater responsibilities as well. Burden sharing in the future will be much more important in this North–South context than in the traditional East–West context.

An economic solution to the debt problem has to embrace both debt management *and* sound economic policies in both industrial and developing countries. In this connection, the Bretton Woods or Marshall Plan policy triad of stable prices, flexible markets, and liberal trade retains its relevance. So also do the Bretton Woods institutions if they address the elements of the policy triad in a comprehensive fashion:

> the IMF concentrating on macroeconomic policy management and *stable prices* in both industrial (multilateral surveillance) and developing (stabilization programs) countries;
>
> the World Bank concentrating on microeconomic (project loans) and structural adjustment (SAL's) loans to create *flexible market* infrastructure and policies; and
>
> the GATT concentrating on reciprocally-negotiated *trade liberalization* among both industrial and developing countries [Nau 1988c].

Although all these elements are at work, to some extent, in the present environment, they must be drawn together in a more visible and sustained way.

The world has no comprehensive forum in which to integrate these policy elements, and anything like a forum of all U.N. countries, such as that envisioned in Global Negotiations (see Chapters 7 and 8), would be counterproductive. There is another way to proceed, however, that also involves the developing countries more directly. After consultation with key developing countries, the industrial country summit could propose a general strategy stressing the postwar policy triad and commission a series of planning activities in the major international economic institutions (IMF, World Bank, and GATT) to adapt these policy directions to the world economy of the 1990s. In the model suggested by the choice-oriented perspective, the summits would provide the broad leadership and supervision for this exercise while tasking international institutions to carry out more specific functions. Each of the major institutions would use this planning exercise to develop new procedures to involve developing countries more equally in decision-making processes.

For example, the group negotiating institutional issues in the Uruguay Round made proposals, in the summer of 1988, to create smaller groups of trade ministers representative of all GATT members to meet at least twice annually to review trade developments and issues before they are submitted to GATT ministerial meetings of all countries (which should be held, the proposals recommend, every 2 or 3 years, instead of once every decade or so as in the past; see *FT*, June 23, 1988). This proposal institutionalizes the informal "Quad-plus" meetings that have been going on among industrial and developing country trade ministers since 1983 and gives key developing countries a new, formal role in the GATT system. Similarly, the IMF and World Bank should experiment with more effective mechanisms that minimize confrontation between industrial and developing countries. Currently, G-10 and G-24 meetings in the IMF exacerbate such confrontations, and IMF Interim Committee meetings, as well as IMF–World Bank Development Committee meetings, are too formalized and ritual-like (with ministers reading set speeches) to be very effective.

Once the multilateral institutions have completed their plans (within, say, a 12-month period), the industrial countries would convene an ad hoc summit meeting with selective heads of state and government of developing countries, along the lines of the Cancun Summit in October 1981, to review the plans and make the highest-level political decisions as to the funding, duration (until another joint summit-level review), and other aspects of the program. The approach outlined here is ambitious, to be sure, but it is practical. It has to be seen as a part of a total effort in the world economic system, not only to deal with debt and development, but also to correct major outstanding macroeconomic and trade imbalances among industrial countries.

Reaching a Consensus in the Cocoon

This study has argued that there are policies that serve the common interest and not just special interests and class politics. The test of such policies is whether, over time, they ultimately benefit a larger number of people, even if they require specific groups of people (i.e., special-interest groups) to make adjustments in the interim.

The major objection raised to U.S. policies of individual freedom and economic competition in the world economy, which this study recommends, is that these policies result in unequal distribution of benefits. Although this debate will not be settled in this study, the historical data from both the U.S. and the world's experience after World War II do not confirm this thesis of unequal benefits. From 1948 to 1973, the data are clear, both in the United States, where the median family income (adjusted for inflation) reached record levels every 1 to 3 years (Levy 1988, p. 17), and between industrial and some 60 middle-income countries where per capita income gaps narrowed on a relative basis (see Chapter 1). During this period liberal, market-oriented policies clearly resulted in less, not more, inequality.

The issue since the early 1970s is whether growth and equity have slowed because liberal policies were abandoned or because new circumstances have made liberal policies no longer relevant or equitable. This study has argued the former. Although there are no conclusive data on this issue, the partial return to market-oriented policies in the United States under the late Carter and subsequent Reagan administrations coincides with some interesting developments. Studies show that the middle class is disappearing in America, but that, contrary to some contentions, the middle class is disappearing into the upper class, not into the lower class. What is more, this pattern was more pronounced under the Reagan administration than under the Carter administration.

Table C-2 shows the distribution of families in the lower, middle, and upper classes from 1969 to 1986.[2] The data clearly show a decline in the lower and middle classes and an increase in the upper class. The middle class declined at about the same rate from 1976 to 1980 as from 1981 to 1986 (57.1 percent to 55.2 percent and 55.2 to 53.0 percent). The lower class did not decline at all from 1976 to 1980 but fell sharply from 34.4 percent in 1981 to 31.7 percent in 1986. The upper class meanwhile grew from 9.7 percent in 1976 to 11.5 percent in 1980 and then shot up dramatically from 11.4 percent in 1981 to 15.3 percent in 1986. Reagan administration policies, therefore, accelerated the movement of both lower- and middle-class families into higher income categories.

If income received is broken up into population quintiles based on individual rather than family incomes, the numbers as shown in Table C-3 suggest growing inequality. These data, however, do not take into account the growing number of two-income families. Moreover, these numbers do not suggest that Reagan administration market-oriented policies have been harder on the poor than interventionist policies before 1981. The percentage of income going to the lower two quintiles decreased more from 1969 to 1980 than from 1981 to 1986, and although income going to the middle quintile declined more in the Reagan years, that going to the highest quintile rose. Thus, even by these

[2]The middle class here is defined as having an income of $20,000–$59,999 at 1986 prices. (See Horrigan and Haugen 1988.)

Table C-2 Distribution of Families in the Lower, Middle, and Upper Classes, 1969–1986

Year	Lower Class (%)	Middle Class (%)	Upper Class (%)
1969	33.7	58.8	7.5
1970	34.3	57.8	7.8
1971	34.9	57.0	8.0
1972	33.1	57.2	9.7
1973	32.1	57.6	10.3
1974	33.1	57.5	9.4
1975	34.6	56.6	8.9
1976	33.1	57.1	9.7
1977	32.8	56.6	10.6
1978	31.8	56.4	11.8
1979	31.8	56.0	12.3
1980	33.2	55.2	11.5
1981	34.4	54.2	11.4
1982	35.0	53.2	11.7
1983	35.4	52.8	11.8
1984	33.8	52.8	13.4
1985	33.3	52.7	14.0
1986	31.7	53.0	15.3

Source: Horrigan and Haugen (1988), p. 9.

distorted numbers, the middle class under Reagan moved more into the upper class, not into the lower class, and the lower class actually suffered a smaller decline of income than it experienced in the 1970s.

If the Reagan administration had followed through on its economic programs, the distribution benefits of market-oriented programs might have been even more evident. Because only \$1 out of every \$5 of non-means-tested entitle-

Table C-3 Income Distribution*

Year	Lowest	2nd	Middle	4th	Highest
1969	5.6	12.4	17.7	23.7	40.6
1974	5.5	12.0	17.5	24.0	41.0
1979	5.2	11.6	17.5	24.1	41.7
1980	5.1	11.6	17.5	24.3	41.6
1981	5.0	11.3	17.4	24.4	41.9
1982	4.7	11.2	17.1	24.3	42.7
1983	4.7	11.1	17.1	24.3	42.8
1984	4.7	11.0	17.0	24.4	42.9
1985	4.6	10.9	16.9	24.2	43.5
1986	4.6	10.8	16.8	24.0	43.7

Source: U.S. Census Bureau.

*Percentage of money income received by each quintile.

ments goes to the poor (see editorial, *NYT*, April 24, 1988), cutting entitlements on a means-tested basis would have distributed the costs of government
retrenchment more evenly to the middle and upper classes. In fact, as this study
has suggested, the real moral shortcoming of the Reagan administration's
budget policies was not the cuts in welfare programs, given the fact that many
of these programs were increasingly ineffective (Murray 1984). These cuts did
not increase poverty in the United States. The poverty rate in 1986, under new
measures that take into account taxes, capital gains, and noncash income, was
only 10.3 percent, lower by one quarter than previously estimated (*WP*, December 28, 1988). The moral failing was the decision not to cut the $4 of
entitlements going to the nonpoor. The middle class failed to step up and
accept its fair share of budget cuts (although it has in effect paid for the
consequences through higher Social Security taxes). Although the decision not
to cut entitlements was a failure of Republican leadership, it was heavily endorsed by the Democratic opposition. If anything, therefore, unequal distribution of benefits to the extent they may have occurred in the United States since
1979 has been more a consequence of politics than of market-oriented economic reforms.

This brief assessment will certainly not settle the debate about the distribution effects of more recent economic policies. The identification of who gets
what, when, where, and how is at the heart of the partisan political debate
driven by special-interest groups. What is clear from this assessment is that the
benefits of growth were spread the widest, both nationally and internationally,
when the Bretton Woods policy triad prevailed from 1947 to 1967. This triad —
price stability, flexible economies, and freer trade — has not applied fully since
1967, although it was partially restored in the 1980s. Income distribution benefits have suffered accordingly. All of this may be blamed on a change in
circumstances, but the argument in this book attributes it to policies. Surprisingly, despite the income distribution record of conservative market-oriented
policies during the first two decades after the war and the total absence of a
comparable record for interventionist command-oriented policies, the myth
persists that market-oriented policies result in greater inequalities.

The Debate Begins

The argument ends at this point but, if this book has done its job properly, the
debate and further research only begin. The choice-oriented perspective outlined here offers an alternative way of looking at America's relationship to the
world economy. This perspective provides an understanding of change from
the standpoint of independent and self-activated participants in the world
economy, who shape their circumstances on the basis of voluntarist choices
and goals and do not simply react to these circumstances on the basis of
predetermined constraints or unintended consequences. At the same time, voluntarist choices of national purpose and economic policies have to meet a
certain test of performance. They have to persuade a majority of participants

in the wider cocoon of social and political forces within a particular national or international society to support these choices and to translate them eventually into official policies within the more immediate bureaucratic policymaking process. Official policies, then, have to produce results, operating within ultimate but not immediate constraints of public acceptability and scarce economic resources. Ultimate constraints of power, wealth, and class influence these choices and outcomes but they do not predetermine them. In the last analysis, individuals make judgments based on their own ideas and preferences for a better political society (the unobserved phenomenon in this study — see Chapter 2). In this way, structural change is possible from within a given set of political and economic structures. Ideas about political community, which are to some extent autonomous, resist and reshape existing political and economic power relationships.

The choice-oriented approach puts issues in both temporal perspective and historical context. It goes beyond the mind-numbing array of special-interest issues that are paraded out daily on Capitol Hill and dominate the news and public policy analysis. According to this approach, for example, the Bretton Woods policy triad was efficient not because price stability was preserved at all costs, or because supply-side incentives existed regardless of the consequences for budgets, or because freer trade was pursued mindlessly even when macroeconomic excesses and structural rigidities created intolerable exchange rate and financial instabilities. It succeeded because all of these standards were pursued together in perspective and moderation vis-à-vis one another. The choice-oriented perspective also adds a historical dimension. The easy references today to the Bretton Woods conference of 1944 as the model for international economic cooperation or the even more ambitious appeals for United Nations conferences to solve the debt and other North–South issues are misleading. The Bretton Woods conference failed. It tried to bridge too many gaps between too many incompatible domestic policies and institutions. United Nations conferences today would fail for the same reason. International economic cooperation requires common political purposes and common economic policy standards, not simply the will to meet and compromise. The content of the Bretton Woods policies came from the Marshall Plan programs of 1948, not the 1944 agreements, although the Bretton Woods conference of 1944 (and U.N. meetings today) helped to sustain a process of cooperation that is clearly a necessary (but not sufficient) condition for substantive agreements.

In the 1990s the postwar world shaped by the Marshall Plan and the Cold War may be in the process of radical change. The question of how societies in central and Eastern Europe will organize themselves politically is once again, as in 1947–1948, the central issue of East–West relations. The transformation of national societies in Western Europe is under way (Europe 1992), and the question has been raised whether Japanese and Western societies, especially American society, differ so fundamentally that they may not be able to continue to coexist in a common liberal international economic order (Prestowitz 1988, van Wolferen 1989). National purpose in many parts of the world is, in short, in flux.

This time, in comparison to 1947 or the 1970s, the movement of national purpose seems to be in the direction of more pluralist societies and more competitive economic rules and institutions. This movement attests to the purposeful leadership of Western societies, and especially the United States, since 1947 and suggests that the primary impulse of human society may be, as this study assumes, the desire to achieve more perfect forms of political and economic fulfillment for the individual. Western standards of democracy and competition appear to resonate with this primary impulse more widely throughout the world than any other standards in this or previous centuries.

But the outcome of current change, however hopeful such change may be, is never certain or guaranteed. From a choice-oriented perspective, there are no technological or historical imperatives. Human consciousness and ideas shape history. Thus, the struggle among ideas about how to improve political society will go on. In the face of non-Western alternatives (e.g., Islam, Confucianism, Shintoism, etc.), Western standards will be continuously tested. As we have argued, the West and the United States do not have to win this contest, in the sense that the whole world must emulate Western values. Alternative standards *can* co-exist; but neither can the West nor the United States lower its standards, particularly as these standards appear to succeed worldwide and offer more hope and prosperity than alternative ones.

Much depends in the future, therefore, on how the West and especially the United States behave. The United States is still the preeminent power in the world economy. The road forward begins with how the United States copes with the more pluralist and competitive world that it has done most to create. Ironically, when Western standards faced eclipse in 1947, at least in central Europe, the United States acted with enlightenment and conviction. Today, when Western standards are spreading as never before, the United States seems trapped in a puzzling debate about its own decline and self-worth, even questioning whether its values of spirited individualism still apply in the modern world (Prestowitz 1988).

When placed in context, as this study suggests, America's decline is a myth. It ignores the convergence of purposes in the world today against which American power is applied. And it ignores the choices that America still has to reaffirm its own purposes and rejuvenate its own power.

America has a choice to reaffirm Western values of human freedom and individual dignity and to reapply the efficient economic policy standards of price stability, flexible domestic markets, and international freer trade of the Bretton Woods era. Only a clear and self-confident American purpose can motivate further economic adjustments among the industrial countries, inspire the progressive integration of developing countries into the postwar international economic system, and guide the delicate and difficult process of economic and political reform in the socialist world. Only a decision to correct American fiscal imbalances and accelerate multilateral trade negotiations can complete the transition back to the efficient policy standards and equitable performance outcomes of the ealier postwar era. If it makes these choices, America has more than ample power to lead the world economy. These choices depend less on special interests and partisan politics than on the crosscutting and broader ideas that inspire social and intellectual coalitions in the wider cocoon of nongovernmental institutions and shape future policies in the political and bureaucratic arena.

In a sense, the debate about America's purpose and policy—how it is conducted and how it is resolved—is the most important element of America's power in the modern world. America leads today, less by sheer size of resources and dominance of international institutions, than by its domestic purposes and procedures, which are widely admired and increasingly emulated around the world. In short, America leads by knowing what it stands for politically and by getting its own house in order economically. The choice-oriented ideas of assertive, but tolerant, national purpose and market-oriented, but equitable, domestic policies are more relevant today than ever before, precisely because America has less relative power to lead by other means. Either America will lead by democratic example and market initiative or it will follow increasingly the purposes and policies of other countries, some of which potentially advocate more centralized forms of both political and economic life in the world community (Garten 1989–1990).

Thinking back, that was the choice America faced in 1947. The choice is less stark today in part because America's purposes and policies after 1947 helped to lead the world toward freer forms of political and economic life. But the choice today is no less real or urgent and this choice, as in 1947, will shape the circumstances of future international relations and decide the fate of liberalism and markets in the 1990s and beyond.

APPENDIX

Table A-1 Economic Policy Indicators, 1947–1987

Countries or Groups of Countries	Macroeconomic Policy			Microeconomic Policy
	Fiscal	Monetary		Public Sector
	Budget Deficit (avg. annual as percent of GDP)	M_1 (avg. annual percent change)	M_2 (avg. annual percent change)	Government Expenditures as pct. of GDP (avg. percentage point change per decade)
	1947–1967	1947–1967	1947–1967	1947–1967
U.S.	−0.2	2.5	5.3	1.6
G-4	−0.7	11.0	12.5	−0.5
G-5	−0.6	9.3	11.0	0.0
G-7	−0.9	9.6	10.7	0.5
Quad	−0.9	9.5	11.2	0.8
DCs	−0.9	5.2	7.9	NA
LDCs	NA	11.5	14.9	NA
	1968–1980	1968–1980 (1977–1979)	1968–1980 (1977–1979)	1968–1980
U.S.	−2.0	6.4 (7.9)	8.8 (9.4)	1.9
G-4	−2.1	11.1 (11.2)	12.9 (10.5)	6.0
G-5	−2.1	10.1 (10.5)	12.0 (10.3)	5.1
G-7	−3.1	11.2 (11.7)	13.3 (12.7)	5.4
Quad	−2.9	10.7 (11.6)	12.8 (12.5)	7.7
DCs	−2.2	9.0 (9.9)	11.4 (11.4)	5.8
LDCs	−3.0	25.0 (28.3)	28.1 (33.8)	4.3

	1981-1987 (1983-1987)	1981-1987 (1980-1982)	1981-1987 (1980-1982)	1981-1987 (1983-1987)
U.S.	-4.6 (-5.0)	9.3 (6.7)	9.0 (6.2)	-0.9 (-4.6)
G-4	-3.6 (-3.4)	9.7 (7.3)	10.5 (10.9)	-1.6 (-4.0)
G-5	-3.8 (-3.7)	9.6 (7.1)	10.2 (10.0)	-1.4 (-4.1)
G-7	-5.2 (-5.3)	10.7 (7.6)	10.0 (10.8)	0.2 (-4.2)
Quad	-6.4 (-6.3)	10.7 (7.1)	11.1 (10.3)	1.4 (-6.0)
DCs	-4.9 (-5.0)	8.7 (6.6)	9.4 (8.9)	0.3 (-3.3)
LDCs	-4.9 (-4.8)	39.1 (30.5)	44.4 (35.0)	3.2 (3.0)

Sources: Budget deficit = Line 80 divided by line 99b (GDP) in IMF, *IFS Yearbook*, various issues. In some cases, for early years, U.N., *Yearbook of National Accounts Statistics*, various issues; and for industrial countries, OECD, *Economic Outlook and Economic Outlook: Historical Statistics 1960–1985 and 1960–1986*.

Govt. expenditure = Expenditures plus lending minus repayment (lines 82 and 83) in IMF, *IFS Yearbook*, various issues. In some cases, for early years, U.N., *Yearbook of National Accounts Statistics*, and Mitchell (1983).

Govt. expenditure as percent of GDP = Lines 82 and 83 divided by line 99b in IMF, *IFS Yearbook*, various issues. In some cases government expenditures series come from OECD, *Economic Outlook: Historical Statistics 1960–1985 and 1960–1986*. Averages calculated only for years for which consistent series exist.

M_1 = Line 34 in IMF, *IFS Yearbook*, various issues.

M_2 = Lines 34 and 35 (quasi-money) in IMF, *IFS Yearbook*, various issues.

Notes: Averages calculated from individual country tables compiled from sources. Averages in columns 1, 2, and 3 do not include all years for every country. Averages in column 4 are standardized on per decade basis.

Definitions: G-4 = G-5 without United States.

G-5 = France, Germany, Japan, United Kingdom, United States.

G-7 = Canada, France, Germany, Italy, Japan, United Kingdom, United States.

Quad = European Community, United States, Canada, and Japan. From 1947 to 1967, EC = 6 (Belgium, France, Germany, Italy, Luxembourg, Netherlands); from 1968 to 1980, EC = 9 (plus U.K., Ireland, and Denmark); from 1981 to 1987, EC = 10 (plus Greece).

DCs = Industrial countries as found in IMF, *IFS Yearbooks*.

LDCs = Developing countries as found in IMF, *IFS Yearbooks*.

Table A–2 Trade Policy Indicators, 1950–1985

Country	NONTARIFF BARRIERS (NTBs)*							Expiration Date‡	TARIFFS§					Percent Change 1965–1985
	1950†	1955†	1960†	1970	1975	1980	1985		1965	1970	1975	1980	1985	
Canada				—— Textiles —→				1991	7.95	5.66	5.46	4.62	3.25	−59.1
						←— Footwear →	1985						
France	53	76	95			←—— Autos ——→		Indefinite	6.05	2.33	1.43	1.07	0.78	−87.1
						Electronics →		Indefinite						
						←— Footwear →		Indefinite						
						←— Motorcycles→		Indefinite						
Germany	47	90	93				←— Autos →	1984	4.64	3.04	2.36	1.81	1.80	−61.3
Italy	54	99	98			←—— Footwear ——→		Indefinite	5.94	4.66	0.57	0.78	0.89	−85.0
					Electronics —→			Indefinite						
				←—— Autos ——→				Indefinite						
Japan		←—— Autos (discretionary licencing) ——→						Indefinite	7.55	7.05	2.96	2.46	2.13	−71.8
						←— Motorcycles ——→		Indefinite						
U.K.	57	84	97			←—— Textiles ——→		1991	5.97	2.76	2.25	2.17	1.45	−75.7
						←—— Autos ——→		Indefinite						
						←— Footwear ——→		Indefinite						
						←— Electronics →		Indefinite						

U.S.		6.75	6.08	3.79	3.08	3.05	−54.8
	Textiles ⟶ 1991						
	Footwear ⟶ 1983						
	⟵ Electronics ⟶ 1984						
	⟵ Autos ⟶ 1989						
	Carbon Steel ⟶						
	⟵ Carbon Steel ⟶ 1989						
	⟵ Specialty Steel ⟶ 1987						
	⟵ Motorcycles ⟶ 1988						
EC							
	Steel ⟶ Indefinite						
	⟵ Textiles ⟶ 1991						
	⟵ Autos ⟶ Indefinite						
	⟵ Electronics ⟶ 1986						
	⟵ Motorcycles ⟶ 1986						
OEEC/OECD	56 84 94	6.79	5.05	3.32	2.43	2.13	−74.9

Sources: Quotas: OBEC (1961), p. 185.
    NTBs: adapted from OECD (1985a), pp. 32–33.
    Tariffs: OECD (1985a), p. 27. 1985 calculated from same sources used in OECD study.

*Includes, in addition to quotas, various restrictive arrangements with various countries, including Basic Price Systems, OMA's, and VER's.
†These columns show percent of private trade freed from quotas (quantitative restrictions) by June 30th of indicated year.
‡Column does not reflect all renewals since 1985.
§Receipts from customs and import duties as percentage of value of imports.

377

Table A-3 Economic Performance Indicators, 1947–1987

Countries or Groups of Countries	Growth	Inflation		Unemployment
	Real GDP (avg. annual percent change)	CPI (avg. annual percent change)	GDP Deflator (avg. annual percent change)	Workforce Unemployed (avg. annual percent)
	1947–1967	1947–1967	1947–1967	1947–1967
U.S.	3.6	2.7	2.9	4.7
G-4	6.6	3.7	4.4	2.0
G-5	6.0	3.5	4.1	2.5
G-7	5.9	3.8	4.7	3.4
Quad	5.3	3.5	4.6	3.8
DCs	4.3	2.5	2.8	3.1
LDCs	5.0	8.6	8.5	NA
	1968–1980	1968–1980	1968–1980	1968–1980
U.S.	2.6	7.2	6.9	5.9
G-4	3.8	8.4	8.6	3.1
G-5	3.5	8.1	8.2	3.7
G-7	3.7	8.5	8.9	4.2
Quad	3.9	8.3	8.7	4.4
DCs	3.3	7.9	7.9	4.4
LDCs	5.7	16.5	17.2	NA

378

	1981–1987 (1983–1987)	1981–1987 (1983–1987)	1981–1987 (1983–1987)	1981–1987 (1983–1987)
U.S.	2.8 (4.0)	4.3 (3.3)	4.6 (3.3)	7.8 (7.5)
G-4	2.8 (2.8)	4.8 (3.4)	4.2 (3.6)	7.8 (8.2)
G-5	2.8 (3.0)	4.7 (3.4)	4.3 (3.6)	7.8 (8.1)
G-7	2.7 (3.1)	5.7 (4.4)	5.5 (4.5)	8.5 (8.8)
Quad	2.1 (2.4)	7.0 (5.5)	7.4 (6.3)	9.1 (9.5)
DCs	2.7 (3.2)	5.2 (3.9)	4.9 (4.0)	8.2 (8.7)
LDCs	1.8 (1.7)	37.0 (40.4)	41.3 (48.4)	NA NA

Sources: Real GDP, CPI, and GDP deflator: From IMF, *IFS Yearbook*, various issues, with some early years (late 1940s and early 1950s) from Mitchell (1981), (1982), (1983); and U.N., *Yearbook of National Accounts Statistics*, various issues.

Unemployment: From U.N., *Statistical Yearbook*, various issues.

Notes: Averages calculated from individual country tables compiled from the preceding sources. They do not include all years for every country.

Table A-4 Economic Policy Shifts after 1967

| Countries or Groups of Countries | Macroeconomic Policy | | | Microeconomic Policy |
| | Fiscal | Monetary | | Public Sector |
	Budget Deficit (avg. annual as percent of GDP)	M_1 (avg. annual percent change)	M_2 (avg. annual percent change)	Government Expenditures as pct. of GDP (avg. percentage point change per decade)
	1961–1967	1961–1967	1961–1967	1961–1967
U.S.	−0.8	3.5	7.5	1.7
G-4	−0.9	11.2	12.6	0.7
G-5	−0.9	9.6	11.6	0.9
G-7	−1.2	10.3	11.6	1.1
Quad	−1.4	9.7	11.1	1.2
DCs	−0.9	6.4	9.4	4.0
LDCs	NA	15.1	16.6	NA
	1968–1974	1968–1974	1968–1974	1968–1974
U.S.	−1.1	6.1	8.3	−1.6
G-4	−0.7	11.2	14.5	4.0
G-5	−0.8	10.2	13.2	3.2
G-7	−1.5	11.4	13.9	3.8
Quad	−1.4	10.5	13.1	4.1
DCs	−1.1	8.9	11.4	4.3
LDCs	−2.6	20.5	28.1	5.7

Sources: Same as Table A-1 in Appendix.

Table A-5 Economic Performance Shifts After 1967

Countries or Groups of Countries	Growth	Inflation		Unemployment
	Real GDP (avg. annual percent change)	CPI (avg. annual percent change)	GDP Deflator (avg. annual percent change)	Workforce Unemployed (avg. annual percent)
	1961–1967	1961–1967	1961–1967	1961–1967
U.S.	4.5	1.7	2.1	5.0
G-4	6.1	3.8	3.6	1.2
G-5	5.8	3.3	3.3	2.0
G-7	5.7	3.3	3.5	2.6
Quad	5.2	3.2	3.5	2.5
DCs	5.0	3.0	3.0	3.1
LDCs	5.2	11.6	8.7	NA
	1968–1974	1968–1974	1968–1974	1968–1974
U.S.	2.6	5.8	6.0	4.9
G-4	5.0	7.3	7.4	2.0
G-5	4.5	7.0	7.1	2.6
G-7	4.7	6.8	7.2	3.1
Quad	5.0	6.8	7.5	3.3
DCs	3.8	5.8	6.9	3.4
LDCs	6.3	11.8	12.4	NA

Sources: Same as Table A-3 in Appendix.

Table A-6 Sector Contribution to Real GNP Growth: Typical vs. Current Recovery

	Annual Rate over First 8 Quarters	
	Typical recovery*	Current recovery†
Real GNP growth (percent change)	5.3	6.0
Sector contribution to GNP growth (percentage points)		
Personal consumption expenditures	3.2	3.3
Durable goods	.9	1.2
Nonresidential fixed investment	.6	1.8
Producers' durable equipment	.5	1.5
Structures	.1	.3
Residential investment	.5	.6
Change in business inventories	.7	1.3
Net exports of goods and services	−.1	−1.3
Exports	.4	.3
Imports	−.4	−1.6
Government purchases of goods and services	.3	.3
Final sales total‡	4.6	4.7

Source: *Economic Report to the President* (February 1985), p. 30.

*Average of recoveries following business cycle troughs in 1954, II; 1958, II; 1961, I; 1970, IV; and 1975, I.

†Calculated from 1982, IV, business cycle trough to 1984, IV; data for 1984, IV, are preliminary.

‡GNP less change in business inventories.

Table A-7 1983–1984 Recovery*

	Past Recoveries		1983–1984 Recovery§
	Typical†	Strongest‡	
Final domestic demand	4.6	5.5	5.9
Contribution of change in inventories#	+0.7	+0.4	+1.4
Total domestic demand	5.4	5.9	7.3
Contribution of net exports#	−0.1	0.0	−1.3
GNP	5.3	5.8	6.0
Personal consumption	5.1	6.0	5.0
Residential construction	13.2	21.5	20.5
Other fixed investment	6.9	7.4	15.0
Producer durables	9.4	10.7	18.6
Public Expenditure	1.2	0.2	1.6
Exports	6.4	8.3	3.7
Imports	8.5	9.2	19.4

Source: Marris (1985), p. 43. © 1987 Institute for International Economics, Washington, DC. Reprinted by permission from *Deficits and the Dollar: The World Economy at Risk*, by Stephen Marris.

*Percentage change in volume, annual rates.

†Average over the first eight quarters of the recoveries starting in 1954, 2; 1958, 2; 1961, 1; 1970, 4; and 1975, 1. The recovery starting in 1949, 4, has been excluded because it was of a different nature, driven by a massive buildup of defense expenditures.

‡Over the eight quarters starting in 1970, 4.

§First eight quarters from 1982, 4, recession trough.

#As percentage of GNP in previous period, annual rate.

Table A-8 Growth Rates of Real GNP Components: Current Expansion vs. Average of Previous Expansions*

	First 3 Years of Expansion		First Six Quarters of Expansion		Second Six Quarters of Expansion	
	Current†	Previous‡	Current†	Previous‡	Current†	Previous‡
Real GNP§	4.5	4.5	6.9	5.2	2.1	3.8
Final sales#	3.8	3.9	4.3	4.1	3.4	3.7
Personal consumption	3.9	4.3	5.2	5.1	2.5	3.4
Gross private domestic investment	17.4	10.6	38.0	15.7	−.1	6.0
Nonresidential fixed investment	11.3	6.4	13.6	6.0	9.0	6.9
Structures	6.8	3.0	4.9	2.6	8.7	3.5
Producers' durable equipment	14.0	8.8	19.1	8.9	9.1	9.1
Residential fixed investment	15.1	9.3	29.2	15.0	2.6	4.2
Exports of goods and services	2.4	8.8	6.0	4.6	−1.1	13.2
Imports of goods and services	14.6	8.6	25.7	10.4	4.5	7.0
Govt. purchases of goods and services	4.0	1.1	1.7	.5	6.3	1.8
Federal	5.9	−1.1	1.2	−1.2	10.8	−.9
State and local	2.4	3.7	2.1	3.1	2.7	4.2

Source: *Economic Report of the President* (February 1986), p. 39.

*Average annual percent change.

†Calculated from 1982, IV, the most recent recession trough.

‡Average of expansions that began in 1954, II; 1961, I; 1970, IV; and 1975, I.

§Real GNP and its components are in 1982 dollars.

#GNP less changes in business inventories.

Table A-9 Economic Policy Shifts After 1980

| Countries or Groups of Countries | Macroeconomic Policy | | | Microeconomic Policy |
| | Fiscal | Monetary | | Public Sector |
	Budget Deficit (avg. annual as percent of GDP)	M₁ (avg. annual percent change)	M₂ (avg. annual percent change)	Government Expenditures as percent of GDP (avg. percentage point change per decade)
	1974–1980	1974–1980 (1977–1979)	1974–1980 (1977–1979)	1974–1980
U.S.	−2.6	6.5 (7.9)	8.9 (9.4)	5.0
G-4	−3.5	10.5 (11.2)	11.4 (10.5)	4.8
G-5	−3.4	9.7 (10.5)	10.9 (10.3)	4.8
G-7	−4.5	10.5 (11.7)	13.1 (12.7)	5.1
Quad	−4.2	10.5 (11.6)	13.0 (12.5)	7.6
DCs	−3.2	9.0 (9.9)	11.4 (11.4)	6.4
LDCs	−2.9	30.4 (28.3)	33.7 (33.8)	2.4
	1981–1987 (1983–1987)	1981–1987 (1980–1982)	1981–1987 (1980–1982)	1981–1987 (1983–1987)
U.S.	−4.6 (−5.0)	9.3 (6.7)	9.0 (6.2)	−0.9 (−4.6)
G-4	−3.6 (−3.4)	9.7 (7.3)	10.5 (10.9)	−1.6 (−4.0)
G-5	−3.8 (−3.7)	9.6 (7.1)	10.2 (10.0)	−1.4 (−4.1)
G-7	−5.2 (−5.3)	10.7 (7.6)	10.0 (10.8)	0.2 (−4.2)
Quad	−6.4 (−6.3)	10.7 (7.1)	11.1 (10.3)	1.4 (−6.0)
DCs	−4.9 (−5.0)	8.7 (6.6)	9.4 (8.9)	0.3 (−3.3)
LDCs	−4.9 (−4.8)	39.1 (30.5)	44.4 (35.0)	3.2 (3.0)

Sources: Same as Table A-1 in Appendix.

Table A-10 Economic Performance Shifts After 1980

Countries or Groups of Countries	Growth	Inflation		Unemployment
	Real GDP (avg. annual percent change)	CPI (avg. annual percent change)	GDP Deflator (avg. annual percent change)	Workforce Unemployed (avg. annual percent)
	1974–1980	1974–1980	1974–1980	1974–1980
U.S.	2.1	9.2	8.2	6.8
G-4	2.4	10.4	10.0	4.0
G-5	2.3	10.1	9.6	4.6
G-7	2.6	11.0	11.1	5.1
Quad	2.7	10.5	10.3	5.9
DCs	2.4	10.1	9.4	5.3
LDCs	5.1	22.6	24.0	NA
	1981–1987 (1983–1987)	1981–1987	1981–1987	1981–1987
U.S.	2.8 (4.0)	4.3	4.6	7.8
G-4	2.8 (2.8)	4.8	4.2	7.8
G-5	2.8 (3.0)	4.7	4.3	7.8
G-7	2.7 (3.1)	5.7	5.5	6.5
Quad	2.1 (2.4)	7.0	7.4	9.1
DCs	2.7 (3.2)	5.2	4.9	8.2
LDCs	1.8 (1.7)	37.0	41.3	NA

Sources: Same as Table A-3 in Appendix.

BIBLIOGRAPHY

Abert, James Goodear, 1969. *Economic Policy and Planning in the Netherlands, 1950–1965*. New Haven, CT: Yale University Press.

Acheson, Dean, 1969. *Present at the Creation*. New York: Norton.

Aggarwal, Vinod K., 1985. *Liberal Protectionism: The International Politics of Organized Textile Trade*. Berkeley, CA: University of California Press.

Aho, C. Michael, and Marc Levinson, 1988. *After Reagan: Confronting the Changed World Economy*. New York: Council on Foreign Relations.

Alker, Hayward R., Jr., Thomas J. Biersteker, and Takashi Inoguchi, 1985. "The Decline of the Superstates: The Rise of a New World Order?" Paper prepared for World Congress on Political Science, International Political Science Association, Paris, 15–20 July.

————, and Thomas J. Biersteker, 1984. "The Dialectics of World Order." *International Studies Quarterly*, vol. 28, no. II (June), pp. 121–142.

Allen, C. G. (ed.), 1977. *Rulers and Governments of the World: 1930–1975*. Volume 3. New York: Bowker Publishing.

Allison, Graham, and Peter Szanton, 1976. *Remaking Foreign Policy: The Organizational Connection*. New York: Basic Books.

Altman, Roger C., 1989. "Awaiting Investment." *Wall Street Journal*, March 10.

Anderson, Martin, 1988. *Revolution*. San Diego, CA: Harcourt Brace Jovanovich.

Anjaria, Shailendra, Naheed Kirmani, and Arne B. Peterson, 1985. "Trade Policy Issues and Developments." Occasional Paper 38, International Monetary Fund, Washington, DC, July.

Aristotle, 1912. *A Treatise on Government* (translated from the Greek by William Ellis). New York: E. P. Dutton and Company.

Arkes, Hadley, 1972. *Bureaucracy, the Marshall Plan, and the National Interest*. Princeton, NJ: Princeton University Press.

Ashford, Douglas E., 1982. *Policy and Politics in France: Living with Uncertainty*. Philadelphia: Temple University Press.

————, 1981. *Policy and Politics in Britain: The Limits of Consensus*. Philadelphia: Temple University Press.

Ashley, Richard K., 1984. "The Poverty of Neorealism." *International Organization*, vol. 38, no. 2 (Spring), pp. 225–287.

Axelrod, Robert, 1984. *The Evolution of Cooperation*. New York: Basic Books.

———, and Robert O. Keohane, 1986. "Achieving Cooperation Under Anarchy: Strategies and Institutions." In *Cooperation Under Anarchy*, ed. Kenneth A. Oye. Princeton, NJ: Princeton University Press, pp. 226–254.

Ayres, Robert L., 1983. *Banking on the Poor: The World Bank and World Poverty*. Cambridge, MA: MIT Press.

Bairoch, Paul, 1982. "International Industrialization Levels from 1750 to 1980." *Journal of European Economic History*, vol. 11, pp. 269–329.

Baldwin, David A., 1985. *Economic Statecraft*. Princeton, NJ: Princeton University Press.

Balassa, Bela, 1983. "Exports, Policy Choices and Economic Growth in Developing Countries After the 1973 Oil Shock." World Bank Development Research Department, DRD Discussion Paper No. 48. Washington, DC: World Bank.

———, Gerardo M. Bueno, Pedro-Pablo Kuczynski, and Mario Henrique Simonsen, 1986. *Toward Renewed Economic Growth in Latin America*. Washington, DC: Institute for International Economics.

Bank for International Settlements (BIS), 1983. *53rd Annual Report*, June 13.

Banks, Arthur S. (ed.), 1988. *Political Handbook of the World, 1986*. Binghamton, NY: CSA Publications.

Bergsten, C. Fred, 1988. *America in the World Economy: A Strategy for the 1990's*. Washington, DC: Institute for International Economics, November.

———, 1986. "America's Unilateralism." In *Conditions for Partnership in International Economic Management*, eds. C. Fred Bergsten, Etienne Davignon, and Isamu Miyazaki. The Triangle Papers 32. New York: The Trilateral Commission.

———, 1985. "The State of the Debate: Reaganomics—The Problem?" *Foreign Policy*, no. 59 (Summer), pp. 132–44.

———, 1981. "The Costs of Reaganomics." *Foreign Policy*, no. 44 (Fall), pp. 24–37.

———, 1975. *Dilemmas of the Dollar*. New York: New York University Press.

Bertsch, Gary K. (ed.), 1988. *Controlling East-West Trade and Technology Transfer*. Durham, NC: Duke University Press.

Beveridge, William H., Sir, 1945. *Full Employment in a Free Society*. New York: W. W. Norton.

Bhagwati, Jagdish N., 1978. *Foreign Trade Regimes and Economic Development: Anatomy and Consequences of Exchange Control Regimes*. Cambridge, MA: Ballinger.

Blanchard, Oliver Jean, 1987. "Reaganomics." *Economic Policy: A European Forum*, vol. 2, no. 2 (October), pp. 17–48.

Blank, Stephen D., 1977. "Britain: the politics of foreign economic policy, the domestic economy and the problem of pluralistic stagnation." In "Between power and plenty: foreign economic policies of advanced industrial states," ed. Peter J. Katzenstein. *International Organization*, special issue, vol. 31, no. 4 (Autumn), pp. 673–723.

Block, Fred L., 1977. *The Origins of International Economic Disorder: A Study of United States International Monetary Policy from World War II to the Present*. Berkeley, CA: University of California Press.

Blum, John Morton, 1967. *From the Morgenthau Diaries: Years of War 1941-1945*. Volume 3. Boston: Houghton Mifflin Company.

Bonn Economic Summit Conference, 1985. "Declaration Towards Sustained Growth and Higher Employment." *Complete Presidential Documents*, pp. 791–795 (May 13).

Borrus, Michael, and John Zysman, 1986. "Japan." In *National Policies for Developing High Technology Industries: International Comparisons*, eds. Francis W. Rushing and Carole Ganz Brown, Boulder, CO: Westview Press, pp. 111–143.

Boskin, Michael J., 1987. *Reagan and the Economy: The Successes, Failures and Unfinished Agenda*. San Francisco: ICS Press.

Bosworth, Barry P., 1984. *Tax Incentives and Economic Growth*. Washington, DC: The Brookings Institution.

Branson, William, 1987. "Discussion." *Economic Policy: A European Forum*, vol. 2, no. 2 (October), pp. 48–52.

Bressand, Albert, 1983. "Mastering the 'Worldeconomy.'" *Foreign Affairs*, vol. 61, no. 4 (Spring), pp. 745–773.

Brock, William E., 1984. "Trade and Debt: The Vital Linkage." *Foreign Affairs*, vol. 52, no. 5 (Summer), pp. 1037–1057.

Brown, William A. Jr., 1950. *The U.S. and the Restoration of World Trade*. Washington, DC: The Brookings Institution.

Bryant, Ralph C., Dale W. Henderson, Gerarld Holtman, Peter Hooper and Steven A. Symansky (eds.), 1988. *Empirical Macroeconomics for Interdependent Economies*. 2 Volumes. Washington, DC: The Brookings Institution.

Brzezinski, Zbigniew, 1983. *Power and Principle*. New York: Farrar, Straus and Giroux.

Buchanan, Allen, 1985. *Ethics, Efficiency and the Market*. Totowa, NJ: Rowman & Allanheld, Publishers.

Buchanan, James M., and Gordon Tullock, 1962. *The Calculus of Consent: Logical Foundations of Constitutional Democracy*. Ann Arbor: University of Michigan Press.

Burns, Arthur F., 1978. *Reflections of an Economic Policy Maker: Speeches and Congressional Statements: 1969–1978*. Washington, DC: American Enterprise Institute for Public Policy Research.

Bush, George (President), 1989. Remarks at Rheingoldhalle, Mainz, Federal Republic of Germany. The White House, Office of the Press Secretary (May 31).

Business Week (weekly issues).

Cagan, Phillip (ed.), 1985. *Essays in Contemporary Economic Problems: The Economy in Deficit 1985*. Washington, DC: American Enterprise Institute.

Calingaert, Michael, 1988. *The 1992 Challenge from Europe: Development of the European Community's Internal Market*. Washington, DC: National Planning Association.

Calleo, David P., 1987. *Beyond American Hegemony: The Future of the Western Alliance*. New York: Basic Books.

_____, 1982. *The Imperious Economy*. Cambridge, MA: Harvard University Press.

_____, and Benjamin M. Rowland, 1973. *America and the World Political Economy: Atlantic Dreams and National Realities*. Bloomington: Indiana University Press.

Caron, François, 1979. *An Economic History of Modern France*. New York: Columbia University Press.

Carr, Edward Hallett, 1964. *The Twenty Years' Crisis 1919–1939*. New York: Harper & Row, Torchback edition (original edition, 1939).

Castle, Eugene W., 1957. *The Great Giveaway: The Realities of Foreign Aid*. Chicago: Henry Regnery Company.

Central Intelligence Agency and Defense Intelligence Agency (CIA and DIA), 1989. "The Soviet Economy in 1988: Gorbachev Changes Course." Report to the Subcommittee on National Security Economics of the Joint Economic Committee, U.S. Congress. April 14. Mimeo.

Chaote, Pat, and Juyne Linger, 1988. "Tailored Trade: Dealing with the World as It Is." *Harvard Business Review* (January–February), pp. 86–93.

Cline, William R., 1989. "The Baker Plan: Progress, Shortcomings and Future Evolution." Paper prepared at the Institute for International Economics. January.

———, 1987. *Mobilizing Bank Lending to Debtor Countries*. Policy Analysis in International Economics, no. 18. Washington, DC: Institute for International Economics, June.

Cohen, Benjamin J., 1986. *In Whose Interest? International Banking and American Foreign Policy*. New Haven, CT: Yale University Press.

Cohen, Stephen S., 1977. *Modern Capitalist Planning: The French Model*. Second edition. Berkeley: University of California Press.

———, and John Zysman, 1987. *Manufacturing Matters: The Myth of the Post-Industrial Economy*. New York: Basic Books.

Collins, Robert M., 1981. *The Business Response to Keynes: 1929-1964*. New York: Columbia University Press.

Comisso, Ellen, and Laura D'Andrea Tyson, 1986. "Power, Purpose and Collective Choice: Economic Strategy in Socialist Countries." *International Organization*, special issue, vol. 40, no. 2 (Spring).

Commission on the Organization of the Government for the Conduct of Foreign Policy (known as the Murphy Commission after its Chairman, Robert D. Murphy), 1975. Volumes 1–7. Washington, DC: Government Printing Office, June.

Congressional Record, 1945. Vol. 91, Part 14, 79th Congress, 1st Session. Jan. 3, 1945–Dec. 21, 1945.

———, 1940. Vol. 86, Part 19, 76th Congress, 3rd Session. Jan. 3, 1940–Jan. 3, 1941.

Conybeare, John A. C., 1984. "Public Goods, Prisoners' Dilemmas and the International Political Economy." *International Studies Quarterly*, vol. 28, no. 1 (March), pp. 5–23.

Cooper, Richard N., 1984. "A Monetary System for the Future." *Foreign Affairs*, vol. 53, no. 1 (Fall), pp. 166–85.

———, 1977. "A New International Economic Order for Mutual Gain." *Foreign Policy*, no. 26 (Spring), pp. 65–140.

———, 1968. *The Economics of Interdependence: Economic Policy in the Atlantic Community*. New York: McGraw-Hill, published for the Council on Foreign Relations.

Cox, Robert W., 1986. "Social Forces, States and World Orders: Beyond International Relations Theory." In *Neorealism and Its Critics*, ed. Robert O. Keohane. New York: Columbia University Press, pp. 204–55.

Day, Alan J., and Henry W. Degenhardt (eds.), 1984. *Political Parties of the World*. Second edition. Detroit, MI: Gale Research Company.

Deaver, Michael K. (with Mickey Herskowitz), 1987. *Behind the Scenes*. New York: William Morrow and Company.

de Larosière, Jacques, 1982. *Restoring Fiscal Discipline: A Vital Element for Economic Recovery*. International Monetary Fund, March 16.

de Menil, George, and Anthony M. Solomon, 1983. *Economic Summitry*. New York: Council on Foreign Relations.

Denison, Edward F., 1962. *The Sources of Economic Growth in the United States*. New York: Committee for Economic Development.

Department of Defense (DOD), 1985. *The Technology Security Program*. A report by Caspar Weinberger, Secretary of Defense, to the 99th Congress, 1st session, February.

Dertouzos, Michael L., Richard K. Lester, Robert M. Solow, and the M.I.T. Commission on Industrial Productivity, 1989. *Made in America: Regaining the Productive Edge*. Cambridge, MA: The M.I.T. Press.

Destler, I. M., 1986. *American Trade Politics: System Under Stress*. Washington, DC: Institute for International Economics, and New York: The Twentieth Century Fund.

_____, 1980. *Making Foreign Economic Policy*. Washington, DC: The Brookings Institution.

_____, Haruhiro Fakui, and Hideo Sato, 1979. *The Textile Wrangle: Conflict in Japanese-American Relations 1969-1971*. Ithaca, NY: Cornell University Press.

de Vries, Margaret Garritsen, 1986. *The IMF in a Changing World 1945-85*. Washington, DC: International Monetary Fund.

Domhoff, G. William, 1990. *The Power Elite and the State: How Policy Is Made in the United States*. New York: Aldine.

Drouin, Marie-Josee, Maurice Ernst, and Jimmy W. Wheeler, 198 . *Western European Adjustment to Structural Economic Problems*. Prepared for the U.S. Department of State, Contract No. 1722-420150 by the Hudson nstitute, 620 Union Drive, Indianapolis, IN 46206 (September 5).

Drucker, Peter F., 1989. *The New Realities*. New York: Harper & R w.

_____, 1986. "The Changed World Economy." *Foreign Affairs*, vol. 64, no. 4 (Spring), pp. 768-92.

Eckes, Alfred E., Jr., 1975. *A Search for Solvency: Bretton Woods and the International Monetary System, 1941-1971*. Austin: University of Texas Press.

Economic Report of the President (ERP), various years. Transmitted to the Congress, together with the Annual Report of the Council of Economic Advisers. Washington, DC: Government Printing Office.

Economist (weekly issues).

Etzioni, Amitai, 1988. *The Moral Dimension: Toward a New Economics*. New York: The Free Press.

_____, 1985. "Encapsulated Competition." *Journal of Post Keynesian Economics*, vol. 7, no. 3 (Spring), pp. 287-302.

Evans, John W., 1971. *The Kennedy Round in American Trade Policy: The Twilight of the GATT?* Cambridge, MA: Harvard University Press.

Evans, Peter B., Dietrich Rueschemeyer, and Theda Skocpol (eds.), 1985. *Bringing the State Back In*. Cambridge, England: Cambridge University Press.

Executive Office of the President (EOP), 1981. *America's New Beginning: A Program for Economic Recovery*. Washington, DC: Government Printing Office.

Fallows, James, 1989. *More Like Us: Making America Great Again*. Boston: Houghton Mifflin.

Federal Reserve Bank, 1986. *Monetary Policy Objectives for 1986*. Summary Report of the Federal Reserve Board (February 19).

Federal Reserve Bank of St. Louis. *Monetary Trends*, various issues.

_____. *U.S. Financial Data*, various issues.

Feldstein, Martin, 1988. "Let the Market Decide." *The Economist* (December 3).

_____, 1987. "Budget Card Tricks and Dollar Levitation." *The Washington Post*, December 1.

_____, and Kathleen Feldstein, 1986. "Foreigners Can't Solve Our Trade Problem." *The Washington Post*, July 8.

Financial Times (FT), daily issues.

Finger, J. Michael, and Andrzej Olechowski (eds.), 1987. *The Uruguay Round: A*

Handbook on the Multilateral Trade Negotiations. Washington, DC: World Bank.

Flemming, J. S., 1978. "The Economic Explanation of Inflation." In *The Political Economy of Inflation*, eds. Fred Hirsch and John H. Goldthorpe. Cambridge, MA: Harvard University Press, pp. 13–37.

Foa, Bruno, 1949. *Monetary Reconstruction in Italy.* New York: King's Crown Press.

Fortune (weekly issues).

Freedenberg, Paul, 1989. "U.S. Export Control Policy." In *The Allies and East-West Economic Relations*, eds. Henry R. Nau and Kevin F. F. Quigley. Washington, DC: U.S.-Japan Economic Agenda (joint project of the Elliott School of International Affairs at the George Washington University and the Carnegie Council on Ethics and International Affairs).

Frieden, Jeffrey A., 1988. "Sectoral conflict and foreign economic policy." *International Organization*, vol. 42, no. 1 (Winter), pp. 59–91.

———, 1987. *Banking on the World: The Politics of American International Finance.* New York: Harper & Row.

Friedman, Benjamin M., 1988. *Day of Reckoning: The Consequences of American Economic Policy Under Reagan and After.* New York: Random House.

Friedman, Milton, 1968. "The Role of Monetary Policy." *American Economic Review*, vol. 58 (March), pp. 1–17.

———, and Rose Friedman, 1980. *Freedom to Choose.* New York: Harcourt Brace Jovanovich.

Frost, Ellen L., 1987. *For Richer, for Poorer: The New U.S.-Japan Relationship.* New York: Council on Foreign Relations.

Fukuyama, Francis, 1989. "The End of History?" *The National Interest*, no. 16 (Summer), pp. 3–19.

Funabashi, Yoichi, 1988. *Managing the Dollar: From the Plaza to the Louvre.* Washington, DC: Institute for International Economics.

Gaddis, John Lewis, 1987. *The Long Peace: Inquiries into the History of the Cold War.* New York: Oxford University Press.

———, 1972. *The United States and the Origins of the Cold War 1941-1947.* New York: Columbia University Press.

Galbraith, Evan G., 1987. *Ambassador in Paris: The Reagan Years.* Washington, DC: Regnery Gateway.

Gardner, Richard N., 1980. *Sterling-Dollar Diplomacy in Current Perspective: The Origins and the Prospects of Our International Economic Order.* New York: Columbia University Press. Expanded edition with revised introduction (original edition, 1956).

Garten, Jeffrey E., 1989-1990. "Japan and Germany: American Concerns," *Foreign Affairs*, vol 68, no. 5 (Winter), pp. 84–102.

Garthoff, Raymond L., 1985. *Detente and Confrontation.* Washington, DC: The Brookings Institution.

General Agreement on Trade and Tariffs (GATT), various years. *International Trade.* Geneva, Switzerland: GATT.

Gibbon, Edward, 1938. *The Rise and Fall of the Roman Empire.* 7 Volumes. London: T. Cadell.

Gilpin, Robert, 1987. *The Political Economy of International Relations.* Princeton, NJ: Princeton University Press.

———, 1986. "The Richness of the Tradition of Political Realism." In *Neorealism and*

Its Critics, ed. Robert O. Keohane. New York: Columbia University Press, pp. 301–21.

————, 1981. *War and Change in World Politics*. Cambridge, England: Cambridge University Press.

————, 1977. "Economic Interdependence and National Security in Historical Perspective." In *Economic Issues and National Security*, eds. Klaus Knorr and Frank N. Trager. Lawrence, KA: Allen Press (for the National Security Education Program), pp. 19–67.

————, 1975. *U.S. Power and the Multinational Cooperation: The Political Economy of Foreign Direct Investment*. New York: Basic Books.

————, 1972. "The Politics of Transnational Economic Relations." In *Transnational Relations and World Politics*, eds. Robert O. Keohane and Joseph S. Nye, Jr. Cambridge, MA: Harvard University Press, pp. 48–70.

Goldstein, Joshua S., 1988. *Long Cycles: Prosperity and War in the Modern Age*. New Haven, CT: Yale University Press.

Goldthorpe, John H., 1978. "The Current Inflation: Towards a Sociological Account." In *The Political Economy of Inflation*, eds. Fred Hirsch and John H. Goldthorpe. Cambridge, MA: Harvard University Press, pp. 186–214.

Gordon, Lincoln, 1984. "Lessons from the Marshall Plan: Successes and Limits." In *The Marshall Plan: A Retrospective*, eds. Stanley Hoffmann and Charles Maier. Boulder, CO: Westview.

Gourevitch, Peter, 1986. *Politics in Hard Times: Comparative Responses to International Economic Crises*. Ithaca, NY: Cornell University Press.

Gowa, Joanne, 1986. "Anarchy, egoism and third images: *The Evolution of Cooperation* and international relations." *International Organization*, vol. 40, no. 1 (Winter), pp. 167–186.

————, 1983. *Closing the Gold Window: Domestic Politics and the End of Bretton Woods*. Ithaca, NY: Cornell University Press.

Greider, William, 1987. *Secrets of the Temple: How the Federal Reserve Runs the Country*. New York: Simon and Schuster.

————, 1981. "The Education of David Stockman." *The Atlantic Monthly* (December), pp. 27–54.

Haas, Ernst B., 1986. "What Is Nationalism and Why Should We Study It?" *International Organization*, vol. 40, no. 3 (Summer), pp. 708–744.

————, 1983. "Words can hurt you; or who said what to whom about regimes." In *International Regimes*, ed. Stephen D. Krasner. Ithaca, NY: Cornell University Press, pp. 23–61.

————, 1975. *The Obsolescence of Regional Integration Theory*. Research Series No. 25, Institution of International Studies. Berkeley: University of California.

————, 1964. *Beyond the Nation-State*. Stanford, CA: Stanford University Press.

Haberler, Gottfried, 1985. "International Issues Raised by Criticisms of the U.S. Budget Deficit." In *Essays in Contemporary Economic Problems: The Economy in Deficit 1985*, ed. Phillip Cagan. Washington, DC: American Enterprise Institute, pp. 121–147.

Haggard, Stephan, 1988. "The Institutional Foundations of Hegemony: Explaining the Reciprocal Trade Agreements Act of 1934." *International Organization*, vol. 42, no. 1 (Winter), pp. 91–121.

————, and Beth A. Simmons, 1987. "Theories of International Regimes." *International Organization*, vol. 41, no. 3 (Summer), pp. 491–517.

Haig, Alexander, Jr., 1984. *Caveat: Realism, Reagan, and Foreign Policy*. New York: Macmillan Publishing Company.

Hall, Peter, 1987. "The Evolution of Economic Policy under Mitterrand." In *The Mitterrand Experiment*, eds. George Ross, Stanley Hoffmann, and Sylvia Malzacher. New York: Oxford University Press, pp. 54–72.

———, 1986. *Governing the Economy: The Politics of State Intervention in Britain and France*. New York: Oxford University Press.

Halperin, Morton H., 1974. *Bureaucratic Politics and Foreign Policy*. Washington, DC: The Brookings Institution.

Harberger, Arnold C. (ed.), 1984. *World Economic Growth: Case Studies of Developed and Developing Nations*. San Francisco: Institute for Contemporary Studies.

Harris, Seymour Edwin, (ed.), 1948. *Foreign Economic Policy for the United States*. Cambridge, MA: Harvard University Press.

Harrod, Roy F., 1971. *The Life of John Maynard Keynes*. New York: Aron.

Hartz, Louis, 1955. *The Liberal Tradition in America*. New York: Harcourt, Brace and World.

Havrylyshyn, Oli, and Iradj Alikani, 1982. "Is There Cause for Export Optimism?" *Weltwirtschaftliches Archiv*, vol. 118, no. 4. Kiel, FRG: Kiel Institute of World Economics, pp. 651–62.

Hayek, Friedrich, 1944. *The Road to Serfdom*. Chicago: University of Chicago Press.

Hellwig, Martin, and Manfred J. M. Neumann, 1987. "Economic Policy in Germany: Was There a Turnaround?" *Economic Policy: A European Forum*, vol. 2, no. 2 (October), pp. 105–141.

Helpman, Elhanan, and Paul R. Krugman, 1985. *Market Structure and Foreign Trade*. Cambridge, MA: M.I.T. Press.

Henning, C. Randall, 1987. *Macroeconomic Diplomacy in the 1980s*. Atlantic Paper No. 65. London: Croom Helm for the Atlantic Institute for International Affairs.

———, 1985. *The Politics of Macroeconomic Conflict and Coordination Among Advanced Capitalist States 1975–1985*. Ph.D. dissertation, The Fletcher School of Law and Diplomacy, Tufts University (April).

Hibbs, Douglas A., Jr., 1987a. *The American Political Economy: Macroeconomics and Electoral Politics in the United States*. Cambridge, MA: Harvard University Press.

———, 1987b. *The Political Economy of Industrial Democracies*. Cambridge, MA: Harvard University Press.

Hillenbrand, Martin J., 1988. "East–West Economic Relations, Export Controls, and Strains in the Alliance." In *Controlling East–West Trade and Technology Transfer: Power, Politics and Policies*, ed. Gary K. Bertsch. Durham, NC: Duke University Press.

Hirsch, Fred, and John H. Goldthorpe (eds.), 1978. *The Political Economy of Inflation*. Cambridge, MA: Harvard University Press.

Hirschman, Albert O., 1980. *National Power and the Structure of Foreign Trade*. Expanded edition. Berkeley: University of California Press (original edition, 1945).

———, 1970. *Exit, Voice and Loyalty: Responses to Decline in Firms, Organizations and States*. Cambridge, MA: Harvard University Press.

Hitchens, Harold L., 1968. "Influences in the Congressional Decision to Pass the Marshall Plan." *Western Political Quarterly*, vol. 21, no. 1 (March), pp. 51–68.

Hogan, Michael J., 1987. *The Marshall Plan: America, Britain and the Reconstruction of Western Europe, 1947-1952*. Cambridge, England: Cambridge University Press.

Horrigan, Michael W., and Steven E. Haugen, 1988. "The declining middle-class thesis: a sensitivity analysis." *Monthly Labor Review* (May), pp. 3-13.

Horsefield, J. Keith, (ed.), 1969. *The International Monetary Fund 1945-1965: Twenty Years of International Monetary Cooperation*. Volume I: Chronicle; Volume II: Analysis; Volume III: Documents. Washington, DC: International Monetary Fund.

House of Representatives, U.S. Congress, 1945. *Bretton Woods Agreement Act*. Committee on Banking and Currency, 79th session, vol. 1, March 7-9, 12-16, 19-23, 1945; vol. 2, April 19-20, 27, 30, May 1-5, 7-11.

Hughes, Thomas L., 1975. "Liberals, Populists and Foreign Policy." *Foreign Policy*, no. 20 (Fall), pp. 97-138.

Hull, Cordell, 1948. *The Memoirs of Cordell Hull*. 2 Volumes. New York: Macmillan Publishing Company.

Huntington, Samuel P., 1988-1989. "The U.S. – Decline or Renewal?", *Foreign Affairs*, Vol. 67, No. 2 (Winter), pp. 76-97.

_____, 1978. "Trade, Technology and Leverage: Economic Diplomacy." *Foreign Policy*, no. 32 (Fall), pp. 63-80.

Hyland, William G., 1986. *Mortal Rivals: Superpower Relations from Nixon to Reagan*. New York: Random House.

Ikenberry, G. John, 1988. *Reasons of State: Oil Politics and the Capacities of American Government*. Ithaca, NY: Cornell University Press.

_____, David A. Lake, and Michael Mastanduno, 1988. "The State and American Foreign Economic Policy." *International Organization*, special issue, vol. 42, no. 1 (Winter), pp. 1-14.

Institute for International Economics (IIE), 1987. *Resolving the Global Economic Crisis: After Wall Street*. A statement by 33 economists from 13 countries. Washington, DC: Institute for International Economics, December.

_____, 1982. *Promoting World Recovery*. A statement on global economic strategy by 26 economists from 14 countries. Washington, DC: Institute for International Economics, December.

International Energy Agency (IEA), 1983. "Meeting of the Governing Board at Ministerial Level," Communiqué and Annex 1, May 8 (mimeo).

International Herald Tribune (IHT), daily issues.

International Monetary Fund (IMFa), various years. *World Economic Outlook*. Washington, DC: International Monetary Fund.

_____, (IMFb), various years. *Direction of Trade Statistics Yearbook*. Washington, DC: IMF.

_____, (IMFc), various issues. *International Financial Statistics: Yearbook*. Washington, DC: IMF.

_____, 1982. *Annual Report 1982*. Washington, DC: IMF.

Ingram, James C., 1983. *International Economics*. New York: John Wiley and Sons.

Isaacson, Walter, and Evan Thomas, 1986. *The Wise Men: Six Friends and the World They Made*. New York: Simon and Schuster.

Jentleson, Bruce W., 1986. *Pipeline Politics: The Complex Political Economy of East-West Energy Trade*. Ithaca, NY: Cornell University Press.

Johnson, Chalmers, 1982. *MITI and the Japanese Miracle: The Growth of Industrial Policy, 1925-1975*. Stanford, CA: Stanford University Press.

Johnson, D. Gale (ed.), 1988. *Agriculture Reform Efforts in Japan and the United*

States. New York: New York University Press, published for the U.S.-Japan Economic Agenda (joint project of the Elliott School of International Relations of The George Washington University and the Carnegie Council on Ethics and International Affairs).

Katzenstein, Peter J., 1987. *Policy and Politics in West Germany*. Philadelphia: Temple University Press.

———, 1985. *Small States in World Markets: Industrial Policy in Europe*. Ithaca, NY: Cornell University Press.

———, (ed.), 1977. "Between power and plenty: foreign economic policies of advanced industrial states." *International Organization*, special issue, vol. 31, no. 4 (Autumn). Also published as book with same title. Madison: University of Wisconsin Press, 1978.

———, 1975. "International Interdependence: Some Long-term Trends and Recent Changes." *International Organization*, vol. 29, no. 4 (Autumn), pp. 1021-1034.

Kendrick, John W., 1989. "Policy Implications of the Slowdown in U.S. Productivity Growth." In *Productivity Growth and the Competitiveness of the U.S. Economy*, ed. Stanley W. Black. Norwell, MA: Kluwer Academic Publications.

Kennedy, Paul, 1987. *The Rise and Fall of the Great Powers*. New York: Random House.

Keohane, Robert O., 1986. "Reciprocity in International Relations." *International Organization*, vol. 40, no. 1 (Winter), pp. 1-29.

———, 1984. *After Hegemony: Cooperation and Discord in the World Political Economy*. Princeton, NJ: Princeton University Press.

———, 1982. "Inflation and the Decline of American Power." In *Political Economy of International and Domestic Monetary Relations*, eds. Raymond E. Lombra and Willard E. Witte. Ames: Iowa State University Press.

———, 1980. "The Theory of Hegemonic Stability and Changes in International Economic Regimes, 1967-1977." In *Changes in the International System*, eds. Ole R. Holsti, Randolph M. Siverson, and Alexander L. George. Boulder, CO: Westview Press, pp. 131-62.

———, 1978. "Economics, Inflation and the Role of the State: Political Implications of the McCracken Report." *World Politics*, vol. 31, no. 1 (October), pp. 108-128.

———, and Joseph S. Nye, 1977. *Power and Interdependence*. Boston: Little, Brown.

Kettl, Donald F., 1986. *Leadership at the Fed*. New Haven, CT: Yale University Press.

Kihwan, Kim and Hwa Soo Chung, 1989. "Korea's Domestic Trade Politics and the Uruguay Round." In *Domestic Trade Politics and the Uruguay Round*, ed. Henry R. Nau. New York: Columbia University Press, pp. 135-167.

Kindleberger, Charles P., 1987. *Marshall Plan Days*. Boston: Allen and Unwin.

———, 1984. "The American Origins of the Marshall Plan: A View from the State Department." In *The Marshall Plan: A Retrospective*, eds. Stanley Hoffmann and Charles Maier. Boulder, CO: Westview, pp. 7-13.

———, 1981. "Dominance and Leadership in the International Economy." *International Studies Quarterly*, vol. 25, no. 3 (June), pp. 242-254.

———, 1973. *The World in Depression 1929-1939*. Berkeley: University of California Press.

Kissinger, Henry (Chairman), 1984. *Report of the National Bipartisan Commission on Central America*. New York: Macmillan Publishing Company.

———, 1982. *Years of Upheaval*. Boston: Little, Brown and Company.

———, 1979. *White House Years*. Boston: Little, Brown and Company.

———, 1976. Speech before the OECD Ministerial Meeting in 1976. *Department of State Bulletin* (July 19).

Knorr, Klaus, and Frank N. Trager (eds.), 1977. *Economic Issues and National Security*. Lawrence, KA: Allen Press, published for the National Security Education Program.

Kolko, Joyce, and Gabriel Kolko, 1972. *The Limits of Power*. New York: Harper & Row.

Korb, Larry, 1989. "Playing the Numbers on the Defense Budget," *The Washington Post*, Outlook Section (December 3).

Kosters, Marvin H., 1986. "Job Changes and Displaced Workers: An Examination of Employment Adjustment Experience." In *Essays in Contemporary Economic Problems: The Impact of the Reagan Program 1986*, ed. Phillip Cagan. Washington, DC: The American Enterprise Institute, pp. 275–306.

Kraft, Joseph, 1984. *The Mexican Rescue*. New York: Group of Thirty.

Krasner, Stephen D., 1987. *Asymmetries in Japanese-American Trade: The Case for Specific Reciprocity*. Policy Papers in International Affairs, no. 32, Institute of International Studies. Berkeley: University of California.

———, 1985. *Structural Conflict: The Third World Against Global Liberalism*. Berkeley: University of California Press.

———, (ed.), 1983. *International Regimes*. Ithaca, NY: Cornell University Press.

———, 1978. *Defending the National Interest*. Princeton, NJ: Princeton University Press.

———, 1977. "U.S. commercial and monetary policy: unraveling the paradox of external strength and internal weakness." In "Between power and plenty: foreign economic policies of advanced industrial states." *International Organization*, ed. Peter Katzenstein, special issue, vol. 31, no. 4 (Autumn), pp. 635–673.

———, 1976. "State Power and the Structure of International Trade." *World Politics*, vol. 28, no. 3 (April), pp. 317–347.

Kratochwil, Friedrich, and John Gerard Ruggie, 1986. "International Organization: A State of the Art or an Art of the State." *International Organization*, vol. 40, no. 4 (Autumn), pp. 753–777.

Krauthammer, Charles, 1984. "On Nuclear Morality." In *Nuclear Arms: Ethics, Strategy, Politics*, ed. R. James Woolsey. San Francisco: Institute for Contemporary Studies.

Kreile, Michael, 1977. "West Germany: the dynamics of expansion." In "Between power and plenty: foreign economic policies of advanced industrial states," ed. Peter J. Katzenstein. *International Organization*, special issue, vol. 31, no. 4 (Autumn), pp. 775–808.

Krieger, Joel, 1986. *Reagan, Thatcher and the Politics of Decline*. New York: Oxford University Press.

Kreuger, Anne O., and Constantine Michalopoulos, 1985. "Developing-Country Trade Policies and the International Economic System." In *Hard Bargaining Ahead: U.S. Trade Policy and Developing Countries*, ed. Ernest H. Preeg. New Brunswick, NJ: Transactions Books, pp. 39–63.

———, 1978. *Foreign Trade Regimes and Economic Development: Liberalization Attempts and Consequences*. Cambridge, MA: Ballinger.

Krugman, Paul R., 1987. "Is Free Trade Passé?" *Economic Perspectives*, vol. 1, no. 2 (Fall), pp. 131–144.

———, (ed.), 1986. *Strategic Trade Policy and the New International Economics*. Cambridge, MA: M.I.T. Press.

———, 1985. "U.S. Macro-Economic Policy and the Developing Countries." In *U.S. Foreign Policy and the Third World: Agenda 1985-86*, eds. John W. Sewell, Richard E. Feinberg, and Valerianna Kallab. New Brunswick, NJ, and Oxford, England: Transaction Books.

Kuznets, Simon, 1966. *Modern Economic Growth: Rate Structure and Spread*. New Haven, CT: Yale University Press.

Laird, Sam, and J. Michael Finger, 1986. "Protection in Developed and Developing Countries: An Overview" (October 21). Obtained from authors at World Bank.

Lake, David, 1984. "Beneath the Commerce of Nations: A Theory of International Economic Structures." *International Studies Quarterly*, vol. 28, pp. 143–170.

Lande, Stephen L., and Craig VanGrasstek, 1986. *The Trade and Tariff Act of 1984: Trade Policy in the Reagan Administration*. Lexington, MA: Lexington Books.

Lawrence, Robert Z., 1984. *Can America Compete?* Washington, DC: The Brookings Institution.

———, and Charles L. Schultze (eds.), 1987. *Barriers to European Growth: A Transatlantic View*. Washington, DC: The Brookings Institution.

Lawson, Fred H., 1983. "Hegemony and International Trade." *International Organization*, vol. 37, no. 2 (Spring), pp. 317–339.

Levy, Frank, 1988. *Dollars and Dreams: The Changing American Income Distribution*. New York: W. W. Norton.

Lewis, W. Arthur, 1984. *The Rate of Growth of the World Economy*. Taipei, ROC: The Institute of Economics, Academia Sinica.

Lieberman, Sima, 1977. *The Growth of European Mixed Economies 1945–1970: A Concise Study of the Economic Evolution of Six Countries*. New York: Schenkman Publishing Company.

Lindblom, Charles E., 1977. *Politics and Markets: The World's Political-Economic Systems*. New York: Basic Books.

London Economic Summit Conference, 1984. Declaration on Democratic Values (June 8); London Economic Declaration, Declaration on East–West Relations and Arms Control, Declaration on International Terrorism (June 9). *Complete Presidential Documents*, pp. 574–578.

Lubar, Robert, 1981. "Reaganizing the Third World." In *Fortune* (November 16), pp. 83–90.

Maier, Charles S., 1981. "The Two Postwar Eras and the Conditions for Stability in Twentieth-Century Western Europe." *American Historical Review*, vol. 86 (Autumn), pp. 327–352.

———, 1978. "The Politics of Inflation in the Twentieth Century." In *The Political Economy of Inflation*, eds. Fred Hirsch and John H. Goldthorpe. Cambridge, MA: Harvard University Press, pp. 37–73.

———, 1977. "The politics of productivity: foundations of American international economic policy after World War II." In "Between power and plenty: foreign economic policies of advanced industrial states," ed. Peter J. Katzenstein. *International Organization*, special issue, vol. 31, no. 4 (Autumn), pp. 607–635.

Makin, John H., 1985. "The Effect of Government Deficits on Capital Formation." In *Essays in Contemporary Economic Problems: The Economy in Deficit 1985*, ed. Phillip Cagan. Washington, DC: American Enterprise Institute, pp. 163–195.

Mandelbaum, Michael, 1985. "The Luck of the President," *Foreign Affairs*, vol. 64, no. 3, pp. 393–413.

Marris, Stephen, 1985. *Deficits and the Dollar: The World Economy at Risk*. Policy Analyses in International Economics, no. 14, Washington, DC: Institute for International Economics (December). Updated edition, 1987.

———, 1984. "East–West Economic Relations: A Longer-Term Perspective," *OECD Observer*, no. 128 (May), pp. 11–18.

———, 1978. "The Politics of Inflation in the Twentieth Century." In *The Political*

Economy of Inflation, eds. Fred Hirsch and John H. Goldthorpe. Cambridge, MA: Harvard University Press, pp. 37–73.

Mastanduno, Michael, 1988. *Between Economics and National Security: CoCom and the Politics of Export Control Policy* (unpublished manuscript).

Matthews, Kent, and Patrick Minford, 1987. "Mrs. Thatcher's Economic Policies 1979–1987." *Economic Policy: A European Forum*, vol. 2, no. 2 (October), pp. 59–93.

McKinnon, Ronald I., 1984. *An International Standard for Monetary Stabilization*. Policy Analysis in International Economics, no. 8. Washington, DC: Institute for International Economics (March).

Mee, Charles L., 1984. *The Marshall Plan: The Launching of the Pax Americana*. New York: Simon and Schuster.

Melton, William C., 1985. *Inside the Fed: Making Monetary Policy*. Homewood, IL: Dow Jones-Irwin.

Milner, Helen V., 1988. "Anarchy and Interdependence." (Unpublished manuscript, September).

_____, 1988. *Resisting Protectionism: Global Industries and the Politics of International Trade*. Princeton, NJ: Princeton University Press.

Milward, Alan S., 1984. *The Reconstruction of Western Europe 1945-51*. Berkeley: University of California Press.

Mitchell, B. R., 1983. *International Historical Statistics: The Americas and Australasia*. Detroit: Gale Research Company.

_____, 1982. *International Historical Statistics: Africa and Asia*. New York: New York University Press.

_____, 1981. *European Historical Statistics—1750-1975*. Second Revised Edition. New York: Facts on File, Inc.

Mitrany, David, 1943. *A Working Peace System*. London: Royal Institute of International Affairs.

Morgenthau, Henry, 1945. "Bretton Woods and International Cooperation." *Foreign Affairs*, vol. 23, no. 2, pp. 182–193.

Morse, Edward L., 1976. *Modernization and the Transformation of International Relations*. New York: The Free Press.

_____, 1973. *Foreign Policy and Interdependence in Gaullist France*. Princeton, NJ: Princeton University Press.

Morris, Robert J., 1989. Testimony before the Subcommittee on International Development, Finance, Trade and Monetary Policy, U.S. House Committee on Banking, Finance and Urban Affairs (November 16).

Moynihan, Daniel Patrick, 1988a. *Came the Revolution: Argument in the Reagan Era*. San Diego: Harcourt Brace Jovanovich.

_____, 1988b. "Debunking the Myth of Decline." *The New York Times Magazine* (June 19).

_____, 1983. "Reagan's Bankrupt Budget." *New Republic*, no. 189 (December 31), pp. 18–21.

Murray, Charles, 1984. *Losing Ground: American Social Policy 1950-1980*. New York: Basic Books.

Myrdal, Gunnar, 1971. *Economic Theory and Underdeveloped Regions*. New York: Harper & Row.

_____, 1968. *Asian Drama: Inquiry into the Poverty of Nations*. Volumes I-III. New York: Pantheon.

National Academy of Sciences (NAS), 1987. *Balancing the National Interest: U.S. National Security Export Controls and Global Economic Competition*. Panel on the Impact of National Security Controls on International Technology Transfer, Committee on Science, Engineering and Public Policy. National Academy of Sciences, National Academy of Engineering, Institute of Medicine. Washington, DC: National Academy Press.

Nau, Henry R., 1989a. "Conclusion: The Three Faces of Change and U.S. Policy." In *The Allies and East-West Economic Relations*, eds. Henry R. Nau and Kevin F. F. Quigley. Washington, DC: U.S.-Japan Economic Agenda (joint project of the Elliott School of International Affairs at the George Washington University and the Carnegie Council on Ethics and International Affairs).

_____, (ed.), 1989b. *Domestic Trade Politics and the Uruguay Round*, New York: Columbia University Press.

_____, 1988a. "Bargaining Barriers: U.S. Policies, the NICs and the Uruguay Round." In *Keeping Pace: U.S. Policies and Global Economic Change*, ed. John Yochelson. Cambridge, MA: Ballinger.

_____, 1988b. "Export Controls and Free Trade: Squaring the Circle in COCOM." In *Controlling East-West Trade and Technology Transfer: Power, Politics and Policies*, ed. Gary K. Bertsch. Durham, NC: Duke University Press.

_____, 1988c. "Policy Comes Before Finance," in *The U.S. Approach to the Latin American Debt Crisis*, FPI Policy Study Groups. Washington, DC: Foreign Policy Institute, SAIS, The John Hopkins University (February).

_____, 1986. "National Policies for High Technology Development and Trade: An International and Comparative Assessment." In *National Policies for Developing High Technology Industries*, eds. Francis W. Rushing and Carole Ganz Brown. Boulder, CO: Westview, pp. 9-31.

_____, 1985a. "The State of the Debate: Reaganomics—The Solution?" *Foreign Policy*, no. 59 (Summer), pp. 144-153.

_____, 1985b. "The NICs in a New Trade Round." In *Hard Bargaining Ahead: U.S. Trade Policy and Developing Countries*, ed. Ernest H. Preeg. New Brunswick, NJ: Transaction Books.

_____, 1984-1985. "Where Reaganomics Works." *Foreign Policy*, no. 57 (Winter), pp. 14-38.

_____, 1984. *International Reaganomics: A Domestic Approach to World Economy*. Significant Issues Series, vol. VI, no. 18, Center for Strategic and International Studies, Georgetown University.

_____, 1981a. "Economics, National Security and Arms Control." In *Arms Control II: A New Approach to International Security*, eds. John H. Barton and Ryukichi Imai. Cambridge, MA: Oelgeschlager, Gunn, and Hain, Publishers, pp. 113-159.

_____, 1981b. "Securing Energy." *The Washington Quarterly*, vol. 4, no. 3 (Summer).

_____, 1980. "The Evolution of U.S. Foreign Policy in Energy: 1973-77." In *International Energy Policy*, eds. Robert Lawrence and Martin Heisler. Lexington, MA: Lexington Books, pp. 37-65.

_____, 1979. "From Integration to Interdependence: Gains, Losses, and Continuing Gaps." *International Organization*, vol. 33, no. 1 (Winter), pp. 119-147.

_____, 1977. "The International Political Economy of Food and Energy." *Journal of International Affairs*, vol. 31, no. 2 (Fall/Winter).

_____, 1976. *Technology Transfer and U.S. Foreign Policy*. New York: Praeger.

_____, 1974. *National Politics and International Technology: Nuclear Reactor Development in Western Europe*. Baltimore: Johns Hopkins Press.

The New York Times (NYT), daily issues.

Newlon, Dan, 1986. "Replication in Empirical Economics." *American Economic Review* (Summer).

Niskanen, William A., 1988. *Reaganomics: An Insider's Account of the Policies and the People*. New York: Oxford University Press.

Nye, Joseph S., Jr., 1988. "Understanding U.S. Strength." *Foreign Policy*, no. 72 (Fall), pp. 105–130.

―――, 1976. "Independence and Interdependence." *Foreign Policy*, no. 22 (Spring), pp. 129–162.

Odell, John S., 1982. *U.S. International Monetary Policy: Markets, Power and Ideas as Sources of Change*. Princeton, NJ: Princeton University Press.

Office of Management and Budget (OMB), 1989. *Mid-Session Review of the Budget*. Washington, DC: Executive Office of the President, July 18.

Oliver, Robert W., 1975. *International Economic Co-operation and the World Bank*. New York: Holmes and Meier.

Olson, Mancur, 1982. *The Rise and Decline of Nations: Economic Growth, Stagflation and Social Rigidities*. New Haven, CT: Yale University Press.

―――, 1965. *The Logic of Collective Action*. Cambridge, MA: Harvard University Press.

Organization for Economic Cooperation and Development (OECD), 1988. *Structural Adjustment and Economic Performance*. Paris: OECD.

―――, 1987a. *National Accounts: Main Aggregates 1970–85*. Paris: OECD.

―――, 1987b. *Economic Outlook*. Paris: OECD (December).

―――, 1987c. *Economic Outlook: Historical Statistics 1960–1985*. Paris: OECD.

―――, 1985a. *Costs and Benefits of Protection*. Paris: OECD.

―――, 1985b. *The Role of the Public Sector: Courses and Consequences of the Growth of Government*, OECD Economic Studies, special issue, No. 4 (Spring).

―――, 1984. *Economic Outlook*. Paris: OECD (July).

―――, 1978. *Public Expenditure Trends*. OECD Studies in Resource Allocation, No. 5. Paris: OECD, June.

―――, 1977. *Towards Full Employment and Price Stability*. Paris: OECD, June.

Organization for European Economic Cooperation (OEEC), 1961. *12th Annual Economic Review*. Paris: OEEC.

Osgood, Robert E., 1953. *Ideals and Self-interest in America's Foreign Relations*. Chicago: University of Chicago Press.

Ottawa Economic Summit Conference, 1981. Declaration Issued at the Conclusion of the Conference. *Complete Presidential Documents*, pp. 574–578 (May 13).

Oye, Kenneth A., Robert J. Lieber, and Donald Rothchild, (eds.), 1987. *Eagle Resurgent?: The Reagan Era in American Foreign Policy*. Boston: Little, Brown and Company.

―――, (ed.), 1986. *Cooperation Under Anarchy*. Princeton, NJ: Princeton University Press.

―――, Robert J. Lieber, and Donald Rothchild, (eds.), 1983. *Eagle Defiant: United States Foreign Policy in the 1980's*. Boston: Little, Brown and Company.

―――, Donald Rothchild, and Robert J. Lieber, (eds.), 1979. *Eagle Entangled: U.S. Foreign Policy in a Complex World*. New York: Longman.

Paarlberg, Robert L., 1976. "Domesticating Global Management." *Foreign Affairs*, vol. 54, no. 3 (April), pp. 563–577.

―――, 1985. *Food Trade and Foreign Policy: India, the Soviet Union and the United States*. Ithaca, NY: Cornell University Press.

Packard, George R., 1987–1988. "The Coming U.S.–Japan Crisis." *Foreign Affairs*, vol. 66, no. 2 (Winter), pp. 348–368.

Panic, M., 1978. "The Origin of Increasing Inflationary Tendencies in Contemporary Society." In *The Political Economy of Inflation*, eds. Fred Hirsch and John H. Goldthorpe. Cambridge, MA: Harvard University Press, pp. 137–160.

Pastor, Robert A., 1980. *Congress and the Politics of U.S. Foreign Economic Policy, 1929–1976*. Berkeley: University of California Press.

Patterson, Gardner, 1966. *Discrimination in International Trade: The Policy Issues*. Princeton, NJ: Princeton University Press.

Peacock, Alan T., and Martin Ricketts, 1978. "The Growth of the Public Sector and Inflation." In *The Political Economy of Inflation*, eds. Fred Hirsch and John H. Goldthorpe. Cambridge, MA: Harvard University Press, pp. 117–137.

Pempel, J. T., 1982. *Policy and Politics in Japan*. Philadelphia: Temple University Press.

———, 1977. "Japanese foreign economic policy: the domestic bases for international behavior." In "Between power and plenty: foreign economic policies of advanced industrial states," ed. Peter J. Katzenstein. *International Organization*, special issue, vol. 31, no. 4 (Autumn), pp. 723–775.

Peterson, Peter G., 1971. *The United States in the Changing World*, vols. I and II. Washington, DC: Government Printing Office (December 27).

Poehl, Karl Otto, 1987. "You Can't Robotize Policymaking," *International Economy*, vol. 1, no. 1 (October/November), pp. 20–28.

Pogue, Forrest C., 1987. *George C. Marshall: Statesman 1945–1959*. New York: Viking.

Pollard, Robert A., 1985. *Economic Security and the Origins of the Cold War, 1945–50*. New York: Columbia University Press.

Polanyi, Karl, 1944. *The Great Transformation*. New York: Ferrar and Rinehart.

Posner, Alan R., 1977. "Italy: dependence and political fragmentation." In "Between power and plenty: foreign economic policies of advanced industrial states." ed. Peter J. Katzenstein. *International Organization*, special issue, vol. 31, no. 4 (Autumn), pp. 809–839.

President's Committee on Foreign Aid (Harriman Committee), 1947. *European Recovery and American Aid*. Washington, DC: Government Printing Office (November 7).

Prestowitz, Clyde V., Jr., 1988. *Trading Places: How We Allowed Japan to Take the Lead*. New York: Basic Books.

Price, Harry Bayard, 1955. *The Marshall Plan and Its Meaning*. Ithaca, NY: Cornell University Press.

Public Opinion Quarterly, 1947, vol. 11, no. 3 (Fall), pp. 494–497; 1947–1948, vol. 11, no. 4 (Winter), pp. 674–676; 1948, vol. 12, no. 1 (Spring), pp. 172–173; 1948, vol. 12, no. 2 (Summer), pp. 364–67.

Putnam, Robert D., and C. Randall Henning, 1989. "The Bonn Summit of 1978: A Case Study in Coordination." In *Can Nations Agree?: Issues in International Economic Cooperation*. eds. Richard N. Cooper, Barry Eichengreen, C. Randall Henning, Gerald Holtman, and Robert D. Putnam. Washington, DC: The Brookings Institution.

———, and Nicholas Bayne, 1984. *Hanging Together: The Seven Power Summits*. Cambridge, MA: Harvard University Press. Revised and enlarged edition, 1987.

Rawls, John, 1971. *A Theory of Justice*. Cambridge, MA: Harvard University Press.

Reagan, Ronald (President), 1981a. Remarks to Meeting of the World Bank Group and International Monetary Fund. *Complete Presidential Documents* (September 29), pp. 1052–1055.

———, 1981b. Remarks to the World Affairs Council of Philadelphia, Bellevue Stratford Hotel, Philadelphia. *Complete Presidential Documents* (October 15), pp. 1137–1144.

———, 1981c. Statement at the Opening of the International Meeting on Cooperation and Development, Cancun, Mexico. *Complete Presidential Documents* (October 22), pp. 1185–1188.

———, (Governor), 1980. "A Strategy of Peace for the 1980's." Television address (October 19).

Rees, David, 1973. *Harry Dexter White: A Study in Paradox*. New York: Coward, McCann and Geoghegan.

Regan, Donald T., 1988. *For the Record: From Wall Street to Washington*. New York: Harcourt Brace Jovanovich.

Reich, Robert B., 1983. "Beyond Free Trade." *Foreign Affairs*, vol. 61, no. 4 (Spring), pp. 773–805.

Report of the National Bipartisan Commission on Central America. Washington, DC: Government Printing Office, January 1984.

Reynolds, Lloyd G., 1983. "The Spread of Economic Growth to the Third World: 1850–1980." *Journal of Economic Literature*, vol. 21, pp. 941–980.

Rhoads, Steven E., 1985. *The Economist's View of the World: Government, Markets, and Public Policy*. Cambridge, England: Cambridge University Press.

Rielly, John, ed., 1987. *American Public Opinion and U.S. Foreign Policy 1987*. Chicago: Chicago Council on Foreign Relations.

Roberts, Paul Craig, 1984. *The Supply-Side Revolution: An Insider's Account of Policymaking in Washington*. Cambridge, MA: Harvard University Press.

Robinson, Roger W., Jr., 1986. "Soviet Cash and Western Banks." *The National Interest*, no. 4 (Summer), pp. 37–45.

Rochester, J. Martin, 1986. "The Rise and Fall of International Organization as a Field of Study." *International Organization*, vol. 40, no. 4 (Autumn), pp. 777–813.

Rosecrance, Richard (ed.), 1976. *America as an Ordinary Country: U.S. Foreign Policy and the Future*. Ithaca, NY: Cornell University Press.

Rosenau, James N., 1986. "Hegemons, Regimes, and Habit-Driven Actors." *International Organization*, vol. 40, no. 4 (Autumn), pp. 849–894.

Rosenberg, Nathan, and Birdzell, L. E., Jr., 1986. *How the West Grew Rich: The Economic Transformation of the Industrial World*. New York: Basic Books.

Ross, George, Stanley Hoffman, and Sylvia Malzacher, 1987. *The Mitterrand Experiment: Continuity and Change in Modern France*. New York: Oxford University Press.

Rostow, Walter W., 1978. *The World Economy: History and Prospect*. Austin: University of Texas Press.

Rothstein, Robert L., 1979. *Global Bargaining: UNCTAD and the Quest for a New International Economic Order*. Princeton, NJ: Princeton University Press.

———, 1977. *The Weak in the World of the Strong: The Developing Countries in the International System*. New York: Columbia University Press.

Ruggie, John Gerard, 1986. "Continuity and Transformation in the World Polity: Toward a Neorealist Synthesis." In *Neorealism and Its Critics*, ed. Robert O. Keohane. New York: Columbia University Press, pp. 131–157.

———, 1983a. *The Antinomies of Interdependence: National Welfare and the International Division of Labor*. New York: Columbia University Press.

———, 1983b. "International Regimes, Transactions, and Change: Embedded Liberalism in the Postwar Economic Order." In *International Regimes*, ed. Stephen D. Krasner. Ithaca, NY: Cornell University Press, pp. 195–233.

Rushing, Francis W., and Carole Ganz Brown (eds.), 1986. *National Policies for Developing High Technology Industries: International Comparisons*. Boulder, CO: Westview Press.

Russett, Bruce, 1985. "The mysterious case of vanishing hegemony; or, is Mark Twain really dead?" *International Organization*, vol. 39, no. 2 (Spring), pp. 207–233.

Schell, Jonathan, 1982. *The Fate of the Earth*. New York: Knopf.

Schelling, Thomas C., 1978. *Micromotives and Macrobehavior*. New York: W. W. Norton.

Schneider, William, 1987. "'Rambo' and Reality: Having It Both Ways." In *Eagle Resurgent: The Reagan Era in American Foreign Policy*, eds. Kenneth A. Oye, Robert J. Lieber, and Donald Rothchild. Boston: Little, Brown and Company, pp. 41–75.

Schott, Jeffrey J., 1989. *More Free Trade Areas?* Washington, DC: Institute for International Economics, May.

Schriftgiesser, Karl, 1960. *Business Comes of Age: The Story of the Committee for Economic Development and its Impact upon the Economic Problems of the United States, 1942-1960*. New York: Harper and Brothers.

Schuker, Stephen A., 1981. "Comment on Charles S. Maier's 'The Two Postwar Eras and the Conditions of Stability in Twentieth-Century Western Europe.'" *American Historical Review*, vol. 86 (April), pp. 353–358.

Shultz, George P., 1985. "National Policies and Global Prosperity." Address before the Woodrow Wilson School of Public and International Affairs, Princeton University, Princeton, New Jersey, April 11. Released by the Bureau of Public Affairs, Department of State, Washington, DC, Current Policy No. 684.

_____, 1981. "Risk, Uncertainty and Foreign Economic Policy." David Davies Memorial Lecture, London (October 27).

_____, 1979. "Light Switch Diplomacy." *Business Week*, May 28, pp. 24–26.

_____, and Kenneth W. Dam, 1977. *Economic Policy Beyond the Headlines*. New York: W. W. Norton.

Schultze, Charles L., 1986. *Other Times, Other Places: Macroeconomic Lessons from U.S. and European History*. Washington, DC: The Brookings Institution.

Senate, U.S. Congress, 1945. *Bretton Woods Agreements Act*. Committee on Banking and Currency, 79th Session, June 12–16, 18–22, 25, 28.

Skocpol, Theda, 1985. "Bringing the State Back In: Strategies of Analysis in Current Research." In *Bringing the State Back In*, eds. Peter Evans, B. Dietrich Rueschemeyer, and Theda Skocpol. Cambridge, England: Cambridge University Press, pp. 3–37.

Slemrod, Joel, 1986. "Taxation and Business Investment." In *Essays in Contemporary Economic Problems: The Impact of the Reagan Program 1986*, ed. Phillip Cagan. Washington, DC: American Enterprise Institute, pp. 45–73.

Smith, Adam, 1976. *The Theory of Moral Sentiments*, ed. D. D. Raphael and A. L. Macfie. Oxford, England: Clarendon Press.

_____, 1937. *The Wealth of Nations*. New York: Random House (Modern Library).

Snidal, Duncan, 1985. "The Limits of Hegemonic Stability Theory." *International Organization*, vol. 39, no. 4 (Autumn), pp. 579–615.

Solomon, Robert, 1982. *The International Monetary System 1945-1981*. New York: Harper & Row.

Somensatto, Eduardo, 1985. "Budget Deficits, Exchange Rates, International Capital Flows and Trade." In *Essays in Contemporary Economic Problems: The Economy in Deficit 1985*, ed. Phillip Cagan. Washington, DC: American Enterprise Institute, pp. 223–281.

Spengler, Oswald, 1932. *The Decline of the West*. One-volume edition. New York: Alfred A. Knopf.

Spindler, J. Andrew, 1984. *The Politics of International Credit: Private Finance and Foreign Policy in Germany and Japan*. Washington, DC: The Brookings Institution.

Sprinkel, Beryl W., 1984. Remarks to the Swiss-American Chamber of Commerce, Zurich Switzerland, February 16.

———, 1981. Testimony before U.S. Congress, Joint Economic Committee, *International Economic Policy*. Hearings, 97th Congress, 1st session, May 4.

Stein, Herbert, 1984. *Presidential Economics: The Making of Economic Policy from Roosevelt to Reagan and Beyond*. New York: Simon and Schuster.

———, 1969. *The Fiscal Revolution in America*. Chicago: University of Chicago Press.

Stockman, David A., 1986. *The Triumph of Politics: Why the Reagan Revolution Failed*. New York: Harper & Row.

Strange, Susan, 1982. "Still an Extraordinary Power: America's Role in a Global Monetary System." In *Political Economy of International and Domestic Monetary Relations*, eds. Raymond E. Lombra and Willard E. Witte. Ames: Iowa State University Press.

Tanzi, Vito, 1985. "The Deficit Experience In Industrial Countries." In *Essays in Contemporary Economic Problems: The Economy in Deficit 1985*, ed. Phillip Cagan. Washington, DC: American Enterprise Institute, pp. 81–121.

Tatom, John A., 1989. "U.S. Investment in the 1980's: The Real Story." Federal Reserve Bank of St. Louis *Review*, vol. 71, no. 2 (March/April), pp. 3–15.

———, 1987. "Will a Weaker Dollar Mean a Stronger Economy?" *Journal of International Money and Finance*, vol. 6, pp. 433–447.

———, 1986. "Domestic vs. International Explanations of Recent U.S. Manufacturing Developments." Federal Reserve Bank of St. Louis *Review*, vol. 68, no. 4 (April), pp. 5–19.

The Times (London), daily issues.

Tolchin, Martin and Susan Tolchin, 1988. *Buying into America: How Foreign Money Is Changing the Face of Our Nation*. New York: Times Books.

Triffin, Robert, 1960. *Gold and the Dollar Crisis: The Future of Convertibility*. New Haven, CT: Yale University Press.

———, 1957. *Europe and the Money Muddle: From Bilateralism to Near-Convertibility, 1947–1956*. New Haven, CT: Yale University Press.

Trilateral Commission Task Force Reports: 1–7, 1977. New York: New York University Press.

Truman, David, 1951. *The Governmental Process: Political Interests and Public Opinion*. New York: Alfred A. Knopf.

United Nations (U.N.), *Yearbook of National Accounts Statistics*, various issues.

———, *Statistical Yearbook*, various issues.

———, *Yearbook of International Trade Statistics*, various years.

UNCTAD, 1987. *Handbook of International Trade and Development: Supplement 1986*. New York: United Nations.

U.S. Department of State, 1947. "European Initiative Essential to European Recovery." Remarks by Secretary of State George C. Marshall at Harvard University Commencement, June 5, 1947. Publication 2882, European Series 25. Washington, DC: Government Printing Office.

Vance, Cyrus R., 1983. *Hard Choices: Four Critical Years in Managing America's Foreign Policy*. New York: Simon and Schuster.

Vandenberg, Arthur H., Jr., (ed.), 1952. *The Private Papers of Senator Vandenberg.* Boston: Houghton Mifflin.

Van der Wee, Herman, 1986. *Prosperity and Upheaval: The World Economy, 1945–1980.* Middlesex, England: Viking.

Van Dormael, Armand, 1978. *Bretton Woods: Birth of a Monetary System.* New York: Holmes & Meier.

Van Hoozer, Terri T., 1988. "Postwar Policymaking: The Shaping of a Consensus in Support of the Marshall Plan." (Unpublished manuscript, April 28).

van Wolferen, Karel, 1989. *The Enigma of Japanese Power.* New York: Alfred A. Knopf.

Vernon, Raymond, 1966. "International Investment and International Trade in the Product Cycle." *Quarterly Journal of Economics* (May), pp. 190–207.

Volcker, Paul, A., 1986a. *Monetary Policy Objectives for 1986.* Testimony of Paul A. Volcker, Chairman, Board of Governors of the Federal Reserve System (February 16).

———, 1986b. Testimony before the Subcommittee on Telecommunications and Finance, U.S. House Committee on Energy and Commerce (April 23).

Walinsky, Louis J., 1982–1983. "Coherent Defense Strategy: The Case for Economic Denial." *Foreign Affairs,* vol. 61, no. 2 (Winter), pp. 272–292.

Wallich, Henry C., 1955. *Mainsprings of the German Revival.* New Haven, CT: Yale University Press.

The Wall Street Journal (WSJ), daily issues.

Walters, Alan A., 1985. *Britain's Economic Renaissance: Margaret Thatcher's Reforms, 1979–1984.* New York: Oxford University Press.

Waltz, Kenneth N., 1979. *Theory of International Politics.* New York: Random House.

Wanniski, Jude, 1978. *The Way the World Works.* New York: Simon and Schuster.

The Washington Post (WP), daily issues.

Wendt, Alexander E., 1987. "The Agent-Structure Problem." *International Organization,* vol. 41, no. 3 (Summer), pp. 335–371.

Wexler, Imanuel, 1983. *The Marshall Plan Revisited: The European Recovery Program in Economic Perspective.* Westport, CT: Greenwood.

White, Harry Dexter, 1945. "The Monetary Fund: Some Criticisms Examined." *Foreign Affairs,* vol. 23, no. 2, pp. 193–210.

Wilcox, Clair, 1949. *A Charter for World Trade.* New York: Macmillan Publishing Company.

Williams, William Appleman, 1962. *The Tragedy of American Diplomacy.* New York: Deli.

Williamson, John, 1983. *The Exchange Rate System.* Policy Analyses in International Economics, No. 5. Washington, DC: Institute for International Economics, September.

Williamson, John, and Marcus H. Miller, 1987. *Targets and Indicators: A Blueprint for the International Coordination of Economic Policy.* Washington, DC: Institute for International Economics, September.

Williamson, Oliver, 1975. *Markets and Heirarchies.* New York: The Free Press.

Winham, Gilbert R., 1986. *International Trade and the Tokyo Round Negotiation.* Princeton, NJ: Princeton University Press.

Woolley, John T., 1984. *Monetary Politics: The Federal Reserve and the Politics of Monetary Policy.* Cambridge, England: Cambridge University Press.

Working Group on Exchange Market Intervention, 1983. *Report on the Working Group on Exchange Market Intervention* (also known as the Jurgensen Report after the chairman of the Group, Philippe Jurgensen, the French financial sherpa). Washington, DC: U.S. Treasury Department, April 29.

World Bank, 1988c. *Trade Liberalization: The Lesson of Experience*. Internal Discussion Paper, Latin America and the Caribbean Regional Series, Report No. IDP 14, April.

_____, 1979b through 1989b. *World Development Report*. New York: Oxford University Press (for the World Bank).

_____, 1981a. *Accelerated Development in Sub-Saharan Africa: An Agenda for Action*. Report No. 3358, August (Bank's Africa Report).

Yamamura, Kozo, 1967. *Economic Policy in Postwar Japan: Growth Versus Economic Democracy*. Berkeley: University of California Press.

Yntema, Theodore and members of the CED Research Staff, 1946. *Jobs and Markets*. Research study for the Committee for Economic Development. New York: McGraw-Hill.

Zysman, John, 1983. *Governments, Markets and Growth: Financial Systems and the Politics of Industrial Change*. Ithaca, NY: Cornell University Press.

INDEX